DEPRESSION AND BRAIN DYSFUNCTION

DEPRESSION AND BRAIN DYSFUNCTION

EDITED BY

Frank G Gilliam MD MPH

Caitlin Tynan Doyle Professor of Neurology; Head, Epilepsy Division, Department of Neurology;

Director, Comprehensive Epilepsy Center, The Neurological Institute of New York, Columbia University, NY, USA

Andres M Kanner MD

Professor of Neurological Sciences and Psychiatry, Rush Medical College;

Director, Laboratory of Electroencephalography and Video-EEG-Telemetry;

Associate Director, Division of Epilepsy and Clinical Neurophysiology and Rush Epilepsy Center,

Rush University Medical Center, Chicago, IL, USA

Yvette I Sheline MD

Professor of Psychiatry, Radiology and Neurology;

Director, Center for Depression and Neuroimaging,

Washington University School of Medicine, St Louis, MO, USA

Taylor & Francis

Taylor & Francis Group

LONDON AND NEW YORK

First published in the United Kingdom in 2006
by Taylor & Francis,
an imprint of the Taylor & Francis Group,
2 Park Square, Milton Park
Abingdon, Oxon OX14 4RN, UK

Tel.: +44 (0) 20 7017 6000
Fax.: +44 (0) 20 7017 6699
E-mail: info.medicine@tandf.co.uk
Website: http://www.tandf.co.uk/medicine

British Library Cataloguing in Publication Data

Data available on application

Library of Congress Cataloging-in-Publication Data

Data available on application

ISBN10 1-84214-214-3
ISBN13 9-78-1-84214-214-1

Distributed in North and South America by

Taylor & Francis
2000 NW Corporate Blvd
Boca Raton, FL 33431, USA

Within Continental USA
Tel.: 800 272 7737; Fax.: 800 374 3401
Outside Continental USA
Tel.: 561 994 0555; Fax.: 561 361 6018
E-mail: orders@crcpress.com

Distributed in the rest of the world by
Thomson Publishing Services
Cheriton House
North Way
Andover, Hampshire SP10 5BE, UK
Tel.: +44 (0) 1264 332424
E-mail: salesorder.tandf@thomsonpublishingservices.co.uk

Composition by Parthenon Publishing
Printed and bound by Antony Rowe Ltd., Chippenham, Wiltshire, UK

Contents

Preface

Depression is a common disorder without geographic, educational, socioeconomic, or racial boundaries. The World Health Organization determined that:

> *Depression is the leading cause of disability as measured by years lived with disability and the fourth leading contributor to the global burden of disease (disability adjusted life years) in 2000. By the year 2020, depression is projected to reach second place of the ranking of disability adjusted life years calculated for all ages, both sexes (www.who.int).*

Advancing our understanding of depression is clearly of utmost importance to the international medical community, but depression is a complex disorder that challenges phenomenological and neurobiological research methodologies.

Two important developments in neuroscience in recent years have served as the primary impetus for this book: (1) major discoveries in the association of depression with monoaminergic system gene polymorphisms, neuronal plasticity and resistance to injury, and also structural and functional neuroimaging abnormalities; and (2) recognition of the increased and early symptoms of depression in common neurological disorders such as Parkinson's disease, epilepsy, Alzheimer's disease, specific stroke syndromes, multiple sclerosis and traumatic brain injury. Hence, the text evolved organically into two sections: the first addressing advances in the neurobiology of depression in persons without other neurological diagnoses, and the second focusing on the results of studies of depression within specific neurological disorders. A premise of the second section is that certain neurological disorders may serve as useful models, or even naturalistic experiments, for the study of the phenomenology and the neurobiology of depression. In order to emphasize the clinical relevance of mood in disorders of brain function, aspects of the epidemiology and management of depression within neurological disorders are also included in these chapters.

In view of the rapid advances in basic and clinical neuroscience of depression, each chapter should be considered as a work-in-progress, and perhaps merely a base from which to begin exploring a specific aspect of depression and brain dysfunction. We sincerely hope that this book will serve as a resource and a catalyst for researchers and clinical practitioners working in the many relevant areas of psychiatry, neurology and neuroscience.

Frank G Gilliam MD, MPH
Andres M Kanner MD
Yvette I Sheline MD

List of contributors

John J Barry MD
Department of Psychiatry
Stanford University School of Medicine
Stanford University Hospital
401 Quarry Road
Stanford, CA 94305
USA

Kevin J Black MD
Departments of Psychiatry, Neurology and Radiology
Washington University School of Medicine and Barnes-Jewish Hospital
660 South Euclid Avenue
St Louis, MO 63110
USA

Li-Shiun Chen MD ScD MPH
Department of Psychiatry
Washington University School of Medicine and Barnes-Jewish Hospital
660 South Euclid Avenue
St Louis, MO 63110
USA

Gabriel A de Erausquin MD PhD
Departments of Psychiatry and Neurology
Washington University School of Medicine and Barnes-Jewish Hospital
660 South Euclid Avenue
St Louis, MO 63110
USA

Andria L Ford MD
Department of Neurology
Washington University School of Medicine
660 South Euclid Avenue
St Louis, MO 63110
USA

Jorge A Franco MD
Department of Psychiatry
39–234 Calle 48
Universidad Autónoma de Bucaramanga
Colombia

Moises Gaviria MD
Division of Neuropsychiatry
Department of Psychiatry
University of Illinois at Chicago
912 South Wood Street
Chicago, IL 60612
and
Christ Hospital Medical Center
4440 West 95th Street
Oak Lawn, IL 60453
USA

Frank G Gilliam MD MPH
Department of Neurology
and
Comprehensive Epilepsy Center
The Neurological Institute of New York
Columbia University
710 West 168th Street
New York, NY 10032
USA

Hrvoje Hećimović MD PhD
Center for Neuroimaging in Epilepsy
Department of Neurology
Washington University School of Medicine
660 South Euclid Avenue
St Louis, MO 63110
USA

Phillip C Jobe PhD
Department of Biomedical and Therapeutic Sciences
University of Illinois College of Medicine at Peoria
Peoria, IL 61656
USA

Andres M Kanner MD
Department of Neurological Sciences
Rush Medical College
Rush Epilepsy Center
Rush University Medical Center
1653 West Congress Parkway
Chicago, IL 60612
USA

Constantine G Lyketsos MD MHS
Division of Geriatric Psychiatry and Neuropsychiatry
Johns Hopkins University
School of Medicine
550 North Broadway, Suite 308
Baltimore, MD 21205
USA

Mark A Mintun MD
Departments of Radiology and Psychiatry
Washington University School of Medicine
660 South Euclid Avenue
St Louis, MO 63110
USA

Robert T Naismith MD
Department of Neurology
Washington University School of Medicine
660 South Euclid Avenue
St Louis, MO 63110
USA

Avni Pandya MD
Department of Family Medicine
Medical College of Georgia
1120 15th Street
Augusta, GA 30912
USA

Paul B Rosenberg MD
Division of Geriatric Psychiatry and Neuropsychiatry
Johns Hopkins University
School of Medicine
550 North Broadway, Suite 308
Baltimore, MD 21205
USA

Juan M Santos MD
Neurological Institute of New York
Columbia University
168th Street
New York, NY 10032
USA

Yvette I Sheline MD
Departments of Psychiatry, Radiology and Neurology
Washington University School of Medicine
660 South Euclid Avenue
St Louis, MO 63110
USA

Andrei G Vlassenko MD PhD
Department of Radiology
Washington University School of Medicine
660 South Euclid Avenue
St Louis, MO 63110
USA

Richard J Weber PhD
Department of Biomedical and Therapeutic Sciences
University of Illinois College of Medicine at Peoria
Peoria, IL 61656
USA

Color plates

The following figures are color versions of selected figures from the main body of the text

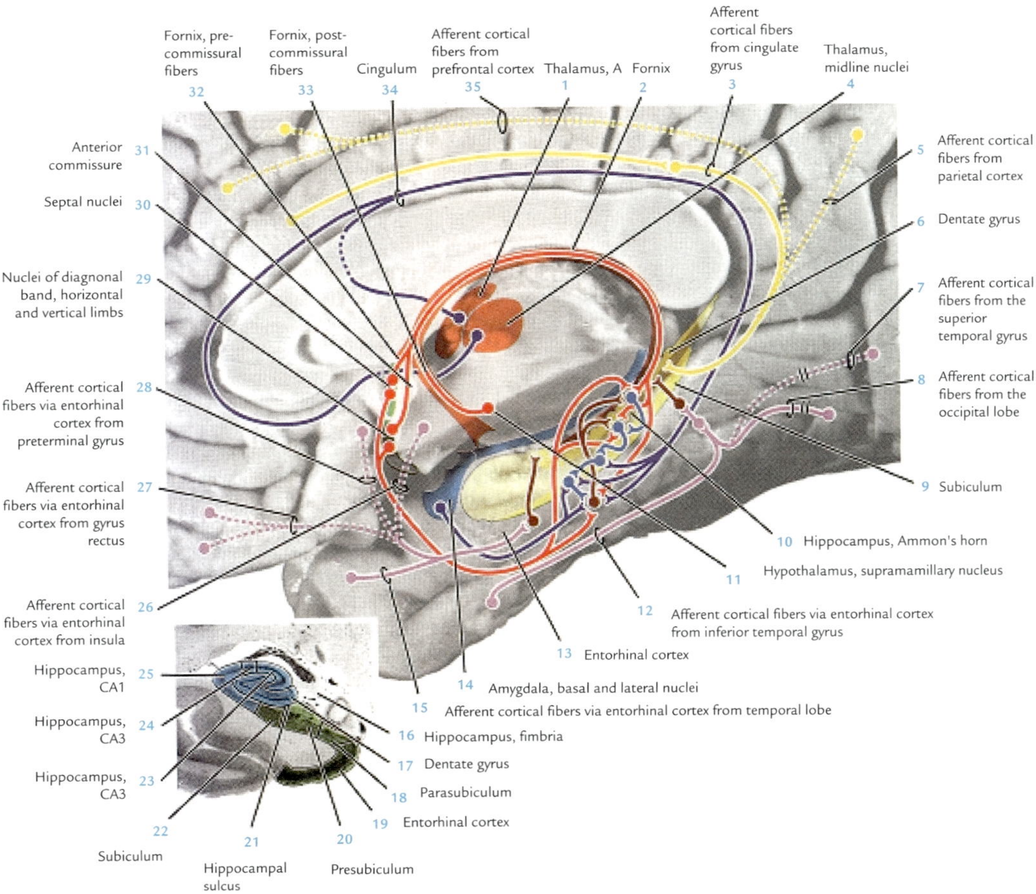

Figure 2.1 Hippocampal afferent pathways. Reproduced from reference 21, with permission

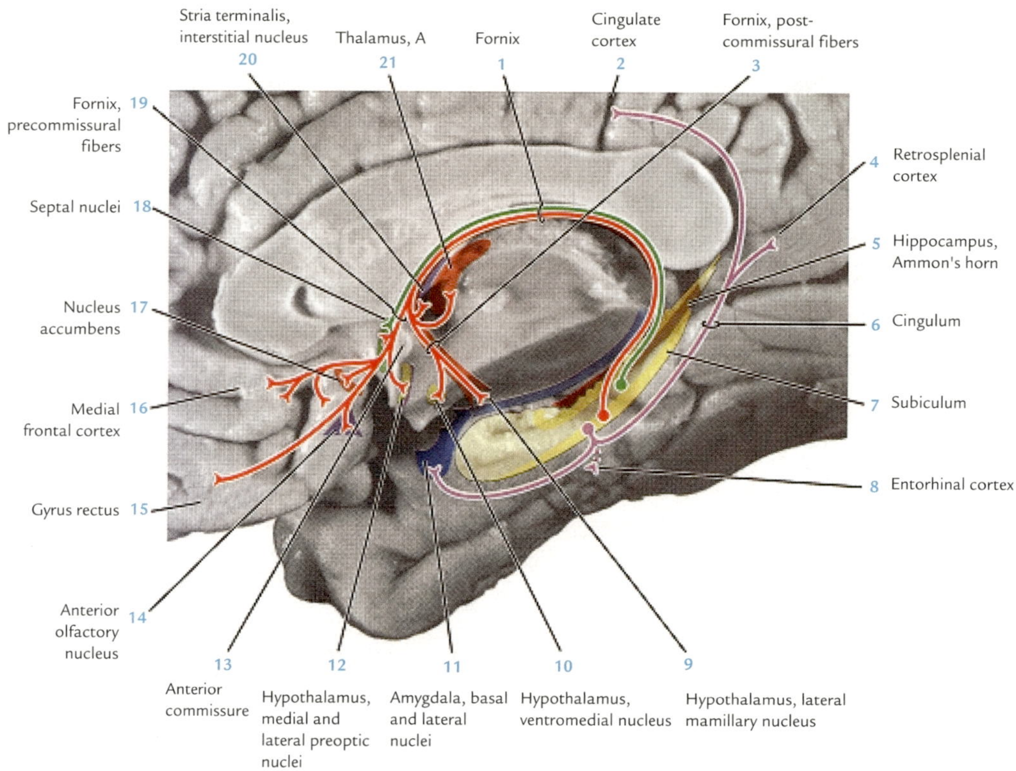

Figure 2.2 Hippocampal efferent pathways. Reproduced from reference 21, with permission

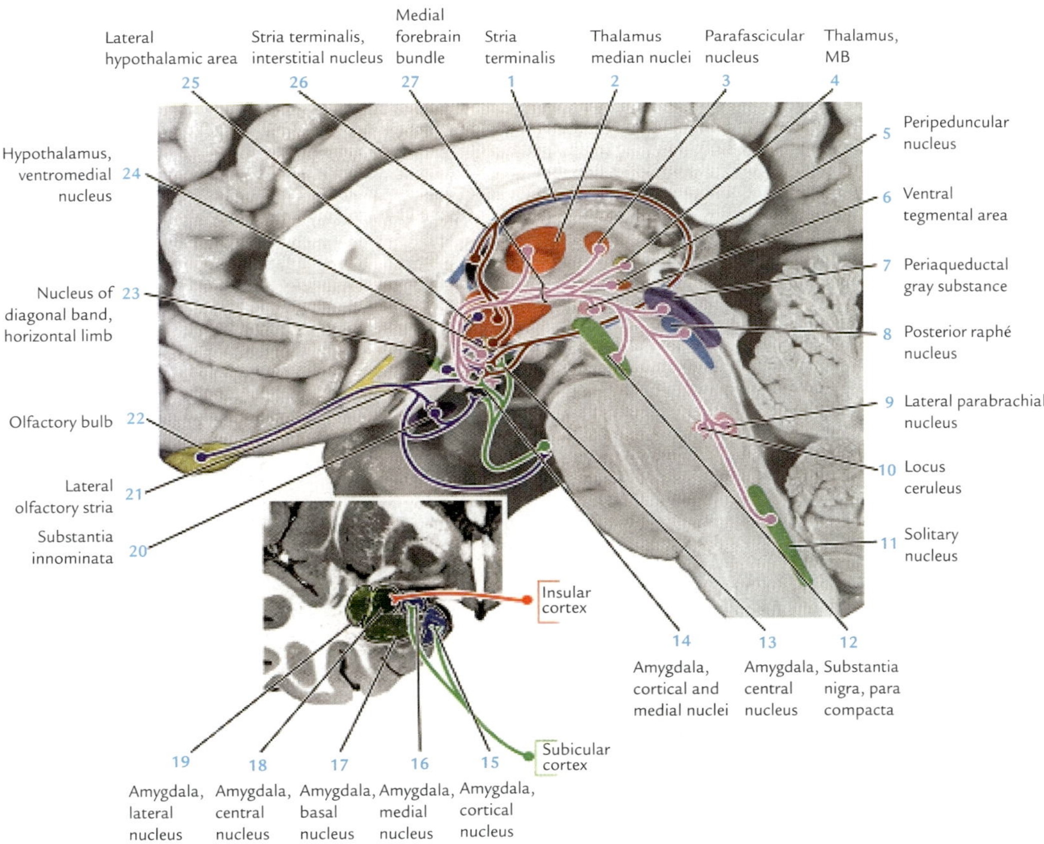

Figure 2.3 Amygdalar afferent pathways. Reproduced from reference 21, with permission

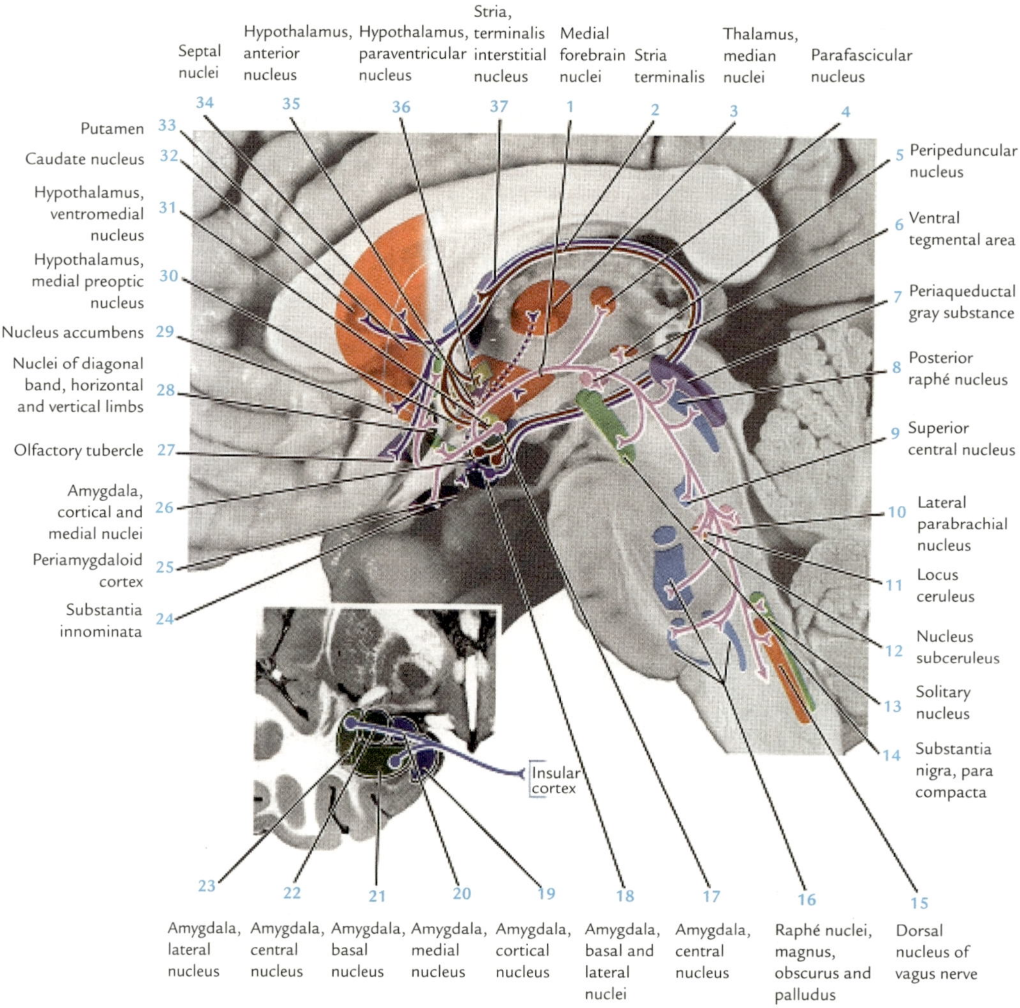

Figure 2.4 Amygdalar efferent pathways. Reproduced from reference 21, with permission

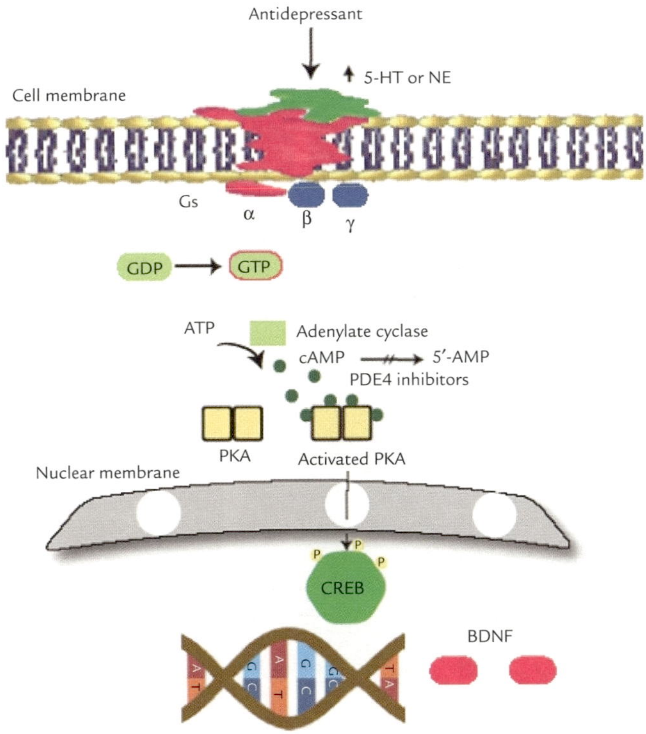

Figure 2.7 Intracellular cAMP signaling pathway. NE, norepinephrine; PKA, protein kinase A; CREB; cAMP response element binding protein; PDE4, phosphodiesterase 4; BDNF, brain-derived neurotrophic factor

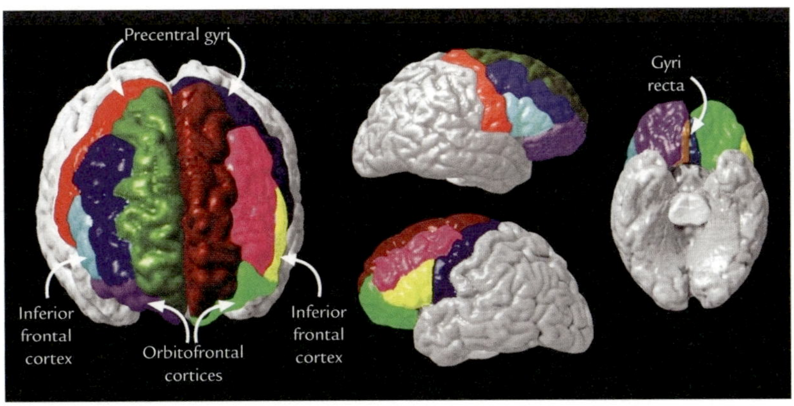

Figure 3.1 Volumetric studies of prefrontal cortex subregions have identified volume decreases in major depression. Reproduced from reference 19, with permission from the *American Journal of Psychiatry*, copyright 2004, American Psychiatric Association

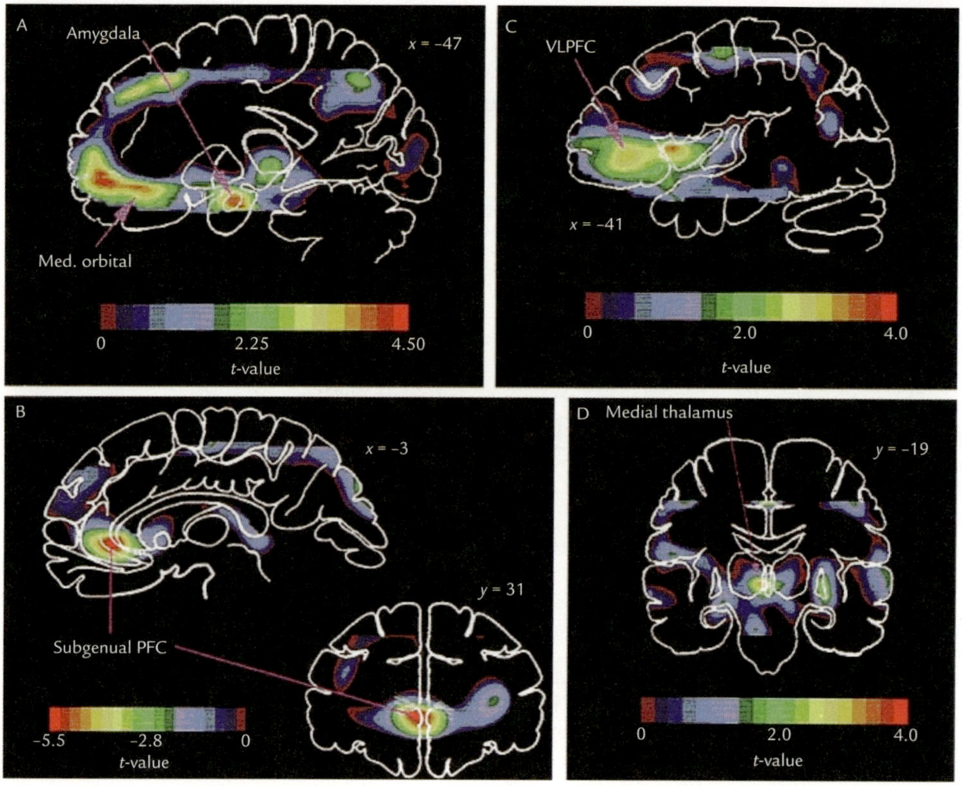

Figure 4.1 (A) Areas of abnormally increased cerebral blood flow (CBF) in familial major depressive disorder (MDD) shown in an image of *t* values, produced by a voxel-by-voxel computation of the unpaired *t* statistic to compare CBF between depressed and control samples. The positive *t* values in this sagittal section at 17 mm left of midline ($x=-17$) show areas of increased CBF in depressives relative to control subjects in the amygdala and medial orbital cortex. Anterior is to the left. (B) Coronal (31 mm anterior to the anterior commissure, or $y=31$) and sagittal (3 mm left of midline, or $x=-3$) sections showing negative voxel *t* values where glucose metabolism is decreased in depressives relative to control subjects. Anterior is to the left. (C) Sagittal *t* image section 41 mm left of midline ($x=-41$) shows an area of increased CBF in depressives relative to healthy control subjects in the left ventrolateral prefrontal cortex (VLPFC), lateral orbital cortex and anterior insula. (D) Coronal *t* image section 19 mm caudal to the anterior commissure (i.e. $y=-19$) showing the area corresponding to increased CBF in the left medial thalamus of depressives relative to control subjects. Reproduced from reference 3: Drevets WC. Neuroimaging studies of mood disorders. *Biol Psychiatry* 2000; 48: 813–829, with permission from the Society of Biological Psychiatry, copyright 2000

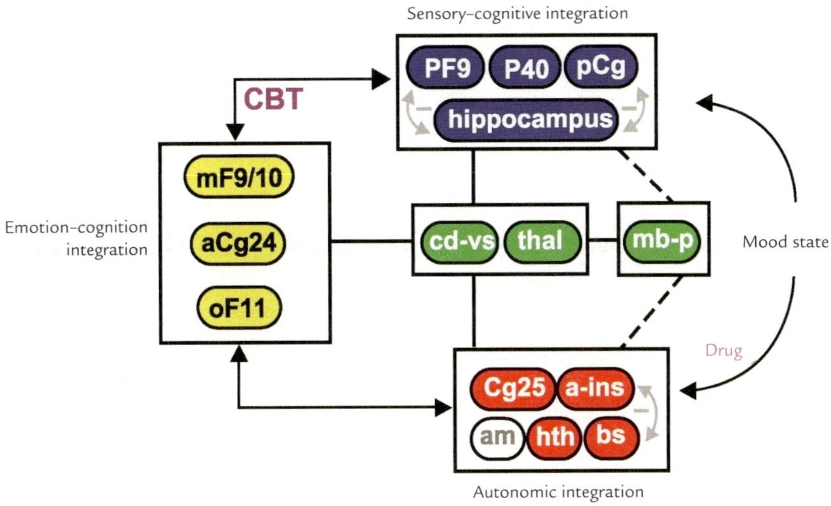

Figure 4.2 Limbic–cortical dysregulation model. Regions with known anatomical interconnections are grouped into four main behavioral compartments. The cortical limbic (dorsal–ventral) segregation within each 'compartment' additionally identifies those brain regions where an inverse relationship is seen across the different PET paradigms. Sadness and depressive illness are both associated with decreases in dorsal neocortical regions (sensory–cognitive compartment) and relative increases in ventral limbic and paralimbic areas (autonomic compartment). The model, in turn, proposes that illness remission occurs when there is appropriate modulation of dysfunctional limbic–cortical interactions (small gray arrows) – an effect facilitated by various forms of treatment (purple). mF, medial prefrontal; aCg, rostral anterior cingulate; oF, orbital frontal; cd-vs, caudate–ventral striatum; thal, thalamus; mb-p, midbrain-pons; Cg25, subgenual cingulate; a-ins, anterior insula; am, amygdala, hth, hypothalamus; bs, brain stem; PF, dorsolateral prefrontal; P, parietal; pCg, posterior cingulate. Numbers: Brodmann designations. Reproduced from reference 6: Seminowicz DA, Mayberg HS, McIntosh AR, *et al.* Limbic–frontal circuitry in major depression: a path modeling metanalysis. *Neuroimage* 2004; 22: 409–418, with permission from Elsevier, copyright 2004

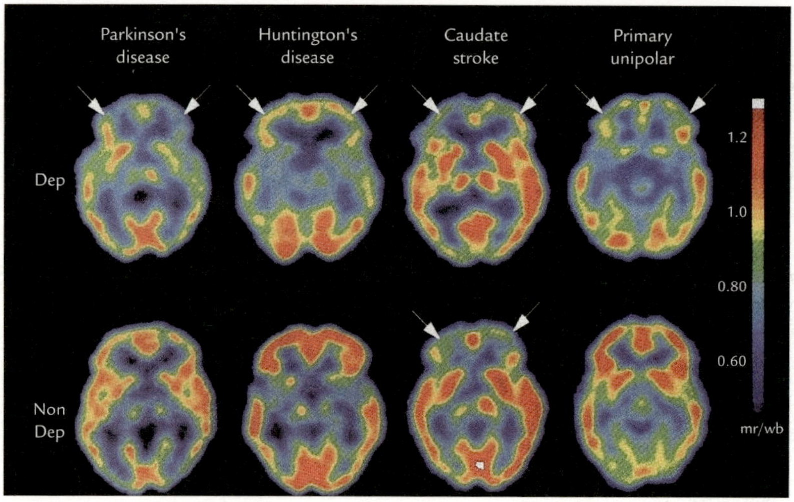

Figure 4.3 ^{18}F-fluorodeoxyglucose (FDG) positron emission tomography (PET) images in primary and secondary depressions – basal ganglia level. There is bilateral frontal lobe hypometabolism in all depressed patients (Dep), independent of disease etiology (arrows, top row). Frontal cortex metabolism is normal in non-depressed patients (Non Dep) except those with strokes (arrows, bottom row). Scale: relative metabolic rate (regional absolute metabolic rate/whole brain metabolic rate). Reproduced from reference 8, with permission from the *American Journal of Geriatric Psychiatry*, copyright 1994, American Psychiatric Association

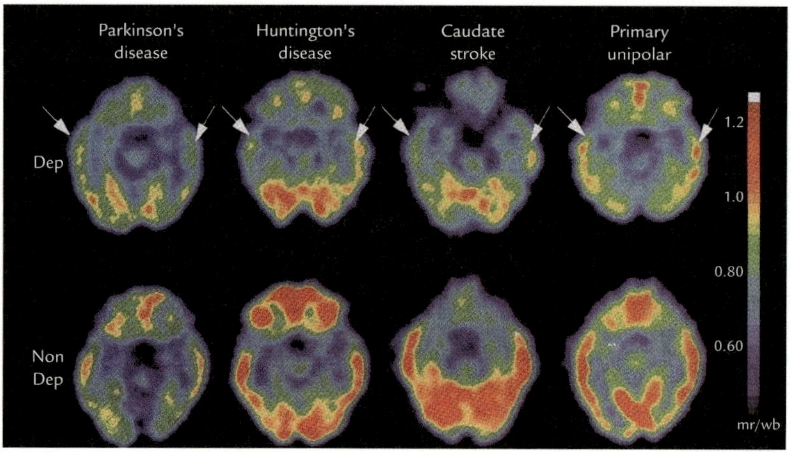

Figure 4.4 ^{18}F-fluorodeoxyglucose (FDG) positron emission tomography (PET) images in primary and secondary depressions – temporal lobe level. There is symmetric bitemporal hypometabolism in all depressed patients (Dep), independent of disease etiology (arrows, top row). Temporal cortex metabolism is normal in all non-depressed patients (Non Dep). Scale: relative metabolic rate (regional absolute metabolic rate/whole brain metabolic rate). Reproduced from reference 8, with permission from the *American Journal of Geriatric Psychiatry*, copyright 1994, American Psychiatric Association

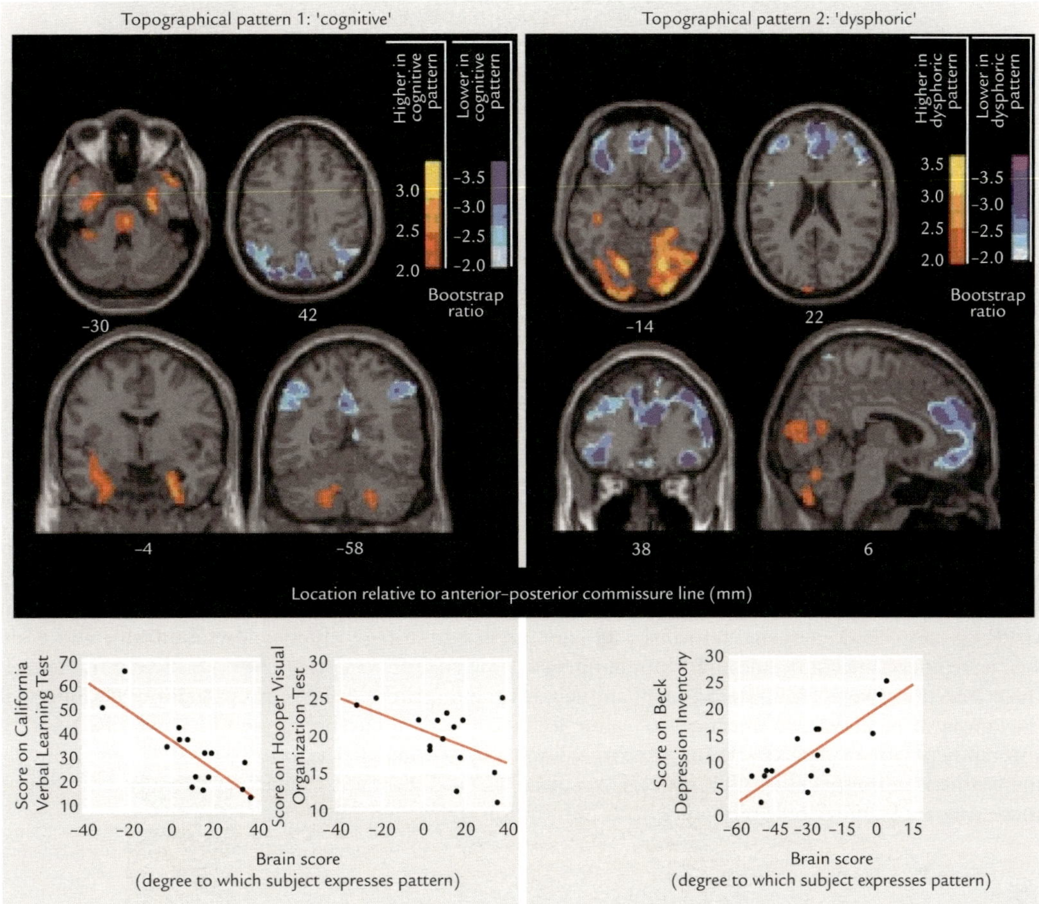

Figure 4.5 Two orthogonal topographic patterns of brain metabolism in 15 patients with Parkinson's disease and doubly dissociated correlations with cognitive and mood measures. The orthogonal topographic patterns were derived from glucose metabolic scans at rest. The bootstrap ratio is the ratio of voxel salience value to its standard error (estimated from 100 bootstrap samples). It is a measure of the stability of high (red scale) and low (blue scale) metabolic values that constitute the topographic pattern. Scatterplots of relationship between behavior and brain score are presented at the bottom row. Brain–behavior correlations for dysphoria and cognition were doubly dissociated, whereas those for the two cognitive measures were similar. Reproduced from reference 56, with permission from the *American Journal of Psychiatry*, copyright 1971, American Psychiatric Association

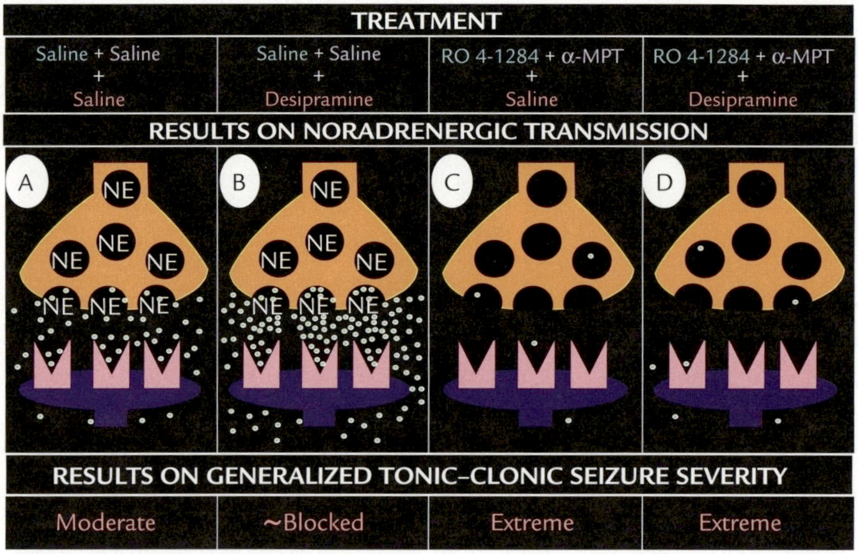

Figure 5.7 Effect of norepinephrine (NE) depletion on the severity of GTCSs induced by sound in GEPR-3s. , noradrenergenic terminals; , noradrenergic storage vesicles; norepinephrine and white dots, norepinephrine molecules; , noradrenergic postsynaptic receptors; , postsynaptic neuron. Three injections were administered to all animals as shown (groups A–D). Group A received saline each time; group B received two injections of saline followed by desipramine; group C received RO 4-1284 (a norepinephrine storage vesicle inactivator) followed by α-methyl-ρ-tyrosine (α-MPT, a norepinephrine synthesis inhibitor) plus saline; group D received RO 4-1284, then, α-MPT, and finally desipramine. These data are summarized from a previous publication (reference 236)

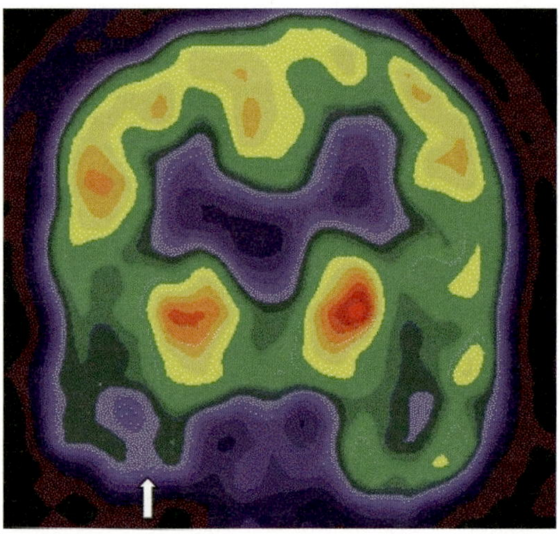

Figure 7.3 ^{18}F-fluorodeoxyglucose–positron emission tomography image of right temporal (white arrow) and right basal ganglia hypometabolism in a patient with refractory right temporal lobe epilepsy. From reference 15, with permission

1

Phenomenology of depression: towards a neuroscientific perspective

Juan M Santos, Li-Shiun Chen and Gabriel A de Erausquin

'The barrier to conspicuous advance in psychiatry has not been stinginess and prejudice on the part of those who decide whether a research project submitted to them should live or die; nor has it been lack of ability among those who are engaged in psychiatric research: it lies in the inherent toughness of the problems.'

Sir Aubrey Lewis, 1963

INTRODUCTION

Overview

Depression is one of the most common disabling disorders worldwide, and has appropriately received extensive attention from both clinical and basic neuroscience investigators. However, a comprehensive definition that includes all aspects of the experience of depression has been elusive. Well known from antiquity by the name of melancholy, it has been a recurrent theme in literature, philosophy and other aspects of human affairs. The depressed person appears to lose capacity in somatic, cognitive, emotional and social spheres. Accurate integration of all of these aspects of depression into a measurable construct is necessary for optimal utilization of advances in research technologies such as genomics and functional neuroimaging.

Depression has been approached in the past as a state of mind, a clinical diagnosis or a syndrome. Its conceptualization as a state of mind offers such an array of problems that it is rendered useless for most research purposes. Considering depression as a diagnosis (a categorical description) has led to elevation of a concept that offers some problems as defined by modern nosological criteria such as DSM-IV or ICD-10. Currently this concept is defined as a particular constellation of symptoms. The selection of the symptoms that define depression has been determined by large field and epidemiological studies, rendering its operational definition reliable, but its validity is less clearly characterized.

Phenomenology offers a means to integrate multiple dimensions of depression into a coherent account. Differentiation of specific dimensions can be approached mathematically, mostly with the help of multivariate statistics, to describe specific patterns of depression in different individuals and different conditions. Thus, a phenomenological perspective may allow detection and description of particular variants of depression, currently grouped together into a single category.

In an era when outstanding contributions in neurobiology (such as genetics and neuroimaging) are being made, and correlations between depression and these findings are being explored, it is of the utmost importance that the endophenotype of depression can be appropriately defined for different types of brain dysfunction, since the pattern of depressive symptoms may be different for different neurological conditions. Not being careful about the ways depression is defined may give undue weight to some criteria, such as the severity of depression, when phenomenological approaches consistently show that subthreshold depression may have a functional impact comparable to that of major depressive disorder, particularly in neurological conditions.

Phenomenological descriptions can be derived from standardized instruments of measurement (BDI, HAM-D, SCAN, CIS, etc.), but can also incorporate other clinical data, such as biological data or course of depression data, which tell us about collateral aspects of depression.

For example, the understanding of depression has been greatly enhanced by considering those dimensions related to time (e.g. relapses, remission, duration of treatment, sequential measurements of quality of life), and this is an area largely underinvestigated in relation to neurological disorders, although some data are available in Parkinson's disease.

The classification of depression remains a topic of discussion. Classifications based purely on clinical criteria have consistency problems. It is possible to say safely that a distinction between melancholic depression (where a categorical model may apply) and non-melancholic depressions can be made. The future of depression subtyping lies in the precise definition of endophenotypes with its associated longitudinal phenomenology.

While depression is said to affect mostly mood, which carries an important emotional effect, its most salient feature is the change in cognition. Thoughts and complex perceptions are negatively biased and there is good evidence that some forms of therapy – such as cognitive behavioral therapy – can raise mood by working on the thoughts associated with depression. Nevertheless, mood, emotions, feelings and thoughts are all related, although the exact way in which they are has not yet been fully identified.

Cultural and social factors may affect the risk of depression or its manifestations, and careful consideration must be given to usage of culturally validated instruments of depression.

Aims

The purpose of this chapter is to provide a basic account of the concept of depression from a phenomenological, also called psychopathological, perspective and to serve as an introduction to the topic of depression in relation to neurological disorders. It is not intended to review comprehensively all the issues related to the phenomenology of depression, a task well beyond the scope of a single chapter, but to equip the reader with some useful and relevant phenomenological considerations.

Indeed, although in the title of this book 'brain dysfunction' is used as a synonym for neurological disorders, the authors would like to emphasize that depression is itself a form of brain dysfunction[1].

Importance of depression in general and other neurological disorders

Depression is among the top five leading causes of disability and disease burden throughout the world[2], and it has been projected that it will be the second most important cause of disability (measured as disability-adjusted life years) worldwide, and the most important one in developing countries by the year 2020. According to estimates by the World Health Organization, there are 121 million people worldwide suffering from depression[3]. This condition is over-represented in practically all of the major types of neurological disorders (stroke, epilepsy, Parkinson's disease, etc.)[4].

The reader is referred to Kupfer and Frank[5] for a review of the impact of depression on medical conditions in general and its costs. These authors hold that, when present, depression increases the cost of treatment of the medical condition as well as the cost of caring for the psychiatric condition itself.

Phenomenology and psychopathology

There has been a longstanding tradition in psychiatry of classifying mental illnesses according to the categories of experiences described by the mentally ill patients. Defining these experiences in a precise way, rendering labeling possible, has been the key element in the progress of phenomenology. The most used nosological classifications, such as the DSM-IV or the ICD-10, are based on the aggregation of symptoms. Each aggregation of symptoms, along with some information about clinical course, defines 'disorders' that are empirically derived and make no reference to etiology or pathophysiology[6,7]. The categorical classification of both DSM-IV and ICD-10[8,9] has inherent problems, which will be discussed below.

According to Sims[10], 'psychopathology is concerned with abnormal experience, cognition and behaviour [...] it describes and categorizes the abnormal experience as recounted by the patient and observed in his behaviour.' Thus, psychopathology can be understood as a description of psychological symptoms and behavior without any added explanation, interpretation or attribution. This reasonably plain method of describing symptoms and syndromes is the basis of classification and diagnosis in psychiatry. Moreover, it is far from being an exhausted mine[11].

Phenomenology can be a misleading term in psychiatry, since its psychiatric meaning is diverse from the philosophical term coined by Husserl. The former was first used by Karl Jaspers, who defined it as an attempt to understand mental or behavioral phenomena, in such a way as to allow a sharing – as much as possible – in the patient's condition, promoting a personal relationship with him or her. Depression has a variety of manifestations, some of them easily observable, such as early morning wakening, while others, such as dysphoria, are essentially subjective. Both kinds of data, observable and subjective, are essential to good phenomenology. Data gathering is achieved, according to Sims[10], by 'precise, insightful and persistent and knowledgeable questioning until the doctor is able to give an account of the patient's subjective experience that the patient recognizes as his own'.

Self-administered questionnaires and often structured and semistructured interviews have the drawback of missing some information, since an adequate phenomenological approach is not possible with them. Many of the criticisms against paper-and-pencil measurements of depression may not be well founded[12], as they provide reliable assessments of diagnosis for many research purposes. On the other hand, in spite of the increasing availability of biological and genetic data, the understanding element of phenomenology remains crucial for the clinical validity of diagnoses[13].

Jaspers and Kurt Schneider are generally credited as the creators of modern psychopathology. The expansion of phenomenology to the Anglo-Saxon world was largely due to the late Frank Fish, whose book *Clinical Psychopathology*[14] influenced the English-speaking countries. Fish, who undertook part of his training in Germany, was able to summarize in a pragmatic manner the accounts of Jaspers and Schneider.

The terms psychopathology and phenomenology are generally used interchangeably, although a purist approach may challenge this view. We endorse this use throughout this chapter. There are other definitions of psychopathology. Ossorio[15] defined it as 'a significant restriction in the ability of an individual to engage in deliberate action and, equivalently, to participate in available social practices.' Ossorio's approach links the description of symptoms directly with their impact on health outcomes, and points towards the correlation of psychopathological dimensions with quality of life indicators. This is an important statement, because it links the experiences (phenomenology) with values (cultural aspects) and health outcomes. Clancy and Eisenberg[16] stated that '[outcomes research] is the study of the results of health services that takes patients' experiences, preferences, and values into account [and it] is intended to provide scientific evidence relating to decisions made by all who participate in health care.'

Wakefield[17] criticized the overinclusiveness of psychopathological criteria and questioned the usefulness of the concept of mental disorder as used in categorical classifications. Vaillant and Vaillant[18] also criticized the shortcomings of a purely atheoretical approach, as used in DSM-IV, pointing out that 'it is reductionistic; it does not emphasize the distinction between state and trait; it is adynamic in that it pays little attention to conflict, adaptation, longitudinal course, and development; and finally, it has consistently sacrificed diagnostic validity on the altar of diagnostic reliability.'

While other branches of medicine may be approached in a far more empirical way, empiricism would be not just highly inadequate and narrow, but also false, as the only way to deal with mental health problems[19]. Phenomenology is in fact the only way so far to approach mental illness in a balanced and realistic way. The avid reader is referred to Harari[19] for an in-depth study of the philosophical, epistemological and evidence-based approach of phenomenology. This author provides a good rebuttal of the purely empirical approach to mental illness. Moreover, phenomenology has given psychiatry a steadiness that other approaches within psychiatry have not been able to provide[20].

A natural expansion of phenomenology towards observation of behavioral changes associated with brain dysfunction is cognitive neuropsychiatry. This discipline can be conceived as 'a systematic and theoretically driven approach in terms of deficits to normal cognitive mechanisms'[21]. On the other hand, stripped of a theory-driven framework, data on cognitive processing and its underlying neural function can be seen as an additional dimension of observable behavior and thus incorporated in an atheoretical phenomenology with a view to description and classification of mental illness. Andreasen[22] proposed a definition of the endophenotype of depression including the use of spectrum and dimensional models (as opposed to categorical ones), cognitive psychology, psychoneurobiology, neurophysiology, neuroimaging, and clinical and lesional studies as the tools to construct a modern phenomenology. This interdisciplinary approach to depression in brain dysfunction forms the basis of an integrative phenomenology.

ETYMOLOGY OF DEPRESSION

The word 'depression' is a re-Latinization (*depressus*) from the French (*depresser*), as are so many other words in English. *Depressare* is related to *deprimere*, which means to press (*premere*) down (the prefix *de-*). This word was used in a psychological context as early as in the 14th century, meaning a state of dejection. During the 20th century it became common to use it to mean a more specific psychological condition.

The multiple meanings of the word depression in general language are illustrated by looking at Webster's dictionary (1990). It defines depression as (1) a hollow or low place, (2) low spirits; dejection, (3) a decrease in force, activity, etc., and (5) a period of reduced business, etc. Synonyms from Roget's International Thesaurus (1992) expand this global definition of depression. There we find that in the same semantic family depression is: (1) downcast, (2) dejection, (3) melancholia, (4) 'the blues', (5) 'in the doldrums', (6) down hearted, and (7) 'moping, melancholy and moonstruck madness'. The key ideas expressed in the word 'depression' are those of lowering, dejection, despondency, subjugation, lessening and sadness.

It is sadness (or sad mood) that has become the central feature of depression. In more recent times the lack of pleasure (anhedonia) has also been considered a core feature of depression[23].

HISTORY OF DEPRESSION (HOW THE CONCEPT OF DEPRESSION CAME TO LIGHT)

Depression has been known since antiquity[24,25]. The Old Testament contains descriptions compatible with the diagnosis of depression such as I Samuel 1, 8 or in Job 3. Mental illness was interpreted from a magic point of view until the ancient Greeks introduced the concept of melancholia and the first mechanical explanatory models[26]. Hippocrates (460–377 BC)[27] is regarded as the distant father of scientific medicine[28]. He considered melancholia in an operational manner: 'repulsion to eat, despondency, insomnia, irritability and restlessness', and pointed to the brain as the source of the problem. Interestingly, he also noted the connection between anxiety and depression in his aphorisms. With the exception of the humor theory, the other questions continue to be hot topics. Melancholia has correlates in practically all cultures[29]. It is interesting that the melancholia concept was taken away from the modern versions of the *Diagnostic and Statistical Manual* only to be claimed back by more recent phenomenological studies.

Galen (AD 131–200), the most important Roman physician and a former slave, abandoned the Hippocratic view and proposed that the heart, and not the brain, was the organ responsible for this malady[30]. However, the Roman Stoics, with their head Seneca, upheld the Hippocratic theory. Lucius Anneus Seneca's *Moral Epistle to Lucilius* is full of what could be called phenomenological examples, and ponders about the influence of heritage and childhood background on the feelings of sadness. Another Roman, Aurelianus, acknowledged aggression as a symptom of depression, particularly in males. The influence of Galen was marked, but the monk–doctors of Europe, such as Casiodorus (490–585), returned to the empirical Hippocratic tradition. The rules of different orders, such as the Hyeronimii monks, prescribed a humanitarian treatment for melancholy and other mental disorders. The invasion of Spain, Portugal and part of Italy by the Arabs brought new ideas[31]. Generally speaking, Islam was also in favor of a humanitarian treatment of psychiatric illnesses. While in the East, in the Baghdad Caliphate, an astringent Muslim theological milieu persecuted those following Hippocrates, in the western Córdoba Caliphate a unique, but short-lived, opportunity of interaction between Christian, Jewish and Muslim cultures flourished. Maimónides, a disciple of Averroes, wrote at length about the topic of depression. His Jewishness was patent in his works about the psychological hygiene of melancholia, which appears to be based on Hippocrates. Avicenna, on the contrary, displayed an eclectic approach towards melancholy, although he also advocated the humor theory. Nevertheless, in both East and West – in Baghdad and Fez, respectively – hospitals for the mentally ill were erected. These experiences of the Islamic civilization undoubtedly influenced Christian Spain. A Spanish Priest, Fr Jofré, created the first modern asylum in Europe in 1409. The humanitarian approach in this first asylum in Valencia was soon propagated to other cities in the Iberian Peninsula, such as the Hospital of Nuestra Señora de Gracia in Zaragoza, which was visited by Philippe Pinnel. This famous French psychiatrist, who would later unchain the mentally ill in the Salpêtrière Hospital in Paris, praised and implemented in France the humanitarian and occupational therapy approach of melancholic patients he had seen in Spain.

During the first half of the 17th century Robert Burton wrote in Oxford, England, his *Anatomy of Melancholy*. It was a masterpiece of synthesis which was well ahead of its time in more than one aspect. Also in the British Isles Henry Maudsley coined the term 'affective disorders' in the 19th century[32]. In America Adolf Meyer fought the great divide between psyche and soma inherited from Descartes. He took one step further in insisting on the biographical, psychological and social influences upon depression, a term he advocated versus melancholia.

Precedents regarding the validity of depression and the difference from normal feelings are found as early as the 6th century. St Isidore of Seville, one of the leading figures of the High Middle Ages, explained melancholia in his encyclopedic work *Etymologies*, taking into account its etiology, typological classification, treatment and social implications. Centuries down the line the nosology of melancholia grew inordinately, making the sharing of knowledge difficult, owing to the disparity in terminology. Jean Étienne Esquirol tried to simplify nosology in the 19th century, but it was not until Kraepelin published careful and detailed descriptions of mental illnesses that modern classifications began to develop. He was the predecessor of phenomenology and Karl Jaspers counted himself in the Kraepelinian school. Jaspers' monumental *General Psychopathology*, written in 1911, is the foundation of modern phenomenology. At the Washington University School of Medicine in St Louis, Robins and Guze[33] continued the neo-Kraepelinian track.

DEFINITION

A depressive disorder is a syndrome (a group of symptoms) associated with a sad mood qualitatively and/or quantitatively (duration and/or intensity) different from normal sadness or the one provoked by specific situations, such as grief. In addition to mood, the full constellation of symptoms affects cognition, behavior and the neurovegetative system. The definition of the syndrome is completed by a significant functional disability (Table 1.1).

The heterogeneity of the disorder has been proposed from consideration of nosological, neurobiological[34] and epidemiological data[35].

Table 1.1 DSM-IV criteria for depressive disorder (with permission from the American Psychiatric Association)

A. Five (or more) of the following symptoms have been present during the same 2-week period and represent a change from previous functioning; at least one of the symptoms is either:

(1) depressed mood or

(2) loss of interest or pleasure

Note: Do not include symptoms that are clearly due to a general medical condition, or mood-incongruent delusions or hallucinations

(1) depressed mood most of the day, nearly every day, as indicated by either subjective report (e.g. feels sad or empty) or observation made by others (e.g. appears tearful). Note: In children and adolescents, can be irritable mood

(2) markedly diminished interest or pleasure in all, or almost all, activities most of the day, nearly every day (as indicated by either subjective account or observation made by others)

(3) significant weight loss when not dieting or weight gain (e.g. a change of more than 5% of body weight in a month), or decrease or increase in appetite nearly every day. Note: In children, consider failure to make expected weight gains

(4) insomnia or hypersomnia nearly every day

(5) psychomotor agitation or retardation nearly every day (observable by others, not merely subjective feelings of restlessness or being slowed down)

(6) fatigue or loss of energy nearly every day

(7) feelings of worthlessness or excessive or inappropriate guilt (which may be delusional) nearly every day (not merely self-reproach or guilt about being sick)

(8) diminished ability to think or concentrate, or indecisiveness, nearly every day (either by subjective account or as observed by others)

(9) recurrent thoughts of death (not just fear of dying), recurrent suicidal ideation without a specific plan, or a suicide attempt or a specific plan for committing suicide

B. The symptoms do not meet the criteria for a mixed episode

C. The symptoms cause clinically significant distress or impairment in social, occupational, or other important areas of functioning

D. The symptoms are not due to the direct physiological effects of a substance (e.g. a drug of abuse, a medication) or a general medical condition (e.g. hypothyroidism)

E. The symptoms are not better accounted for by bereavement (i.e. after the loss of a loved one); the symptoms persist for longer than 2 months or are characterized by marked functional impairment, morbid preoccupation with worthlessness, suicidal ideation, psychotic symptoms, or psychomotor retardation

Alternatively, depression has been described as a mood state, a disorder or a disease[36]. Since, in the case of neurological disorders, depression occurs in the context of different, and well-defined, etiologies, we propose that a phenomenological approach may be more useful for research purposes in this setting.

It is reasonable to expect, given the syndromic nature of the concept of depression, that its sensitivity may be high, but not as high as its specificity. Thus, the major advantage of studying the manifestations of depression in neurological disorders with phenomenological rigor may turn out to be an understanding of the relationship of specific symptom dimensions to well-known mechanisms of disease, adding specificity to the diagnosis.

Although intuitively one may think of depression as a disease affecting primarily the emotional sphere of the individual, it is rather mood that is affected. We briefly review the relevant relationship between thought and emotions and mood.

CRITICAL APPROACH TO THE CONCEPT OF DEPRESSION (LIMITATIONS OF EXISTING NOSOLOGICAL CLASSIFICATIONS)

Until the second half of the 20th century, the word 'melancholy' was used to describe depressive syndromes. Kraepelin replaced the term melancholy by depression in his description of manic-depressive illness, and set the boundaries between this condition and dementia praecox (later known as schizophrenia) on the basis of course and prognosis. Some authors, such as Fleming, anticipated this distinction in the mid-19th century. Eugen Bleuler rejected Kraepelin's emphasis on prognosis and preferred the classification on the basis of aggregation of symptoms. Modern ICD-10 and DSM-IV classifications try to integrate both Kraepelin and Bleuler contributions in an operational definition. DSM-IV and ICD-10 emerged from the Feighner Criteria (St Louis, 1972), and the Research and Diagnostic Criteria (RDC). Our current concept of depression is based on these atheoretic (no weight given to etiological considerations) and descriptive classifications.

From a nosological perspective other types of classification, such as histopathological or etiopathogenic, are better than purely descriptive clinical classifications, since the former can explain the structure of the concept. Yet, since the pathogenesis of depression is unknown and its pathology is ill-defined, the obvious challenge in depression is linking its phenomenological (i.e. purely descriptive) and clinical (i.e. operational) aspects with emerging neurobiological and genetic data. Some of the advances in the understanding of the neural basis of depression have been reviewed by Davidson *et al.*[37], who suggested a 'potentially very fruitful approach to subtyping that does not rely on the descriptive nosology of psychiatric diagnosis but rather is based upon more objective characterization of the specific affective deficits in patients with mood disorders'.

The current concept of depression is based on the addition of unspecific symptoms. There are certainly some criteria which are not symptoms *per se*, such as the decrease in performance and the duration of symptoms, which are also taken into account. Moreover, DSM-IV defines some subcategories in a more precise way, such as melancholy, seasonal affective disorder and atypical depression. Moreover, it may well be the case that each of the symptoms of depression has a different neurobiological foundation and is mediated by specific circuits[38].

Therefore, depression is a heterogeneous diagnostic category. Regardless of the point of view (clinical, symptomatological, etiological, course, phenomenological, etc.) its heterogeneity is well acknowledged.

In their classical paper of 1970, Robins and Guze[33] proposed criteria for the validity of psychiatric diagnoses, which included: (1) clinical description; (2) laboratory studies; (3) delimitation from other disorders; (4) follow-up studies; and (5) family studies. They insisted that a diagnosis be rendered true by its validity and the provision of information regarding prognosis and treatment. Kendell and Jablensky[39] stated that a diagnostic category is valid only if: (1) the defining characteristics constitute a syndrome separated from neighboring syndromes and normality by a zone of rarity (if it has been shown to be a discreet entity with natural boundaries); or (2) the category is defined by a physiological, anatomical, chromosomal, or molecular abnormality. Depression remains weak as an entity either by the Robins and Guze criteria[33] or by the Kendell and Jablensky criteria[39].

Andreasen[40] has proposed that clinical manifestations (and epidemiological considerations) are not enough to establish a mental disorder as a valid disease construct. Rather, clinical description of symptoms and diagnoses should be linked to their neural substrates to achieve validity (see Chapter 2), which is the task undertaken in this book. A diagnosis should be backed up by a deep database including genetics, molecular biology, neurochemistry, neuroanatomy, neurophysiology and cognitive neuroscience in addition to clinical dimensions and outcome parameters[40]. Kendell[41] advocated a polydiagnostic approach of such deep databases where multiple sets of criteria could be applied to the same subjects. The one with the highest validity is chosen. McGuffin and Farmer[42] suggested that 'research data [be] collected in such a way as to allow the application of competing definitions of the disorder'. They proposed 'decomposing diagnostic criteria into their component systems' in a flexible fashion, where the patient meets the criteria, rather than the clinician or researcher having to allocate a particular dimension previously defined in a rigid way. This approach may be useful for overcoming problems such as transcultural variations of depression or overlapping of categories mentioned later in this chapter.

Categoric and dichotomic models of depression (e.g. endogenous depression versus neurotic depression) lack sufficient evidence[43]. A possible exception is melancholic depression, which probably fits into a categorical model[44]. Some authors[45] have summarized the data indicating the clinical instability of non-melancholic depression diagnoses. Non-melancholic depressions are defined by criteria not sufficiently specific to differentiate them from other disorders. In patients with medical conditions, the use of standard criteria is more problematic, given the overlap of symptoms between the medical condition and depression itself or the secondary effects of drugs. The diagnostic category of 'depression due to a general medical condition' used in DSM-IV is a recognition of a statistical association between the underlying disease and depressive symptomatology, but without presumption of knowledge about the pathogenesis.

The system used by the DSM and ICD fits disorders into clearly defined categories. Validation depends primarily on ascertainment of prognosis and response to treatment. If the diagnosis can predict these issues, then it is valid. Nevertheless, the predictive value of depression is relatively poor. The DSM-IV based the operational definition of depression on field trials that evaluated diagnostic reliability and thresholds for criteria[46], and which have been replicated[47,48]. In spite of the criticism previously stated, operational criteria of depression have improved the diagnostic precision and the inter-rater reliability.

Some of the boundaries of the concept of major depression, such as with mixed anxiety–depressive disorder or depressive personality disorder[49], are far from being fully resolved. This is not just a question of co-morbidity, but has profound prognostic implications, since both anxiety and disordered personality function are amongst the most important predictors of outcome of depression[50,51]. Another poorly defined boundary is the delimitation with dysthymia[46,52,53]. Co-morbidity between depression and dysthymia is common (62% in relation to a current depressive episode and 80% in relation to life). Up to 30% of those diagnosed with depression have antecedents of dysthymia, and current dysthymia increases the risk for major depressive disorder. Kendler *et al.*[54] studied symptoms of depression in a sample of more than 1000 female twins for 12 months and, using latent class analysis techniques, showed that atypical depression and typical depression were segregated by the symptom clusters, dividing the latter into mild and severe subgroups. Furthermore, these two diagnostic categories also differed in course and familial aggregation. These studies supported the validity of major depressive disorder as an entity.

A third problematic boundary is the threshold of severity and duration of depression, which is particularly relevant in patients with neurological disorders who often do not reach the threshold for major depressive disorder but nonetheless exhibit a substantial clinical and functional deterioration. These subjects are referred to as experiencing subthreshold depression, subsyndromal depression, minor depression, analog depression or a depressive equivalent, although there is a degree of controversy related to the boundaries of these terms. Subthreshold depression would be perhaps more accurately referred to as subthreshold depressive symptoms (the existence of depressive symptoms not reaching criteria for major depressive disorder). At this level of severity, the boundaries with normality might be blurred, and inter-rater reliability decreases with less severe symptoms[55]. There is a growing consensus[56–58] supporting the continuity between subthreshold depression symptoms and syndromal depression (or DSM-IV major depressive disorder, MDD). Solomon *et al.*[59] also favored a continuum using a taxometric approach, because subjects with subthreshold depression have an elevated risk of future episodes of MDD, increased frequency of family history of affective disorders and shortened rapid eye movement (REM) sleep latency as observed in MDD[60]. REM sleep latency appears to differentiate personality/character 'depressions' and anxiety conditions from subthreshold/clinical depression[60]. Nevertheless, Enns *et al.*[56] have suggested that subthreshold depression may be associated with personality traits, more specifically neuroticism, and a history of parental overcontrol in childhood. It seems, however, that people suffering from subthreshold depression have less anxiety than those suffering from MDD[61,62]. Neurovegetative symptoms may be absent in subthreshold depression[58], but this finding has been contradicted[60].

The strong likelihood of progression from subthreshold depression into major depressive disorder (particulary in patients with a family history), or the partial relapse of depressive disorder into subthreshold depression, are well known[58,63,64], but the reported rates are variable. Given the high prevalence of depression in most neurological disorders, subthreshold depression in this population should be considered a risk factor for MDD. Furthermore, pre-existing depressive symptoms may be either a herald or a risk factor for conditions such as Parkinson's disease (see Chapter 9) or stroke[65]. Recovery from depression should be assumed only if there is a return to asymptomatic status, but not if there is persistent subthreshold depression[66,67]. Subthreshold depression may require most attention when it appears in adolescence, because in adult life it is associated with increased risk of suicidal behaviors, depression and anxiety[68]. It is noteworthy that there are no major differences in health outcomes or quality of life measures, particularly those related to socioeconomic burden, between subthreshold depression and MDD[69,70]. Subthreshold depression is an impairing condition that causes household strain, social irritability and financial strain, and is associated with an increased requirement of disability and welfare benefits[71]. Minor depression is quite a persistent and stable condition[58,72], but it is also easily treatable[73,74].

The presence of subthreshold depression in patients with neurological disorders should warrant clinical attention. Indeed, subthreshold depressive symptoms are extremely frequent in patients with neurological disorders. For example, Wiegartz *et al.*[75] found that 25% of their patients with epilepsy suffered from minor depression.

Methodological inconsistencies have been noted across studies of subthreshold depression, including problems with gender distribution and qualitatively distinct subtypes of depression[59]. Enns *et al.*[56] cautioned about the personality and developmental variables that could represent important areas of discontinuity between clinical and subthreshold depression. In a systematic review of prospective studies, Cuijpers and Smit[64] found that the concept of subthreshold depression was too broad, and a consensus about its definition is deeply needed.

In summary, the DSM-IV or ICD-10 definition of depression is a diagnosis highly reliable but with low validity. This weakens its diagnostic stability. The problem of the DSM-IV validity is not limited to depression. In the years to come, thanks to the faithfully collected operationalized data from interview protocols in large samples – particularly those with follow-up data sets – we will be in a much better position to explore the effects of combining and splitting diagnoses, omitting criteria or re-weighting them, and of choosing altered algorithms with respect to age at onset, number of symptoms and duration of episodes[76]. Here, some of the criteria of Robins and Guze[33] and Kendell[77] may still be relevant. The operational concept of depression also lacks precision, because the syndrome described appears to include a variety of conditions. Accuracy will improve by defining the specific clinical picture, including those aspects of the longitudinal phenomenology of depression in those settings (such as neurological diseases) where a particular endophenotype can be defined. Indeed, a phenomenological approach to depression using multivariate statistics in patients with neurological disorders of known etiology and pathogenesis may help to develop some hierarchical classification, and to identify specific types of depression with increased validity.

The subjective experience of depression

This critical approach to the concept of depression would be incomplete if we did not point out the increasing number of authors who claim an integration of subjective experiences in shaping modern phenomenology and psychopathology. A forceful example of such criticism was put forth by Alexopoulos[78], who writes that 'exclusive reliance on the experimental process produces incomplete knowledge in basic, clinical, or psychosocial fields. For this reason, pragmatic–social processes have been and will continue to exert major influence on [...] scientific initiatives'. He further argued that 'experimentally sound scientific findings have only relative value' and that 'awareness of the influence of nonscientific processes on [...] scientific endeavors can promote and guide cross-fertilization of the sciences of psychopathology.' Korab-Karpowicz[79], along similar lines, states 'the objectivity of scientific research is not free from subjectivity. Objective, scientific knowledge is based on indifference, the state of mind that constitutes scientific attitude. [...]; to look at things as if they were mere objects is a way of relating to them from a certain perspective and can lead to their being known only in part'. One example of normally overlooked experiences of depression is the slowed subjective flow of time in patients with depression[80].

As it stands, the current operational concept of depression is broad enough to include most of the depressions, yet it does not adequately capture the subjective quality of some of the criteria for depression (such as depressed mood, as it differs from normal sadness). Qualitative approaches using adjective checklists, such as the Depression Adjective Check List (DACL), as a way to investigate them have been used[81]. Subjective well-being would be the opposite of low mood. Coyne[82], studying college students, suggested that distress may also be related to depression and not just to anxiety, although distress is empirically and conceptually different from depression in DSM-III-R. As an example of the well-researched influences of personality and culture on subjective well-being, the reader is referred to the excellent review by Diener *et al.*[83], which illustrates the complexity of mood and the need to use holistic approaches to study it. Cultural and other environmental influences may also offer an explanation for the increased rate of depression in industrialized societies.

To add weight to the complexity of the experience of depression and its modulation, we might mention the impact of religion on the behavior and outcome of depressed patients. Dervic *et al.*[84] found that depressed patients with religious affiliation had less frequency of completed and attempted suicide, both in themselves and in their families. Some authors also suggest that the emotional regulation techniques of traditional religions have a positive impact on mood[85,86]. The search for meaning within depression and the experiences associated with it can be justified from the perspective of dynamic psychotherapy[87,88] as well as from the interface between philosophy and psychology[89]. Von Morstein[89] proposed that depression could be understood as a failure to address basic philosophical questions satisfactorily, particularly in response to alienating experiences (*grenszituation*). This is just one amongst many examples of the profound impact of depression on the human being for which the scientific model may not be fully equipped. A phenomenological account of depression without a brief mention of these boundaries would be incomplete.

EMOTIONS AND MOODS

Emotions can be defined as phasic changes in the subject involving distinct, integrated psychophysiological response systems[90]. Any given emotion may involve a number of differentiable response systems including a prototypic form of expression, typically facial, which clinically is characterized as the affect, a pattern of autonomic changes, a distinct subjective state involving pleasure or displeasure (the emotion proper, or feeling), and an overt behavior[90,91]. Emotions are usually a response to events in the environment, to thought content, or to homeostatic or other bodily changes.

Moods, on the other hand, are tonic, or background, subjective states on which emotions are superimposed. As opposed to the fleeting nature of neurovegetative changes that participate in emotions, moods involve circadian or ultradian changes in function (such as sleep architecture, feeding patterns, or reproductive behavior and function). It has been noted that 'every emotion tends to prolong itself into a mood', and conversely 'every emotion tends to entail a mood change'[92–94]. The tonic state imposed by moods would then operate as a threshold for triggering a particular set of emotions[92,95–97]. Accordingly, DSM-IV defines mood as 'a pervasive and sustained emotion that colours the perception of the world'. It is a global and prolonged prevailing state or disposition and it influences both psychological and somatic experiences.

Table 1.2 points out some of the differences between mood and affect, and Table 1.3 shows differences between feelings and mood.

The overt behavior associated with emotional states defines the affect, a term used to describe differentiated specific feelings directed towards objects (we understand as object the target of feelings, usually a person, although it can be a thing or an idea). The type or quality of affect is the predominant emotion expressed. Happiness, surprise, interest, sadness, shame, disgust, fear/anxiety, anger and contentment are quoted as the principal types. Other parameters, such as range, intensity, stability and congruence are necessary to render a complete description of affect.

Classical theories of emotion attributed the experience of emotions to bodily changes, but this view has recently come into question[98]. Modern emotion research links emotions with cognitive factors[99,100]. From this perspective the cognitive interpretation of the physiological change is what constitutes the experience of emotion. There is evidence that cognitions affect physiological responses to stress[101], particularly in survival-threatening situations, but

Table 1.2 Differences between mood and affect

	Mood	*Affect*
Emotional/feeling state	Internal	Visible, external
Objectivity/subjectivity	Subjective	Objective
Temporal dimension	Longitudinal (steady along time)	Transversal (at a given moment, fluctuating)

Table 1.3 Differences between mood and feelings (from reference 94, with permission)

	Mood	*Feelings*
Duration	Sustained, long	Short
Course	More or less permanent	Unstable
Relationship to life events and environment	Autonomous	Linked to life events and circumstances
Starting and ending	Without apparent reason	Response to life events
Tone	Deep	More 'superficial' or 'higher'
Psychopathology	No direction Directed towards self No intentionality Basic and primary experience	Generally directed More social Usually teleology evident Mediated by psychological processes
Co-existence	(of different moods) rare – only mixed and rapid cycling states	(of different feelings) norm

the impact of cognition has not been studied in the social context in which most emotions normally occur[102].

Whereas cognitive appraisal is paramount to understand the genesis and nature of emotions, emotions themselves exert a powerful influence on cognition. Emotional status affects attention, perception, memory, learning, decision-making and sociability[95]. On the other hand, emotional intensity is regulated by selective attention[103].

The complex and somewhat overinclusive nature of the concept of emotion is manifested by a classification of feelings, exposed systematically by Weitbrecht[104], who proposed a hierarchy at the bottom of which lie feelings related to bodily experiences, and topped by those associated with spiritual notions (Figure 1.1).

Normal and pathological sadness

Sadness is a common emotion, often triggered by unpleasant or adverse conditions. The core issue behind sadness is loss. While normal sadness is the expression of a feeling, pathological sadness is related to mood. The differential diagnosis between both of them is not always easy. Historically pathological sadness has been known as melancholia or 'vital' sadness. Pathological sadness may, and often does, appear without an identifiable trigger. Alternatively, pathological sadness may have a precipitating factor but the disproportion (either in intensity or duration) between the cause and the sadness is obvious. Pathological sadness may also be qualitatively different from normal sadness, and may be perceived by the sufferer with a distinct flavor commonly not fully understood by others.

According to the DSM-IV, a major depressive episode that occurs in response to a psychosocial stressor is distinguished from adjustment disorder with depressed mood simply by a categorical count of symptoms, which do not reach the specified threshold in the latter. On

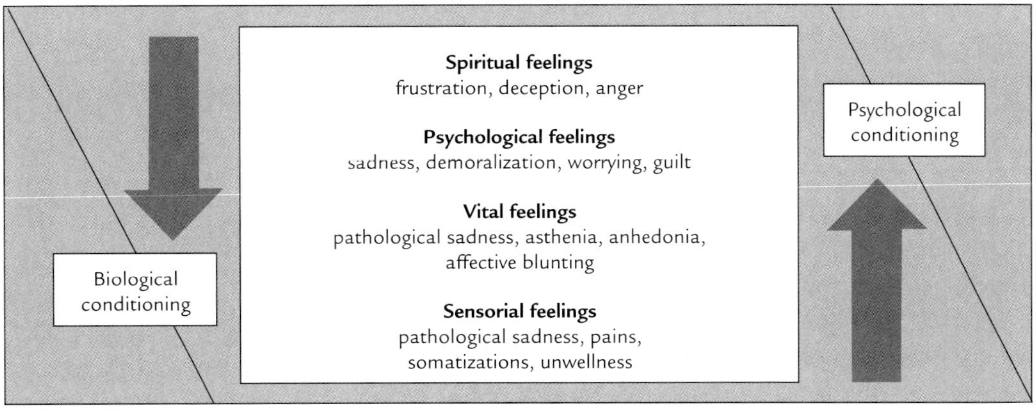

Figure 1.1 Schneider's hierarchy of 'feelings' (modified from reference 94, with permission)

the other hand, if the precipitating factor is the loss of a loved one, even if depressive symptoms are of sufficient duration and number to reach the threshold for a major depressive episode, they are considered normal and attributed to bereavement unless they persist for more than 2 months, cause marked functional impairment, or include severe symptoms such as morbid preoccupation with worthlessness, suicidal ideation, psychotic symptoms, or psychomotor retardation. A detailed account of the varieties of grief experience from a phenomenological perspective has been given by Bonanno and Kaltman[105].

Mood in depression

Depressed mood is described in a variety of negative ways, but always conveys a subjective experience of misery. It may or may not improve substantially in circumstances when ordinary sadness would be alleviated, but it is important to insist that the patient suffering from depression experiences his mood as different from ordinary sadness, either in degree or in quality[106]. Sadness is frequently accompanied by irritability or anxiety; the former may be the prevalent mood in children and adolescents, as well as in adult males.

The clinical range of mood in depression is wide. There are patients with a pervasive sadness that permeates every aspect of their lives and precludes them from any enjoyment, a state called anhedonia. In others, sadness may be less prominent, reactivity is still present and anxiety may be the most noticeable symptom; they may have difficulties seeking pleasure, but can still experience it in some favorable settings.

Cognitive impairments of depression

Although generalized impairment in most cognitive domains can be seen in acute depression, selective attention appears to be prominently impaired, and may predict response to treatment, remission of symptoms and risk of relapse[107].

Depressed patients show deficits in automatic attentional processing, memory scanning and memory span, which altogether suggest a reduced speed of information processing in automatic subtasks[108].

Airaksinen *et al.*[109] found that the type of cognitive impairment varied according to the depression subtype, so that patients with major depression and mixed anxiety–depressive disorder exhibited significant memory dysfunction, whereas individuals with dysthymia showed pronounced difficulties in mental flexibility. Minor depression did not affect cognitive performance. Verbal fluency and perceptual–motor speed were not affected by depression. This indicates that specific types of depression may carry specific types of neuropsychological dysfunction. However, the interpretation of experimental data is limited because of the paucity of normative data on the cognitive aspects of depression[110].

The available neurobiological evidence suggests that there is a different pattern of neural response for emotion than for mood, and another one for thought. Although traditionally emotions have been seen as contrary to thought, certain emotional processes can influence complex perceptions and cognition, including interpersonal judgements. This regulation of interpersonal interaction by emotions has been proposed as the basis for empathy, a quality often lost in depression, giving rise to a dysregulation leading to anger and aggression[111]. The difficulty in proving such a line of causation from thought to emotion or vice versa has been acknowledged by different authors[112]. The concept of 'core affect' has been proposed as a framework for such an explanatory model[113]. Core affect would be the combination of mood and emotion at any given time. Whereas it is easy to see, from the previous discussion, that core affect would have an effect on cognition, it is not easy to see what heuristic advantage the concept has over the analytical distinction proposed by classic phenomenology.

Beck's cognitive model of depression

Aaron Beck proposed a simple model to explain the relationship of mood, emotions and thought in a clinical context[114–116]. According to this model, feelings do not occur in a vacuum, but instead follow a perception, image, interpretation, or attribution. In other words, feelings follow from the thoughts or cognitions we have about our situation. The model assumes that the individual's consciousness is rooted in his cognitions. These cognitions are the result of a synthesis of internal and external stimuli that reflect the conception of the individual about his self, his past, his future and his surrounding environment. Emotions, mood and behavior are determined by conscious thoughts[117]. Hence, modification of the thoughts will influence mood and behavior.

Beck proposed that the core of depression is the presence of cognitive negative distortions, whether related to the self, to the present or to the future. This negative attributional habit is ingrained in the schema, a cognitive general representation and interpretational approach to the world which selectively filters perceptual input. The cognitive behavioral general model in depression is illustrated in Figure 1.2.

Cognitive distortions give rise to negative thoughts. Amongst the most frequently found cognitive distortions are: arbitrary inference (drawing a conclusion lacking evidence to support it); selective abstraction (focusing on particular aspects of a given situation with obvious ignorance of other aspects or 'the big picture'); overgeneralization (drawing a conclusion based on a single fact or incident); or personalization (relating negative external events to oneself instead of to external circumstances). According to Beck, negative core beliefs (or dysfunctional beliefs) are particular types of cognition preceding depression and

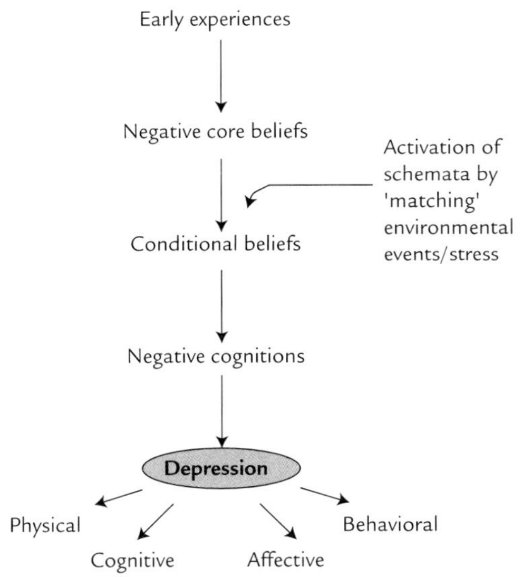

Figure 1.2 General model in cognitive behavior therapy

predisposing to it. Negative cognition or negative automatic thoughts are conditioned by the cognitive distortions and can be defined as the immediate evaluation a person makes of his situation. There is a continuous interplay within this model summarized in Figure 1.3: symptoms arise from distorted thinking (cognition) and this distorted thinking impacts on mood, behavior ('the way in which one thinks about a situation or attributes meaning leads to a set behavior or affect') and bodily function. In turn, all these four aspects influence each other.

There are some major advantages to the cognitive behavior therapy (CBT) model. First, it allows bridging between subjective (psychological issues) and objective (behavioral) issues. Second, it is internally consistent and validated by clinical data[118]. Third, it is consistent with experimental data; in fact, there are cognitive biases in recall of negative information in depression[119]. Fourth, it has proven therapeutic efficacy in depression[120–123]. Adding evidence in favor of CBT, Goldapple *et al.*[124] showed, using positron emission tomography (PET) that CBT modulated specific areas in the frontolimbic network (known to be involved in depression), which are different from those modulated by paroxetine.

There are authors[125] who have questioned the efficacy of CBT because of the heterogeneity of groups in whom it has been studied and the difficulties in providing adequate control therapies. Peselow *et al.*[126] indicated the difficulties in proving the negative core beliefs during clinical recovery, since they tend to remit. Some researchers[127] have also reviewed the weaknesses of the cognitive vulnerability model and have suggested that theory-guided high-risk experiments be designed to address this thorny issue. The ultimate difficulty in proving the cognitive model of depression lies in the inherent difficulty of elucidating whether these cognitive biases are a consequence or a cause of depression. Some critics[128] have pointed out that 'longitudinal assessment of originally euthymic subjects appears the best method to investigate any cognitive risk to depression and the significance of diathesis–stress models' as the best way to overcome this problem. The question is whether these cognitive biases are an epiphenomenon of depression or rather a cause of it.

We would like to highlight that CBT has deeply influenced our understanding of the phenomenology of depression. Some of the phenomenological instruments used to measure depression, in particular the Beck's Depression Inventory (BDI), are derived from the cognitive–behavioral conceptual model of depression.

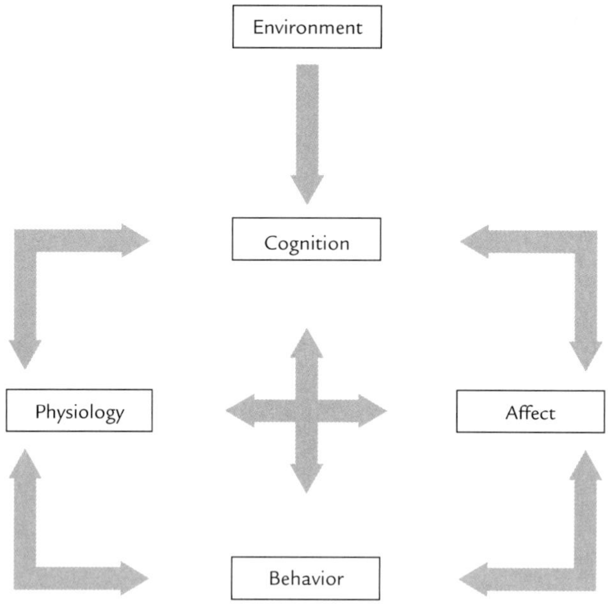

Figure 1.3 Interactions within the cognitive behavior therapy (CBT) model

CATEGORIES AND DIMENSIONS (IS DEPRESSION A DISEASE?)

We have reviewed pertinent aspects of the concept of depression. The rest of the chapter is devoted to some specific problems and issues related to the phenomenology of depression. We start this second part of this chapter by reviewing the categorical and dimensional approaches to depression, an area with important repercussions in phenomenology.

Mental illness description and diagnosis can be approached from either a categorical or a dimensional perspective. In the categorical model there are separate and mutually exclusive groups of subjects within each classificatory class or group. When applied to diagnostic subtypes of depression the categorical view emphasizes the clustering of symptoms in characteristic combinations or complexes. These symptom complexes will in turn define a class. Alternatively, description may be assumed to follow a continuum along, for example, dimensions of severity or duration. In this case symptom complexes can co-exist in the same person. One of the difficulties with the categorical approach used in the most widely used classifications of mental illness, DSM and ICD, is that with each successive version the number of categories has increased without gain in the specificity or validity of each class. In addition, Nathan and Langenbucher[129] said that 'as diagnoses proliferated, the frequency of co-morbidity increased, causing clinicians to ask whether co-morbidity represents the co-occurrence of two or more mental disorders or a single disorder that has simply been labelled in different ways'.

Haslam[130] reviewed the whole spectrum of psychiatric disorders using Meehl's 'taxometrics'.

This procedure allows a mathematical model to be created that is inclined towards neither categorical nor dimensional approaches but is able to validate a latent class if adequate statistical power is present. With the use of this model, melancholia favors a categorical approach but a dimensional model tends to be favored for depression in general. Strictly speaking, the dimensional approach produces a much more accurate description of reality (such as response to treatment or functional outcome), particularly when the time dimension is taken into account[131]. However, there are advantages to the categorical approach, such as its easier handling for statistics, epidemiology, diagnostic and treatment algorithms and health policy. The weakness of the categorical approach is in the validity of the construct.

We have summarized the main differences between these two approaches in Table 1.4.

When multivariate statistical techniques are used, depression symptom severity measures such as the Hamilton Depression Rating Scale (HDRS, also HAM-D) tend to have a relatively reliable factor structure, although the specific factor architecture may not be totally stable in different studies[132]. Factors extracted from the transformation of raw scale data are continuous variables representing underlying dimensions that need to be interpreted at face value. For example, a factor that receives contributions from the variance in endorsing questions about the lack of enjoyment of food, social contacts and other pleasurable activities would be consistent with a dimension representing normal hedonic pleasure to anhedonia. These transformed variables, or factors, tend to follow normal distributions and can be treated with parametric tools. Raw data may or may not follow parametric distributions, but when they do they may correlate with categorical classifications; for example, in some studies those depression subtypes associated with lower scores in the BDI are the non-recurrent, seasonal and atypical ones[133].

Table 1.4 Differences between the categorical and dimensional models (modified from reference 94, with permission)

	Categorical	*Dimensional*
Type	Qualitative	Quantitative
Statistics	Cluster analysis	Factor analysis
Information processing	Dichotomous	Quantitative
Classifying procedure	Ideal types (prototypes)	Natural grouping of symptoms
Classifying approach	Exclusive	Complementary
Diagnostic categories	Dependent	Independent
Associated problems	Simple Elevated co-morbidity Atypical cases Mixed forms	Complex No standard instruments
Inspiring philosophy	Platonic	Aristotelian

Figure 1.4 provides a general algorithm from published data[48,134], which suggests selections of statistical tools of classification based on the nature of the raw data. The specific test to be chosen also depends on the availability of normal (parametric) data, the size of the sample and the question that needs to be addressed. The greatest virtue of multivariate approaches is that they allow the use of richer phenomenological sets and, depending on the technique chosen, they may permit atheoretical classification of cases into what are assumed to be naturally occurring groups. On the other hand, when small samples are used, factor analysis may yield unstable solutions rendering face validity inadequate, as was rightly pointed out by Streiner[135].

Problems with the nature of data (for example, an excessive number of individual variables compared with the sample size, or non-parametric distributions) can sometimes be addressed by mathematical treatment of the raw data. A number of data reduction strategies have been used in published reports, but a detailed discussion is beyond the scope of this chapter. It is important to note, however, that regardless of the particular techniques used to analyze the source data, a general consensus in the literature indicates that there is a relatively reliable factor structure in data obtained from severity scales of depression in community samples. On the other hand, the natural segregation of cases in subcategories, or latent structure, of community samples of patients with depression remains an area of open research. Part of the problem may arise from the nature of the symptoms, which may occur within a continuum that includes normal behavior. There is a growing

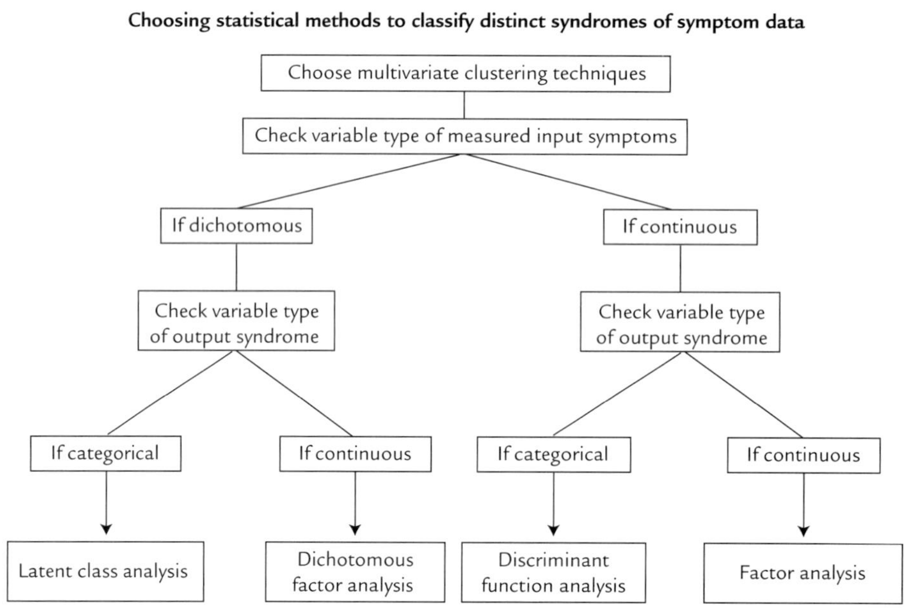

Figure 1.4 Algorithm of statistical techniques to deal with phenomenological approaches to depression (from Chen, unpublished, with permission)

body of evidence that supports this statement in non-melancholic depression[48,59,136], whereas melancholia, the psychotic disorders and the eating disorders appear to have sharper separations from other disorders and from normal controls, at least in statistical terms[130]. It is possible, then, that a dimensional (continuous) conceptualization of the depressive traits 'as part of an overarching model of personality structure, rather than a discreet diagnostic entity' may be most appropriate[137].

There are certain types of specific problem, such as the overlap between depressive personality disorder and dysthymia, which can be solved only through a dimensional approach. Another problem that warrants further research is the continuity between dysthymia and subthreshold depression. While Ryder *et al.*[137] did not find sufficient evidence in their review to make the depressive personality disorder a category with operationalized criteria, they stressed that a major differential diagnosis key between this condition and dysthymia is the absence of somatic and vegetative features in the former and their presence in the latter.

Depressive symptomatology is over-represented in a variety of neurological disorders, including Parkinson's disease, stroke, multiple sclerosis and epilepsy. Research in depression in these patient populations will be significantly affected by the selected conceptual framework. Indeed, if depressive symptomatology occurs along a continuum that reflects underlying pathological processes, then there is a risk and possibly an error in the dichotomization of samples between 'depressed' and 'non-depressed' patients at an arbitrary level of symptom severity or duration. Alternatively, it would be more appropriate to use factorially derived dimensions from instruments such as the HAM-D or the BDI to reflect the impact of underlying pathology on mood, and to evaluate more precisely the treatment impact in clinical trials[138].

For an extensive review of different data 'slimming' techniques, the reader is referred to Bryant and Yarnold[139]. For a comprehensive review of the BDI, the most widely used instrument worldwide for depression, see Richter *et al.*[140].

In Figure 1.5 we provide an algorithm to deal with the statistical approaches to explore the validity of a diagnosis and its utility.

A dimensional approach to depression may help to refine neurobiological and genetic aspects of depression. A good example of this is the identification of those dimensions of depression that seem to be transmitted within families regardless of the presence or absence of threshold criteria for categorical diagnosis. Korszun *et al.*[141] found four factors of depression (mood symptoms and psychomotor retardation; anxiety; psychomotor agitation; guilt, and suicidality) with positive family correlation in the largest study of this kind conducted to date. Interestingly, other aspects, such as sleep biological markers, were not found to have this positive correlation.

Similarly, some authors are intelligently linking clinical symptoms of depression not just to specific areas of the brain, as measured by functional neuroimaging, but also to the circuit of relevant areas to depression treated as a whole[142,143]. It has recently been proposed that a similar approach might be used to answer questions posed by specific types of depression associated with neurological disorders[144]. Indeed, a dimensional phenomenological approach combined with neuroimaging findings may throw light upon questions such as whether depression in epilepsy is different from depressions in community samples, while

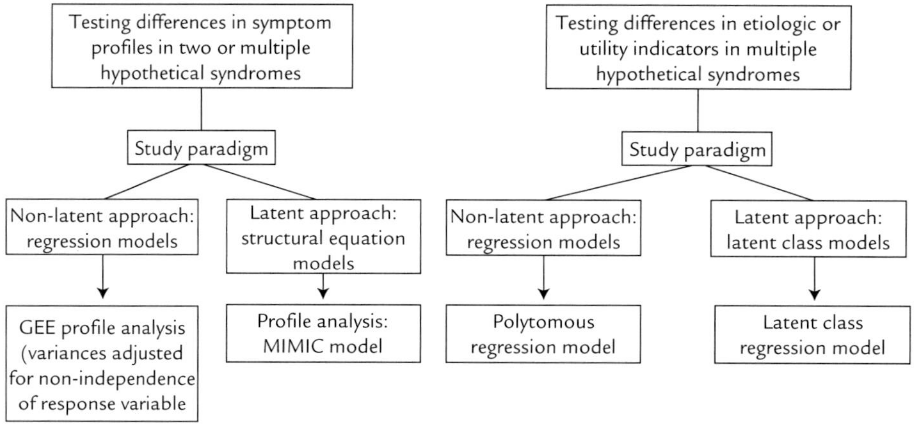

Figure 1.5 Statistical methods to examine diagnostic validity or utility. MIMIC, multiple indicator, multiple causes (from Chen, unpublished data, with permission)

adding an extra layer of evidence by linking phenomenology to specific brain networks.

CLASSIFICATION OF DEPRESSION (WHERE IS THE SAFE GROUND?)

There have been a number of conceptual models for the classification of depression throughout history. An attempt to instill some order in the treacherous waters of categorical classification has been made by Peralta and Cuesta[94], using a clinically eclectic approach (Table 1.5).

This approach, however useful, gives rise to a number of criticisms, some of them harsh[145]. While, in the last quarter of the 20th century, a unitarian approach to depression with attention to its severity was clearly prevalent, some of the old controversy between dualistic approaches (e.g. endogenous versus reactive) remained. In the next few paragraphs those subtypes of depression with more empirical support are briefly reviewed.

Melancholic depression seems a separable clinical entity[146–149]. Parker[145] has suggested three separate subtypes of depression: psychotic, melancholic and non-melancholic depression (Figure 1.6). The main characteristics of the melancholic subtype of depression are the very prominent psychomotor disturbance – such as feeling slowed down – consummatory anhedonia and non-reactive mood[150]. This type has been found to respond well to antidepressants and electroconvulsive therapy (ECT).

While Paykel[146] based his conclusions about melancholia on the results of cluster analysis, Parker[145] incorporated neuroendocrinological markers (direct correlation with dexamethasone non-suppression) into the criteria for classification and extrapolated his findings into an animal model of the melancholic type of depression in primates using a lesion of the basal ganglia–prefrontal circuit. This author advocates a dimensional approach of the more heterogeneous non-melancholic group. The psychotic subtype has also been validated in other samples[151].

Table 1.5 Categorical classification of depressions (from reference 94, with permission)

1. According to etiology
    a. Primary
        i. Endogenous
            1. with hereditary factors
            2. without hereditary factors
        ii. Reactive (to psychosocial stressors)
    b. Secondary (to biological or medical factors)
2. According to the presence of certain specific factors
    a. Seasonal
    b. Hormonal
        i. Catamenial
        ii. Postpartum
        iii. Postmenopausal
3. According to the presence of melancholic syndrome
    a. Melancholic
    b. Non-melancholic
        i. Neurotic (anxious)
        ii. Dysthymic
        iii. Atypical
4. According to clinical profile
    a. Simple melancholy
        i. Agitated depression
        ii. Inhibited (stuporous) depression
        iii. Obsessive depression
    b. Vital depression
        i. Anergic
        ii. Vegetative (masked)
    c. Depression with depersonalization symptoms
    d. Pyschotic depression
5. According to its course
    a. Recurrent depressive disorder (monopolar depression)
    b. Bipolar depression (includes cyclothymia)
    c. Chronic (dysthymia)
    d. Treatment-refractory depression
6. According to the severity
    a. Mild
    b. Moderate
    c. Severe
        i. With psychotic symptoms
        ii. Without psychotic symptoms
7. According to age of onset
    a. Child depression
    b. Adult onset
    c. Elderly depression

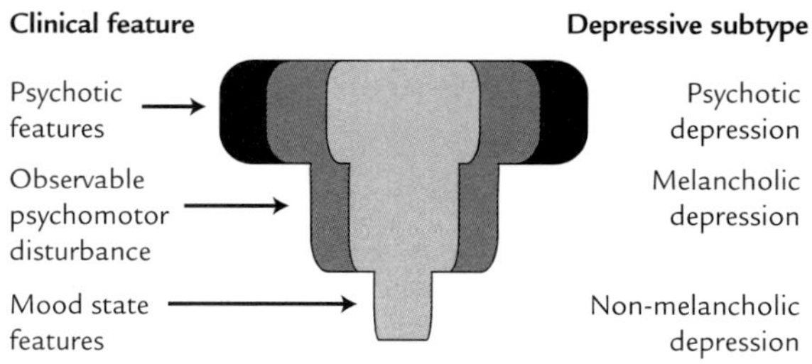

Figure 1.6 Relationships between empirically validated forms of depression (from reference 145, with permission)

Melancholia is a concept worth further explanation. In the second half of the 20th century melancholia was associated with endogenicity, i.e. a form of depression more independent of external events than other types of depression, although this has been challenged by some[149]. Melancholia has a stronger heritability and tends to be more severe than non-melancholic depressions[152,153]. There is ample evidence that melancholic depression responds better to tricyclic antidepressants than to selective serotonin reuptake inhibitors (SSRIs)[154], while the opposite may be true for non-melancholic depression[155].

Atypical depression is a syndrome characterized by mood reactivity to positive events (DSM-IV), in association with increased appetite or weight gain, hypersomnia, leaden paralysis (extreme fatigue and heaviness in the limbs), and a longstanding pattern of extreme sensitivity to perceived interpersonal rejection. Some authors add pronounced anxiety[156] and yet others consider atypical depression as part of the anxiety spectrum[157,158]. Liebowitz[156] also insisted on the fluctuation of the level of depression in atypical depression. Atypical depression seems to respond better to monoamine oxidase inhibitors (MAOIs) than to other drugs[159].

Coryell *et al.*[160] followed a large number of patients diagnosed with depressive disorder. They studied the stability of some subtypes of depression and found that patients with psychotic, agitated/retarded, or endogenous subtypes of the disorder in a given episode would manifest the same subtype in subsequent episodes. The greatest stability was for the psychotic subtype. On the other hand, some authors have questioned the stability of melancholic syndrome[161].

In other large epidemiological data sets, segregation of subcategories does not occur in a consistent fashion. Sullivan *et al.*[162,163] studied a large sample of female twins using DSM-IV criteria (categorical approach) and latent class analysis (LCA) to conclude that there was evidence of a typical MDD as defined by the DSM-IV, but also of atypical MDD and 'minor' (or subthreshold) depression. Using a prospective longitudinal phenomenological approach (explained later), with 1920 adults from the Baltimore Epidemiological Catchment Area (ECA) study, Chen *et al.*[134] concluded after applying LCA that there are four patterns of depressive phenomenology: anhedonic, suicidal, psychomotor and severely depressed subtypes. The validity of these classes is further confirmed by their usefulness in predicting the course of illness. The anhedonic type had a later onset, fewer recurrent episodes and stronger family association. These authors postulated that this subtype may be related to vascular depression[145]. The psychomotor subtype identified with LCA is similar to the melancholic type identified in other studies already cited.

Genetics may also shed light on depression subtypes. Willeit *et al.*[164] found that the expression of the serotonin transporter promoter gene was related to the type of depression: long allele for the melancholic depression and atypical for the short allele. Not only genetics, but also neuroimaging offers increased possibilities to understand the classification of depression. Fountoulakis *et al.*[165] found different distinctive patterns in regional cerebral blood flow using single photon emission computed tomography (SPECT) in atypical depressions as compared to the melancholic and non-melancholic ones.

The studies mentioned in this section all support the concept of depression as a heterogeneous condition with at least two major subclasses: melancholic and non-melancholic.

Anxiety and depression are closely related. This relationship between the concepts and diagnosis of anxiety and depression is succinctly reviewed by Nathan and Langenbucher[129]. Anhedonia, cognitive distortions related to failure and loss, and dysphoric mood may all be more specific of depression[166,167]. Nonetheless, a clinical entity combining both symptom clusters (anxious depression) appears to be empirically supported[168].

THE PROBLEM OF SUBTHRESHOLD DEPRESSION

The problem of setting an arbitrary threshold of severity of depressive symptoms has received some attention from the literature, particularly focusing on those patients who have some degree of functional impairment but do not reach the predetermined level of diagnosis. These patients are referred to as experiencing subthreshold depression, subsyndromal depression, minor depression, analog depression or a depressive equivalent, although there is a degree of controversy related to the boundaries of these terms. The subthreshold depression would be perhaps more accurately referred to as subthreshold depressive symptoms, i.e. the existence of depressive symptoms not reaching criteria for major depressive disorder.

There is a growing consensus[56–58] supporting the continuity between subthreshold depression symptoms and syndromal depression (MDD). Enns *et al.*[56] have also suggested that subthreshold depression may be associated with personality traits, more specifically neuroticism, and a history of parental overcontrol in childhood. Solomon *et al.*[59] favor a continuum view using a taxometric approach, because subjects with subthreshold depression have an elevated risk of future episodes of MDD, increased frequency of family history of affective disorders or shortened REM latency. Akiskal *et al.*[60] found a similar polysomnographic pattern in subthreshold depression and in clinical depression. Moreover, it allowed them to differentiate personality/character 'depressions' and anxiety conditions from subthreshold/clinical depression. It seems, though, that people suffering from subthreshold depression are not so prone to develop anxiety as those suffering from depressive disorder[61]. There is some controversy about whether there are fewer neurovegetative symptoms in subthreshold depression[58] or about the same as in MDD[60].

The likelihood of progression from subthreshold depression into major depressive disorder, or the partial relapse of depressive disorder into subthreshold depression, is well known[58,64]. These last authors conducted a systematic review of prospective studies of subthreshold depression to find an increased incidence of MDD amongst the sufferers from subthreshold depression. There is, however, a large variation in the incidence rates of MDD in subthreshold depression. An increased incidence of major depressive disorder in high-risk groups with subthreshold depression, such as those with other risk factors, such as family history, has been reported[63]. Given the high prevalence of depression in most neurological disorders, subthreshold depression in brain dysfunction should be considered a high risk. Furthermore, pre-existing depressive symptoms may either be a herald or a risk factor for conditions such as Parkinson's disease (see Chapter 9) or stroke[65]. Some authors advocate that recovery from depression should only be assumed if there is return to asymptomatic status, but not if there is subthreshold depression[67]. Others[66] have conducted large prospective studies and found that the chronicity of MDD is often associated with the presence of subthreshold depression. Subthreshold depression may be of extreme interest when it appears in adolescence, since the risk of suicidal behaviors, depression and anxiety is markedly increased in adult life[68]. This last point is relevant to those types of brain dysfunction with teen onset.

Furthermore, there are no major differences in health outcomes and quality of life measures, particularly those related to socioeconomic burden, between subthreshold depression and MDD[69]. Subthreshold depression is an important public health problem due to its high prevalence and psychosocial impairment[70]. It can therefore be considered a very impairing condition affecting areas such as household strain, social irritability and financial strain. Increased disability and over-representation of welfare benefits in subthreshold depression has been reported[71]. As social dysfunction and disability are high in subthreshold depression, this should warrant attention to studying subsyndromal depression in brain dysfunction. Minor depression is quite a persistent and stable condition[58,72]. Subthreshold depression is a treatable condition and some studies using SSRIs have demonstrated improvement in clinical and health outcomes[73].

Interestingly, subthreshold depressive symptoms are extremely frequent in patients with neurological disorders. For example, Wiegartz *et al.*[75] found that 25% of the patients with epilepsy suffered from minor depression. Co-morbidity of subthreshold depression with anxiety is also noticeable, although it is less than at diagnostic level[62]. Subthreshold depression co-morbidity can only be properly researched using a dimensional model.

Solomon *et al.*[59], however, point out some methodological inconsistencies across studies of subthreshold depression. For these authors there are problems with the gender distribution of this category and the fact that there might be qualitatively distinct subtypes of depression. Enns *et al.*[56] cautioned about the personality and developmental variables that could represent important areas of discontinuity between clinical and subthreshold depression. In a systematic review of prospective studies Cuijpers and Smit[64] found that the concept of subthreshold depression was too broad and a consensus about its definition is needed.

Although this remark may exceed the scope of a chapter about the phenomenology of depression, it is worth noting that subthreshold depression can be treated with relatively minimal investment of time and resources, as a recent randomized trial has shown[74].

LONGITUDINAL PHENOMENOLOGY OF DEPRESSION

The concept of longitudinal phenomenology emerges from considering the time dimension of the disorder. The definition of a phenotype, e.g. depression in refractory epilepsy, is not complete until the natural history (time dimension) has been considered, including age of onset, recurrence (chronicity) and duration of episodes. Analysis of such data has served to clarify the heterogeneity of depression in previous studies[48], highlighting the bimodality of its onset. In a well-designed prospective study using ECA data these authors reported the difference in cumulative incidence between men and women. When the weighted cumulative incidence was calculated, that in women in the 18–29-year age group was nearly triple that of the men of the same age. In the next age segment (30–44 years) the difference persisted two-fold but, interestingly, there were no significant differences in cumulative incidence from 45 years onwards. This indicates that the highest risk period for men is the young adult and middle years, while for women it is at a younger age. The peak incidence is about 30 years for women and 40 for men, with a second peak at approximately 55 in both sexes. These authors could not find any statistically significant effect of marital status or years of educational attainment upon the incidence of depression.

An interesting finding related to subthreshold symptoms indicates that these can be present for an average of 1 year in the form of a prodrome to a major depressive episode[48]. Dysphoria was the earliest herald of a major depressive disorder in this study (median 5 years) and interestingly suicidal thoughts and behavior could antedate the episode of MDD by 2 years.

Jackson *et al.*[169] published a systematic review of available longitudinal data and reported that early symptoms of depression were inconsistently present, yet mood change followed by psychomotor symptoms were the subjective experiences most frequently recalled. Interestingly, most residual symptoms of depression also occur in the prodromal phase and they are associated with neuroendocrinological and polysomnographic correlates[170]. Accordingly, residual symptoms have been interpreted as prodromal to further episodes of disease[170].

Fewer than half of the lifetime cases will ever experience a recurrence; the duration of episodes appears to be shorter with repeated instances of disease, but is not affected by marital status or educational level. Likewise, the interval between episodes appears to shorten with successive recurrences. Recovery rate 5 years after an episode of MDD approximates 80%. The recurrence rate is similar for men and women, as are the episode durations and rates of recovery. In the study by Eaton *et al.*, the duration of episodes was shorter in the later than the first episodes by approximately 11 weeks[48]. In another study[171], the mean episode duration was 16 weeks, with severe or very severe role impairment in nearly 60% of those suffering from depression.

Angst *et al.*[172] have analyzed the recurrence of major depression in samples of hospitalized patients for over 20 years. The data generated on the impact of depression showed that up to 20% of the adult lifetime of those suffering from recurrent MDD is spent in illness. Judd *et al.*[66] had previously reported that patients with MDD at successive follow-up exhibited active symptoms 42% of the time and, conversely, that more than half of the 12 years following their initial episode was spent with some degree of affective symptoms, whether in obvious illness or below the severity cut-off point for diagnosis, yet with unfavorable outcomes. Similar findings, with slightly higher rates of recurrence and chronicity, were reported by Uebelacker *et al.*[173].

Chen *et al.*[174] looked at the Baltimore Epidemiological Catchment Area Study data set and subsequent follow-up data up to 13 years, to address the heterogeneity of depressive syndromes from a phenomenological point of view. They focused on MDD, subthreshold depression, dysthymia and double depression (dysthymia with superimposed MDD). They found that transitions amongst these categories were frequent. Indeed, using dimensional measures they found that some showed some homogeneity amongst these categories – using a dimensional approach – and added evidence to the continuum model of subthreshold depression and MDD, although a subgroup with subthreshold depression did not progress to MDD.

There are other important conclusions from this study worth mentioning. Family history was associated with all of them, although somewhat less with dysthymia, but life events were more closely associated with subthreshold depression. The double depression was strongly associated with female gender, more than with MDD. In fact, contrary to other studies, female gender was not found to be a significant risk factor for MDD. The greater severity of the depression also seemed to be associated with female gender. There seemed to be a heritability of the subthreshold depression that warrants further investigation. The outcomes of subthreshold depression were found to be as bad, if not worse, than those of MDD. The highest levels of impairment were achieved by those suffering from double depression. The age of onset was different for these four categories.

This study also highlights that it is only through longitudinal phenomenology that a refined epidemiology and study of health outcomes can take place. We have previously mentioned the importance of deep databases in relation to depression in brain dysfunction. Longitudinal phenomenology, including time dimension, onset, duration of episodes, recovery, recurrence and life chart, play a key role.

Longitudinal phenomenology has to take into account life events and life stresses. Ideally, these types of data should be included in research and be part of the same database where clinical data or biological elements, such as neuroimaging or genetics, are kept. In an excellent review of the effects of life stresses upon affective disorders, mostly focused on controlled studies, Paykel[175] found that life stresses showed only moderate specificity. Relapse and remission of depression were also impacted by these issues, including social successes. Life events, however, did not seem to have the same impact on melancholic as on non-melancholic depression. Paykel also pointed out that a drawback of many of these studies was their retrospective nature. He stated that life events were relevant both in the onset of new depression and in subsequent episodes. Other authors[176,177] have found that the relative weight of life events is higher in the new-onset depressions than in relapses, which is in tune with other studies of enhanced sensitization in subsequent episodes of depression[178].

Chen *et al.*[134] found that stressful life events were associated with mild and moderate depression, but not with severe MDD. Female gender, as in the previous study, was associated with severe MDD, but not with mild or moderate cases. In this study they could not find a 'natural cut-off' point for the number of symptoms of depression, but rather a continuous spectrum of intensity or severity of depression.

One of the most promising areas of longitudinal phenomenology is the integration of this discipline with genetics. In a remarkable article by Caspi *et al.*[179], these authors investigated prospectively and longitudinally the effects of life events on a birth cohort. They found that those individuals with one (s/l) or two copies (s/s) of the short allele of the serotonin transporter gene were much more prone to develop major depression and to attempt suicide than those homozygous for the long allele (l/l), as is illustrated in Figure 1.7. This study has served to refine the diathesis-stress models of depression in a proven gene–environment interaction.

While Caspi *et al.*[179] used subjects below the age of 30, other authors have tried to replicate this study in a larger sample with a wider age range[180]. They were unable to replicate Caspi's gene–environment interaction findings, although they did find a significant impact of life events on depression.

A holistic longitudinal phenomenology should also include social aspects often not sufficiently considered, such as family disruption and residential stability in childhood[181], occupational stress[182] or caring partners[183]. This last issue is particularly relevant in neurological conditions.

A caveat is in order regarding life events. One should not necessarily assume life events are *per se* the cause or precipitants of depression. It is often the case that it is the consequences of depression that precipitate life events. Therefore, careful timing should be recorded; this is, together with recall bias and forgetting, the major drawback of retrospective studies.

There seems to be a paucity of data in the studies of co-morbidity of depression in neurological disorders, although it is now widely accepted that co-morbidity should be included

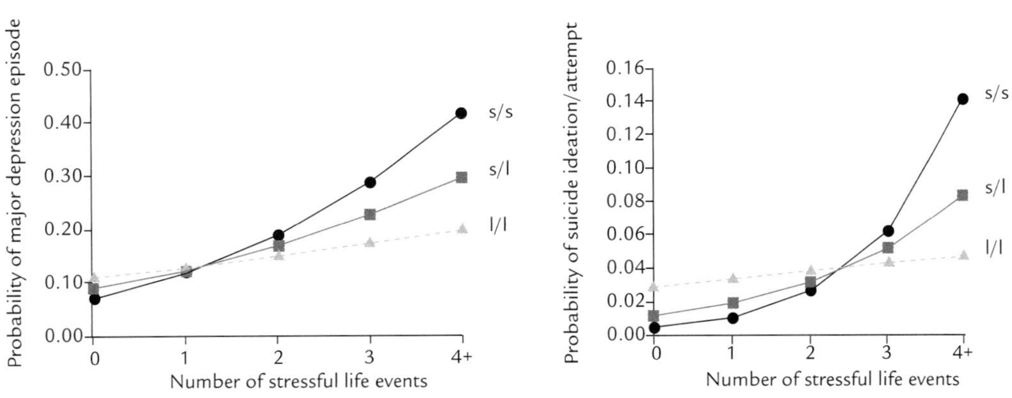

Figure 1.7 Probability of major depressive disorder and suicidality related to the number of stressful life events according to the genetics of the 5-HT transporter gene (one s/l or two copies s/s of the short allele of the 5-HT transporter gene or homozygous for the long allele l/l) (from reference 177, with permission)

in the design of clinical trials. The co-morbidity of depression and anxiety is well known and frequent, and this should be addressed in future research on depression in brain dysfunction. Merikangas *et al.*[184] found, however, that the diagnosis of depression carried a strong stability, as did the even more persistent syndrome of co-morbid anxiety and depression. In this paper anxiety alone was not as stable and had frequent transitions with depression. Kessler *et al.*[171] rarely found MDD alone in their National Comorbidity Survey Replication Study. In their paper more than 70% of the MDD cases fulfilled the criteria for at least one more DSM-IV co-morbid diagnosis. Hence there is the importance of both approaching depression in brain dysfunction with a wider view and the use of instruments (such as the SCAN) that may allow identification of other co-morbid psychiatric disorders.

While this longitudinal phenomenological approach (including onset, duration of illness, number of recurrences and life events) is paramount from a diagnostic point of view, some researchers[185] have found these aspects not to have much impact upon response to pharmacological treatment of depression after a careful review of all data available.

A classification integrating all evidence has not yet reached consensus, but there have been some proposals (Chen *et al.*, unpublished data, Table 1.6).

A word must be said here about the accuracy of recall of symptoms in depression. Wells and Horwood[186] have shown that the recall of key symptoms in depression is well below 25% by former depression patients, which questions all possible retrospective studies where an added layer of information is intended. There is

Table 1.6 Proposal of classification of depression based on current available evidence (from Chen *et al.*, unpublished, with permission)

Classification base	*Subtype(s)*
Longitudinal course	Unipolar vs. bipolar Single vs. recurrent Primary vs. secondary Recurrent brief depression Dysthymia
Symptomatology	Psychotic Minor depression Melancholic Atypical
Etiology or risk factors	Bereavement Substance-induced depression Depression due to general medical condition Pure, spectrum, sporadic Vascular depression
Biological factors	Norepinephrine or serotonin depression Dexamethasone Suppression Test

probably no substitute in this regard for longitudinal prospective studies.

INTRODUCTION TO THE TRANSCULTURAL INTERPRETATION OF DEPRESSION

Depression seems to be a universally recognizable condition[187]. However, even a cursory look at prevalence rates of mental illness across different minorities and countries imposes a consideration of transcultural issues[188]. The wide range of variation in current (Table 1.7) and lifetime (Table 1.8) rates of depression may be indicative of multiple differences in the manifestation of illness, its diagnosis or its assessment and reporting. The relative contribution of each of these factors remains to be established.

It seems reasonable that wherever possible depression studies in brain dysfunction should be adjusted for race and immigrant status. In the USA, for example, Black women have lower rates of depression and suicide than White women, while immigrants exhibit lower rates of depression than their descendants[189].

Table 1.7 Transcultural variation in the prevalence of depression (from reference 192, with permission)

Center	*Country*	*Current depression* (%)
Santiago	Chile	29.5
Rio de Janeiro	Brazil	15.8
Paris	France	13.7
Manchester	UK	16.9
Groningen	The Netherlands	15.9
Mainz	Germany	11.2
Bangalore	India	9.1
Athens	Greece	6.4
Berlin	Germany	6.1
Ankara	Turkey	11.6
Seattle, WA	USA	6.3
Verona	Italy	4.7
Nagasaki	Japan	2.6
Shanghai	China	4.0
Ibadan	Nigeria	4.2
Total		10.4
Men		6.8
Women		12.4

Table 1.8 Transcultural variation in lifetime rate, onset and sex ratio of depression (from reference 192, with permission)

	Lifetime rate *% (SE)*	*Female : male ratio* *Mean (SE)*	*Age at onset* (years) *Mean (SE)*
USA	5.2 (0.24)	2.6(0.11)	25.6 (0.30)
Canada (Edmonton)	9.6 (0.60)	1.9 (0.13)	24.8 (0.52)
Puerto Rico	4.3 (0.59)	1.8 (0.29)	29.5 (1.19)
France (Paris)	16.4 (1.16)	2.1 (0.16)	29.2 (0.52)
Germany	9.2 (1.50)	3.1 (0.39)	29.7 (1.18)
Italy (Florence)	12.4 (1.33)	3.0 (0.26)	34.8 (1.12)
Lebanon (Beirut)	19.0 (1.76)	1.6 (0.19)	25.2 (1.00)
Taiwan	1.5 (0.12)	1.6 (0.17)	29.3 (0.88)
Korea	2.9 (0.24)	2.0 (0.18)	29.3 (0.88)
New Zealand (Christchurch)	11.6 (0.96)	2.1 (0.18)	27.3 (0.58)

The development of cultural competence skills, beyond some partial knowledge about a particular culture (with the risk of stereotyping) and the 'dos and don'ts' within that culture, is necessary in cross-cultural interactions[190].

For a review of specific culturally bound diagnosis related to depression and anxiety, please see reference 191.

Different expressions of depression in different cultures

A priori, the assumption that depression as conceptualized by a prevalent biomedical psychiatry model is valid throughout the world presents many problems[193]. Indeed, cultural background is likely to influence the expression of depression, whether psychological, emotional or physical. Cross-cultural studies have shown that the same range of symptoms of depression and anxiety can be identified in all cultures, but the question remains as to whether the way affective disorders are conceptualized in the Western tradition remains adequate in other cultural settings[192]. It may be preferable to accept existing conceptualizations and terminology rather than imposing a Western model loaded with the influence of stereotypes. For example, the typical depressed patient in the USA is said to be a middle-aged woman who has multiple losses or an unsatisfactory marriage, whereas in Japan, it is a man in his thirties who has just received a promotion and is working too hard[194].

Cultural factors in relation to depression play an important role. Depression in developing countries has an increased prevalence of somatic symptoms[195]. Goodman[196] wrote about the expression of distress in psychological terms as something linked to modern Western thinking related to Cartesian splitting between mind and body. Excess somatization may confound the statistics of depression. According to Patel[197], 'somatic symptoms are also the commonest presenting features of depression in industrialised societies [and] the classic psychological symptoms of depression can often be elicited. Acute presentations of depression are more likely to be somatic; as the illness becomes chronic the patient re-evaluates the illness and becomes more likely to present with psychological symptoms'.

A classical example of the impact of the cultural setting in the manifestations of depression is the study by Escobar[198] in Puerto Ricans. He used structured interviews only to replicate what the ECA study had previously found: Puerto Ricans tend to score higher in somatization indexes[199,200] in Puerto Rico, but not in New York, where these indexes decreased. An excellent insight into the problem was given by Guarnaccia *et al.*[201]. They controlled the somatization parameters using an index for the '*ataque de nervios*'. The '*ataque de nervios'* is a culturally validated syndrome which is considered an accepted way to express distress in some subgroups of the Hispanic community.

Depression is less likely to present with cognitive manifestations in Chinese culture and predictably is more likely to be expressed somatically[202]. Stigma and political influences have created a myth of 'weakness of character' associated with depression but also a culturally higher threshold for it. The intimate connection between Chinese *weltanschauung* and language has created plenty of metaphors of physical content to express distress. This high level of somatization leads to diagnoses such as neurasthenia or '*shenjing shuairo*' to express it. Parker *et al.*[202] stated that 'Western influences on Chinese society and on the detection and identification of depression are likely to have modified the expression of depressive illness quite sharply since the early 1980s'. There are protective factors against depression in Chinese culture, such as a widely accepted stoic philosophy, strong extended family networks and less urbanization. 'In many parts of Chinese society, the experience of depression is physical rather than psychological. Many depressed Chinese people do not report feeling sad, but rather express boredom, discomfort, feelings of inner pressure, and symptoms of pain, dizziness, and fatigue'[189]. The picture is even more complicated in the Chinese immigrants in the USA, for example.

It is because of such transcultural differences in the clinical manifestations of depression that some authors[203] have spoken about the 'category fallacy'. By this they mean that a particular validated construct in culture A is inappropriately applied to culture B. One way to control for the effect of culture in phenomenology, when a dimensional approach is used to study depression, is to measure the 'factorial invariance'[204], that is to say, the equivalence of factor structure amongst cultures, races, minorities, etc.

In summary, the key aspect of the impact of culture on the phenomenology of depression is that the expression of its symptoms may vary in different settings, whereas the precipitant factors and response to treatment appear to remain relatively invariant. Not only is there a cultural variation in which distress is expressed, but culture may influence the categorization of mental conditions[205]. The way verbal or emotional distress is expressed in Western societies is not the same as in other societies[206,207]. Moreover, the difference is not just in the expression of distress, but also in its reporting, both by patients and clinicians. Westermeyer[208] emphasized the social psychology impact on diagnosis, since attitudes, beliefs, historical background and other factors may influence diagnosis.

Impact of social factors on depression

If ethnic factors are important, sociocultural factors are not less crucial. For example, even in the same developed country, such as Great Britain, there is a huge difference in the clinical expression of depression by upper-middle-class Oxfordians from the University community compared with a population of depressed patients in the working-class areas of Glasgow (Santos, unpublished observation). Amongst social factors the socioeconomic status has to be taken into account. Poverty may involve associated distress related to violence or substance abuse in some inner-city environments in America[209]. These associated psychosocial stressors cannot be overlooked in depression research and clinical practice. Depression is nearly twice as common among the poor as among the wealthy[189]. However, that relation between poverty and depression is still controversial, and some researchers have found that rising out from poverty does not necessarily mean reducing the rates of depression[210].

In general, less developed countries and rural settings are better able to tolerate or accommodate some forms of mental illness, such as depression. Attitudes and beliefs are unquestionably linked to culture and these, in turn, influence mental illness[208].

Pérez-Urdániz *et al.*[211] have explained in detail the transition from traditional societies to modern societies and acknowledged its impact in mental health parameters. They postulated that this change is partly responsible for the increased prevalence of depression, anxiety and personality disorders in recent times.

If poverty has been traditionally associated with mental illness for different reasons, the culture of affluence is not devoid of problems of its own. There is some evidence that very affluent children may be more at risk of depression. A recent review by Luthar[212] quoted excessive pressures to achieve and emotional isolation from parents. In adults, Luthar reported that an inordinate emphasis on material success, with its negative impact on close relationships and supportive networks plus the emphasis on autonomy and control, may be the most important reasons for increased risk of depression in affluent segments of the population. This author[212] also analyzed gender-specific factors, such as the working-maternity status of very affluent women.

Problems in transcultural psychiatry of depression

Somatization

As was said previously, the depressive disorder has cognitive symptoms/dimensions related to the expression of distress. These symptoms are particularly sensitive to culture. There are, on the other hand, somatic symptoms that are perhaps less likely to be so. However, those dimensions related to somatization pose a particular problem. Bhugra and Mastrogianni[192] stated that 'somatic symptoms are a prominent feature of depression and it is hopeful that contemporary cultural psychiatry identifies them more as culturally determined 'somatic metaphors' than part of a 'somatisation' process'. These same authors call our attention to the fact that globalization is affecting the patterns of recognition of depression across the world. They point to global mass media, urbanization and industrialization as the vectors changing the way depression is perceived and understood by both the public and professionals in nominally very different cultures.

Grief

Grief, mourning and bereavement offer difficulties in the differential diagnosis of clinical depression. There are cultures where more outward expression of psychological pain is allowed and they even accommodate for abnormalities of perception related to the deceased person. In contrast, Western societies do not only isolate death as a phenomenon (in hospitals, away from family, early burial, etc.) but also encapsulate it psychologically[213,214]. One has to be very aware of these cultural differences[215] to make a correct differential diagnosis under these circumstances.

Diagnostic reliability and related issues

In those multicultural societies, such as the USA, or in recent years, Western Europe, there is a risk that minorities may not be reached adequately. Identifying and obtaining samples from these minority populations is far from easy, but it is nonetheless essential to interpret epidemiological data adequately.

Using instruments not culturally validated to measure depression is another source of bias. King[216] denounced the low rate of depression in Afro-Americans held by previous studies for these and other reasons.

There are many subtle factors to be taken into consideration. For example, Shaw *et al.*[217] found that the prevalence of depression in Afro-Caribbeans in Northern England was higher than in Caucasians, but this was significant only for women. Given these kinds of data it is crucial to refine the diagnostic tools used and to replicate studies done in other settings and communities. In their highly recommendable review of the topic, Bhugra and Mastrogianni[192] advocated taking into account the narratives and explanatory models of mental disorders.

The inter-rater reliability (kappa, κ) of different mood symptoms has not been studied as much as psychotic symptoms across cultures[218]. Some dimensions, especially those with a more psychological component, have more transcultural variability. Interestingly, background information such as the previous social–labor history, previous functioning or personality traits, exhibit a great degree of transcultural variation. Depression with psychotic features demonstrates a high transcultural agreement. This serves to illustrate that health outcome measures also have to be culturally validated, and that no deep databases of depression in brain dysfunction can be constructed without health outcome measures. Separation of psychosocial functioning from the neurobiology of depression leads to poor linking of many relevant data.

Specific language and expression issues

Malgady *et al.*[218] used expressions of anger to identify depression and anxiety symptoms in first- and second-generation Puerto Ricans in New York. These expressions correlated positively with depression and anxiety symptoms and heightened the diagnostic sensitivity in this population. This is an example of how culture can be used to create instruments that are more adequate for identifying depression.

In those situations in which minorities may use their own language, symptom assessment in epidemiological surveys may be adversely affected. Cultural issues or difficulties in understanding a different language may induce individuals from minorities to endorse (or not) questions not completely understood. For instance, there are languages that do not have a clear word to express depression. Whenever the interviewer and the subject belong to different cultural subgroups, the former may be subjected to bias which may be uncontrollable or difficult to control. The use of interpreters does not solve the problem immediately. The interpreters are generally good for facts, but not so much for emotions[208]. It is widely recommended that they should translate literally, rather than in a free fashion. The interviewer also needs to be familiar with the non-verbal aspects of communication in different cultures, and the interpreter may give hints about these issues.

The problem of stereotyping runs parallel with the problem of language and vocabulary. Jones and Gray[219] reviewed the underdiagnosis of depression in Afro-Americans as compared with Caucasians, not just by White but also by Black clinicians. Social and cultural distance from the doctors has been quoted as an explanation, but so has stereotyping of Afro-Americans. Here the differences in vocabulary and expression of distress may play a role.

Interestingly, when adjusting for migration components, the acculturation of the sample can be measured as a proxy by fluency in the new language. It appears that acculturated individuals are more likely to be depressed[220]. Some authors[221] argue that in a large sample of Asian adults this is due to increased self-attention, a phenomenon that occurs as part of acculturation, and that this also explains why the more acculturation within their sample, the higher the prevalence of affective disorders and the lower the expression of somatic complaints.

Suicide

Race also plays a role in suicidality. Multiple studies have shown that some communities are more at risk than others, such as Asian women in the UK, or non-White Hispanics and Caucasian women in the USA. Other groups, such as Hispanic-Americans, are less at risk, while the suicidality statistics in Blacks is increasing[188,192].

Population genetics and biological factors

Cultural aspects of depression are also relevant to brain dysfunction. For instance, Trimble and Krishnamoorthy[222] illustrate the extent to which transcultural factors can modify the neuropsychiatric presentations of epilepsy. While epilepsy unfortunately carries a degree of stigma and decreased quality of life[223], there are some cultures, such as Navajo Indian-Americans, where epilepsy is highly regarded as providing high status within the group[224]. It is not entirely clear in the latter case whether improved social status protects epileptic patients from depression, but inevitably some health outcomes would be expected to be better.

Thinking of ethnic differences in psychiatry as a purely cultural and anthropological issue is a great mistake, because biological and genetic factors may be involved[192,225]. There are biological differences amongst different races, ranging between sleep, neuroendocrinology, pharmacokinetics, pharmacodynamics and pharmacogenetics. For instance, Afro-Americans with depression show alterations in sleep studies that differ from those seen in their Caucasian counterparts, particularly in depth of sleep and phasic REM measures[226].

Co-morbidity

To add complexity to this issue, depression may be masked by other behavioral factors, such as substance and/or alcohol abuse, in some cultural settings. There are important efforts aimed at extricating these masking factors, such as the National Epidemiologic Survey on Alcohol and Related Conditions (NESARC), which is the largest co-morbidity survey of its kind conducted to date ($n = 43\,093$).

Transcultural issues should not obscure attention to the individual

Ballenger *et al.*[194] said individual differences are as important as cultural differences, if not more. According to them the emphasis should be more on the individual than on the culture. There are no rules of thumb to identify the cultural factors. Often it will not be possible to diagnose depression in a single session, and sufficient time should be allocated for either the clinician or the researcher.

Undoubtedly 'rethinking psychiatry diagnosis in ways that account for, even embrace, the diversity of culture and the singularity of the individual'[227] can be overwhelming on account of the multiple diagnostic possibilities. There is still a long way to go in integrating the cultural aspects of depression. Most researchers in transcultural psychiatry share the view of Kirmayer[191], for whom 'conventional psychiatric research is ill-suited to explore the cultural meaning of illness narratives to a checklist of symptoms and signs of disorder'.

Practical aspects in transcultural psychiatry of depression

We will finish this section by trying to summarize a few practical issues based on references 192 and 197.

- The clinical presentation of depression in all cultures includes multiple somatic symptoms of chronic duration; psychological symptoms are important for diagnosis and can be easily elicited in most patients.
- Different explanatory models of depression between patient and researcher/clinician should be taken into account.
- It might be necessary to spend more time with a patient from a different culture.
- Awareness of culturally appropriate terminology for depression is a useful way of bridging the gap between lay and biomedical models of illness, and may help improve levels of recognition and treatment compliance.
- The use of somatic metaphors in particular is highly relevant, because of some cultural variation of clinical dimensions or the reluctance in some minorities to disclose depressive symptoms; however, somatic diagnosis has to be considered, especially in the elderly population.
- If there is a language and cultural barrier that precludes good communication, the use of a bicultural interpreter is the best option.
- The non-verbal aspects of communication need to be addressed.
- Culturally validated instruments of depression should be used in both research and clinical practice.
- There is a need to replicate studies of depression in different cultural settings.

REFERENCES

1. Kanner A. Is major depression a neurologic disorder with psychiatric symptoms? Epilepsy Behav 2004; 5: 636–644.
2. Murray CJ, López AD. Alternative projections of mortality and disability by cause 1990–2020: Global Burden of Disease Study. Lancet 1997; 349: 1498–1504.

3. Anonymous. Energizing depression research [Editorial]. Nature Neurosci 2004; 8: 787.

4. Kanner AM, Barry JJ. The impact of mood disorders in neurological diseases: should neurologists be concerned? Epilepsy Behav 2003; 4: S3–S13.

5. Kupfer DJ, Frank E. Comorbidity in depression. Acta Psychiatr Scand 2003; 418: S57–S60.

6. Frances AJ, Widiger TA, Pincus HA. The development of DSM-IV. Arch Gen Psychiatry 1989; 46: 373–375.

7. Nathan PE. DSM-IV and its antecedents: enhancing syndromal diagnosis. In: Barron JW, ed. Making Diagnosis Meaningful: Enhancing Evaluation and Treatment of Psychological Disorders. Washington, DC: American Psychological Association, 1998.

8. Sadler JZ, Hulgus YF, Agich GJ. On values in recent American psychiatric classification. J Med Philos 1994; 19: 261-277.

9. Spitzer RL. Values and assumptions in the development of DSM-III and DSM-III-R: An insider's perspective and a belated response to Sadler, Hulgus and Agich's 'On values in recent American psychiatric classification'. J Nerv Ment Dis 2001; 189: 351–359.

10. Sims A. Symptoms in the Mind: An Introduction to Descriptive Psychopathology, 3rd edn. London: Saunders, 2003.

11. Huber G. Psychopathologie – eine versiegende Quelle? Ein kritischer Kommentar. Fortsch Neurol Psychiatrie 2002; 70: 393–402.

12. Weary G, Edwards JA, Jacobson J. Depression research methodologies: a reply. J Pers Soc Psychol 1995; 68: 885–891.

13. Van Staden CW. Diagnostic understanding and diagnostic explanation in psychiatry. Curr Opin Psychiatry 2003; 16: 667–671.

14. Hamilton M, ed. Fish's Clinical Psychopathology: Signs and Symptoms in Psychiatry. Bristol: John Wright and Sons, 1974.

15. Ossorio P. Pathology. In: Ossorio P, ed. Essays on Clinical Topics. Ann Arbor: Descriptive Psychology Press, 1997.

16. Clancy CM, Eisenberg JM. Outcomes research: measuring the end results of health care. Science 1998; 282: 245–246.

17. Wakefield JC. Normal inability versus pathological disability: why Ossorio's definition of mental disorder is not sufficient. Clin Psychol Sci Pract 1997; 4: 249–258.

18. Vaillant GE, Vaillant LM. The role of ego mechanisms of defense in the diagnosis of personality disorders. In: Barron JW, ed. Making Diagnosis Meaningful: Enhancing Evaluation and Treatment of Psychological Disorders. Washington, DC: American Psychological Association, 1998.

19. Harari E. Whose evidence? Lessons from the philosophy of science and the epistemology of medicine. Aust NZ J Psychiatry 2001; 35: 724–730.

20. Glas G. History of psychiatric patients and symptoms. Curr Opin Psychiatry 2003; 16: 661–665.

21. Halligan PW, David AS. Cognitive neuropsychiatry: towards a scientific psychopathology. Nature Rev Neurosci 2001; 2: 209–215.

22. Andreasen NC. Linking mind and brain in the study of mental illnesses: a project for a scientific psychopathology. Science 1997; 275: 1586–1593.

23. Klein D. Endogenomorphic depression. Arch Gen Psychiatry 1974; 4: 447–449.

24. Jackson S. Historia de la melancolía y la depresión. Madrid: Turner, 1989.

25. López Piñero JM. Historia de la Medicina. Madrid: Historia 16, 1990.

26. Riera J. Historia, Medicina y Sociedad. Madrid: Pirámide, 1985.

27. Hipócrates. Tratados médicos. Barcelona: Planeta DeAgostini, 1990.

28. Alsina J. Los orígenes helénicos de la medicina occidental. Barcelona: Guadarrama, 1982.

29. Berrios G, Porter R, eds. A History of Clinical Psychiatry: The Origin and History of Psychiatric Diseases. Continuum International. London: Athlone, 1995.

30. García Ballester L. Galeno. Madrid: Guadarrama, 1972.

31. Dieckhöfer K. El desarrollo de la Psiquiatría en España: elementos históricos y culturales. Madrid: Gredos, 1984.

32. Berrios G. The History of Mental Symptoms: Descriptive Psychopathology since the Nineteenth Century. Cambridge: Cambridge University Press, 1996.

33. Robins E, Guze SB. Establishment of diagnostic validity in psychiatric illness: its application to schizophrenia. Am J Psychiatry 1970; 126: 983–987.

34. Nestler EJ, Barrot M, DiLeone RJ, et al. Neurobiology of depression. Neuron 2002; 28: 13–25.

35. De Almeida Fleck MP, Simon G, Herrman H, et al. Major depression and its correlates in primary care settings in six countries: 9-month follow-up study. Br J Psychiatry 2005; 186: 41–47.

36. Parker G. Modern diagnostic concepts of the affective disorders. Acta Psychiatr Scand 2003; 418: S24–S28.

37. Davidson RJ, Pizzagalli D, Nitschke JB, Putnam K. Depression: perspectives from affective neuroscience. Annu Rev Psychol 2002; 53: 545–576.

38. Stahl SM. An emerging, neurobiologically based therapeutic strategy for the modern psychopharmacologist. J Clin Psychiatry 2003; 64: 1145–1146.

39. Kendell R, Jablensky A. Distinguishing between the validity and utility of psychiatric diagnoses. Am J Psychiatry 2003; 160: 4–12.

40. Andreasen NC. The validation of psychiatric diagnosis: new models and approaches. Am J Psychiatry 1995; 152: 161–162.

41. Kendell RE. The choice of diagnostic criteria for biological research. Arch Gen Psychiatry 1982; 39: 1334–1339.

42. McGuffin P, Farmer A. Polydiagnostic approaches to measuring and classifying psychopathology. Am J Med Genet (Neuropsychiatry Genet) 2001; 105: 39–41.

43. Andreasen NC, Scheftner W, Reich T, et al. The validation of the concept of endogenous depression. A family study approach. Arch Gen Psychiatry 1986; 43: 246–251.

44. Vallejo J. Trastornos depresivos. In: Vallejo J, ed. Introducción a la psicopatología y psiquiatría, 4th edn. Barcelona: Masson, 1998.

45. Gastó C. Depresiones no melancólicas. In: Vallejo J, Gastó C, eds. Trastornos afectivos: ansiedad y depresión. Barcelona: Salvat, 1990.

46. Keller MB, Klein DN, Hirschfeld RM, et al. Results of the DSM-IV mood disorders

field trial. Am J Psychiatry 1995; 152: 843–849.

47. Eaton WW, Regier DA, Locke BZ, Taube CA. The Epidemiologic Catchment Area Program of the National Institute of Mental Health. Public Health Rep 1981; 96: 319-325.

48. Eaton WW, Anthony JC, Gallo J, et al. Natural history of Diagnostic Interview Schedule/DSM-IV major depression: The Baltimore Epidemiological Catchment Area follow-up. Arch Gen Psychiatry 1997; 54: 993–999.

49. Von Zerssen D. Development of an integrated model of personality, personality disorders and severe axis I disorders, with special reference to major affective disorders. J Affect Disord 2002; 68: 143–158.

50. Parker G, Wilhelm K, Mitchell P, Gladstone G. Predictors of 1-year outcome in depression. Aust NZ J Psychiatry 2000; 34: 56–64.

51. Clark LA, Vittengl J, Kraft D, Jarrett RB. Separate personality traits from states to predict depression. J Pers Disord 2003; 17: 152–172.

52. Clayton P, Guze S, Cloninger RC, Martin R. Unipolar depression: diagnostic inconsistency and its implications. J Affect Disord 1992; 26: 111–126.

53. Kocsis J, Markowitz J, Prien R. Comorbidity of dysthymic disorder. In: Maser J, Cloninger C, eds. Comorbidity of Mood and Anxiety Disorders. Washington, DC: American Psychiatric Press, 1990.

54. Kendler KS, Eaves LJ, Walters EE, et al. The identification and validation of distinct depressive syndromes in a population-based sample of female twins. Arch Gen Psychiatry 1996; 53: 391–399.

55. Ramos Brieva J. Instrumentos de medida actual de la depresión. In: Ayuso JL, Saiz J, eds. Depresión: visión actual. Madrid: Aula Médica, 1997.

56. Enns MW, Cox BJ, Borger SC. Correlates of analogue and clinical depression: a further test of the phenomenological continuity hypothesis. J Affect Disord 2001; 66: 175–183.

57. Lavretsky H, Kumar A. Clinically significant non-major depression: old concepts, new insights. Am J Geriatr Psychiatry 2002; 10: 239–255.

58. Rapaport MH, Judd LL, Schettler PJ, et al. A descriptive analysis of minor depression. Am J Psychiatry 2002; 159: 637–643.

59. Solomon A, Haaga DAF, Arnow BA. Is clinical depression distinct from subthreshold depressive symptoms? A review of the continuity issue in depression research. J Nerv Ment Dis 2001; 189: 498–506.

60. Akiskal HS, Judd LL, Gillin JC, Lemmi H. Subthreshold depressions: clinical and polysomnographic validation of dysthymic, residual and masked forms. J Affect Disord 1997; 45: 53–63.

61. Sherbourne CD, Wells KB, Hays RD, et al. Subthreshold depression and depressive disorder: clinical characteristics of general and mental health specialty outpatients. Am J Psychiatry 1994; 151: 1777–1784.

62. Preisig M, Merikangas KR, Angst J. Clinical significance and comorbidity of subthreshold depression and anxiety in the community. Acta Psychiatr Scand 2001; 104: 96–103.

63. Clarke GN, Hornbrook M, Lynch F, et al. A randomized trial of a group cognitive intervention for preventing depression in adolescent offspring of depressed parents. Arch Gen Psychiatry 2001; 58: 1127–1134.

64. Cuijpers P, Smit F. Subthreshold depression as a risk indicator for major depressive disorder: a systematic review of prospective studies. Acta Psychiatr Scand 2004; 109: 325–331.

65. Jonas BS, Mussolino ME. Symptoms of depression as a prospective risk factor for stroke. Psychosom Med 2000; 62: 463–471.

66. Judd LL, Akiskal HS, Maser JD, et al. A prospective 12 years study subsyndromal and syndromal depressive symptoms in unipolar major depressive disorders. Arch Gen Psychiatry 1998; 55: 694–700.

67. Judd LL. The role of residual subthreshold depressive symptoms in early episode relapse in unipolar major depressive disorder [Letter]. Arch Gen Psychiatry 1999; 56: 764–765.

68. Fergusson DM, Horwood LJ, Ridder EM, Beautrais AL. Subthreshold depression in adolescence and mental health outcomes in adulthood. Arch Gen Psychiatry 2005; 62: 66–72.

69. Judd LL, Paulus MP, Wells KB, Rapaport MH. Socioeconomic burden of subsyndromal depressive symptoms and major depression in a sample of the general population. Am J Psychiatry 1996; 153: 1411–1417.

70. Judd LL, Schettler PJ, Akiskal HS. The prevalence, clinical relevance, and public health significance of subthreshold depression. Psychiatry Clin North Am 2002; 25: 685–698.

71. Judd LL, Rapaport MH, Paulus MH, Brown JL. Subsyndromal symptomatic depression: a new mood disorder? J Clin Psychiatry 1994; 55: 18–28.

72. Kennedy N, Abbott R, Paykel ES. Longitudinal syndromal and sub-syndromal symptoms after severe depression: 10 year follow-up. Br J Psychiatry 2004; 184: 330–336.

73. Rapaport MH, Judd LL. Minor depressive disorder and subsyndromal depressive symptoms: functional impairment and response to treatment. J Affect Disord 1998; 48: 227–232.

74. Godelief RW, Willemse M, Smit F, et al. Minimal-contact psychotherapy for subthreshold depression in primary care: Randomised trial. Br J Psychiatry 2004; 185: 416–421.

75. Wiegartz P, Seidenberg M, Woodard A, et al. Co-morbid psychiatric disorder in chronic epilepsy: recognition and etiology of depression. Neurology 1999; 53 (Suppl 2): 3–8.

76. Robins LN. Using survey results to improve the validity of the standard psychiatric nomenclature. Arch Gen Psychiatry 2004; 61: 1188–1194.

77. Kendell RE. Clinical validity. In: Robins LN, Barrett JE, eds. The Validity of Psychiatric Diagnosis. New York: Raven Press, 1989.

78. Alexopoulos GS. On the 'infallibility' of psychopathology and its implications for action [Editorial]. Am J Psychiatry 2004; 161: 2151–2154.

79. Korab-Karpowicz WJ. Knowing beyond science: What can we know and how can we know? Humanitas 2002; 15: 60–73.

80. Bschor T, Ising M, Bauer M, et al. Time experience and time judgment in major depression, mania and healthy subjects. A controlled study of 93 subjects. Acta Psychiatr Scand 2004; 109: 222–229.

81. Lubin B. Adjective checklists for the measurement of depression. Arch Gen Psychiatry 1965; 12: 57–62.

82. Coyne JC. Self-reported distress: analog or ersatz depression? Psychol Bull 1994; 116: 29–45.

83. Diener E, Oishi S, Lucas RE. Personality, culture, and subjective well-being: emotional and cognitive evaluations of life. Annu Rev Psychol 2003; 54: 403–425.

84. Dervic K, Oquendo MA, Grunebaum MF, et al. Religious affiliation and suicide attempt. Am J Psychiatry 2004; 161: 2303–2308.

85. Watts FN. Psychological and religious perspectives on emotion. Int J Psychol Religion 1996; 6: 71–87.

86. Schimmel S. The Seven Deadly Sins: Jewish, Christian, and Classical Reflections on Human Psychology. New York: Oxford University Press, 1997.

87. Frankl VE. Man's Search for Meaning: An Introduction to Logotherapy. New York: Washington Square Press, 1963.

88. Frankl VE. The Doctor and the Soul: From Psychotherapy to Logotherapy. New York: Vintage Books, 1973.

89. Von Morstein P. Anxiety and depression: a philosophical investigation. Radical Psychology (1999) 1. Downloaded as http://www.geocities.com/Athens/Forum/5914/Anxiety.html on 8 December 2004.

90. Watson D, Clark LA. Affects separable and inseparable on the hierarchical arrangement of the negative affects. J Pers Soc Psychol 1992; 62: 489–505.

91. Eysenck MW. Simply Psychology, 2nd edn. Hove: Psychology Press, 2002.

92. Fridja NH. Universal antecedents exist, and they are interesting. In: Ekman P, Davidson RJ, eds. The Nature of Emotion Fundamental Questions. Oxford: Oxford University Press, 1994.

93. Crespo Blanco JM. Depresión mayor. In: Roca Bennasar M, ed. Trastornos del Humor. Madrid: Panamericana, 1999.

94. Peralta V, Cuesta MJ. Psicopatología y clasificación de los trastornos depresivos. Anales Sis San Navarra 2002; 25 (Suppl 3): 7–20.

95. Dolan RJ. Emotion, cognition and behavior. Science 2002; 298: 1191–1194.

96. Tellegen A, Watson D, Clark LA. On the dimensional and hierarchical structure of affect. Psychol Sci 2003; 10: 297–303.

97. Green DP, Salovey P. In what sense are positive and negative affect independent? Psychol Sci 2003; 298: 304–306.

98. Bermond B, Nieuwenhuyse B, Fasotti L, Schwerman J. Spinal cord lesions, peripheral feedback and intensities of emotional feelings. Cogn Emotion 1991; 5: 201–220.

99. Schachter S, Singer JE. Cognitive, social and physiological determinants of an emotional state. Psychol Rev 1962; 69: 379–399.

100. Lazarus RS. Thoughts on the relations between emotion and cognition. Am Psychologist 1982; 37: 1019–1024.

101. Steptoe A, Vogele C. Are stress responses influenced by cognitive appraisal? An experimental comparison of coping strategies. Br J Psychol 1986; 77: 243–255.

102. Parkinson B, Manstead ASR. Appraisal as a cause of emotion. In: Clark MS, ed. Review of Personality and Social Psychology. New York: Sage, 1992.

103. Compton RJ. The interface between emotion and attention: a review of evidence from psychology and neuroscience. Behav Cogn Neurosci Rev 2003; 2: 115–129.

104. Weitbrecht HJ. Manual de Psiquiatría. Madrid: Gredos, 1969.

105. Bonanno GA, Kaltman S. The varieties of grief experience. Clin Psychol Rev 2001; 21: 705–734.

106. Ogrodniczuk JS, Piper WE, Joyce AS, et al. Differentiating symptoms of complicated grief and depression among psychiatric outpatients. Can J Psychiatry 2003; 48: 87–93.

107. Majer M, Ising M, Künzel H, et al. Impaired divided attention predicts delayed response and risk to relapse in subjects with depressive disorders. Psychol Med 2004; 34: 1453–1463.

108. Den Hartog HM, Derix MMA, Van Bemmel AL, et al. Cognitive functioning in young and middle-aged unmedicated outpatients with major depression: testing the effort and cognitive speed hypotheses. Psychol Med 2003; 33: 1443–1451.

109. Airaksinen E, Larsson M, Lundberg I, Forsell Y. Cognitive functions in depressive disorders: evidence from a population-based study. Psychol Med 2004; 34: 83–91.

110. Dozois DJ, Covin R, Brinker JK. Normative data on cognitive measures of depression. J Consult Clin Psychol 2003; 71: 71–80.

111. Davidson RJ, Putnam KM, Larson CL. Dysfunction in the neural circuitry of emotion regulation – a possible prelude to violence. Science 2000; 289: 591–594.

112. Baron RA, Byrne D. Social Psychology: Understanding Human Interaction, 6th edn. Boston: Allyn & Bacon, 1991.

113. Russell JA. Core affect and the psychological construction of emotion. Psychol Rev 2003; 110: 145–172.

114. Beck AT. Depression: Clinical, Experimental and Theoretical Aspects. New York: Harper and Row, 1967.

115. Beck AT. Cognitive Therapy and the Emotional Disorders. New York: International Universities Press, 1976.

116. Beck AT. Cognitive models of depression. J Cogn Psychother 1987; 1: 2–38.

117. Franco Lovaco B, Rodríguez Vega B, Cañas de Paz F. Psicoterapia en los trastornos del humor. In: Roca Bennasar M, ed. Trastornos del humor. Barcelona: Panamericana, 1999.

118. Clark LA, Steer RA, Beck AT. Common and specific dimensions of self reported anxiety and depression: implications for the cognitive and tripartite models. J Abnorm Psychol 1994; 103: 645–654.

119. Lloyd GG, Lishman WA. Effects of depression on the speed of recall of pleasant and unpleasant experiences. Psychol Med 1975; 5: 173–180.

120. Robinson LA, Berman JS, Neimeyer RA. Psychotherapy for the treatment of depression: a comprehensive review of controlled outcome research. Psychol Bull 1990; 108: 30–49.

121. Stuart S, Bowers WA. Cognitive therapy with inpatients: review and meta-analysis. J Cogn Psychother 1995; 9: 85–92.

122. Barkham M, Hardy GE. Counselling and interpersonal therapies for depression: towards securing an evidence base. Br Med Bull 2001; 57: 115–132.

123. Paykel ES, Scott J, Teasdale JD, et al. Prevention of relapse by cognitive therapy in residual depression: a controlled trial. Arch Gen Psychiatry 1999; 56: 829–835.

124. Goldapple K, Segal Z, Garson C, et al. Modulation of cortical–limbic pathways in major depression: treatment-specific effects of cognitive behavior therapy. Arch Gen Psychiatry 2004; 61: 34–41.

125. Parker G, Roy K, Eyers K. Cognitive behavior therapy for depression? Choose horses for courses. Am J Psychiatry 2003; 160: 825–834.

126. Peselow ED, Robins C, Block P, et al. Dysfunctional attitudes in depressed patients before and after clinical treatment and in normal control subjects. Am J Psychiatry 1990; 147: 439–444.

127. Just N, Abramson LY, Alloy LB. Remitted depression studies as tests of the cognitive vulnerability hypotheses of depression onset: a critique and conceptual analysis. Clin Psychol Rev 2001; 21: 63–83.

128. Gladstone G, Parker G. Depressogenic cognitive schemas: enduring beliefs or mood state artefacts? Aust NZ J Psychiatry 2001; 35: 210–216.

129. Nathan PE, Langenbucher JW. Psychopathology: description and classification. Annu Rev Psychol 1999; 50: 79–107.

130. Haslam N. Categorical versus dimensional models of mental disorder: the taxometric evidence. Aust NZ J Psychiatry 2003; 37: 696–704.

131. Peralta V, Cuesta MJ, Giraldo C, et al. Classifying psychotic disorders issues regarding categorial vs. dimensional approaches and time frame to assess symptoms. Eur Arch Psychiatry Clin Neurosci 2002; 252: 12–18.

132. O'Brien KP, Glaudin V. Factorial structure and factor reliability of the Hamilton Rating Scale for Depression. Acta Psychiatr Scand 1988; 78: 113–120.

133. Thase ME. Treatment of severe depression. J Clin Psychiatry 2000; 61 (Suppl 1): 17–25.

134. Chen LS, Eaton WW, Gallo JJ, Nestadt G. Understanding the heterogeneity of depression through the triad of symptoms, course and risk factors: a longitudinal, population-based study. J Affect Disord 2000; 59: 1–11.

135. Streiner DL. Figuring out factors: the use and misuse of factor analysis. Can J Psychiatry 1994; 39: 135–140.

136. Beach SHR, Amir N. Is depression taxonic, dimensional or both? J Abnorm Psychol 2003; 112: 228–236.

137. Ryder AG, Bagby RM, Schuller DR. The overlap of depressive personality disorder and dysthymia: a categorical problem with a dimensional solution. Harvard Rev Psychiatry 2002; 10: 337–352.

138. Van Praag HM, Asnis GM, Kahn RS, et al. Nosological tunnel vision in biological psychiatry. A plea for functional psychopathology. Ann NY Acad Sci 1990; 600: 501–510.

139. Bryant FB, Yarnold PR. Principal-components analysis and exploratory and confirmatory factor analysis. In: Grimm LG, Yarnold PR, eds. Reading and Understanding Multivariate Statistics. Washington, DC: American Psychological Association, 1995.

140. Richter P, Werner J, Heerlein A, et al. On the validity of the Beck Depression Inventory. A review. Psychopathology 1998; 31: 160–168.

141. Korszun A, Moskvina V, Brewster S, et al. Familiality of symptom dimensions in depression. Arch Gen Psychiatry 2004; 61: 468–474.

142. Mayberg HS. Positron emission tomography imaging in depression: a neural systems perspective. Neuroimag Clin North Am 2003; 13: 805–815.

143. Seminowicz DA, Mayberg HS, McIntosh AR, et al. Limbic–frontal circuitry in major depression: a path modeling metanalysis. Neuroimage 2004; 22: 409–418.

144. Gilliam FG, Santos J, Vahle V, et al. Depression in epilepsy: ignoring clinical expression of neuronal network dysfunction? Epilepsia 2004; 45 (Suppl 2): 28–33.

145. Parker G. Classifying depression: should paradigms lost be regained? Am J Psychiatry 2000; 157: 1195–1203.

146. Paykel ES. Classification of depressed patients: a cluster analysis derived grouping. Br J Psychiatry 1971; 118: 275–288.

147. Parker G, Hadzi-Pavlovic D. Melancholia: a disorder of movement and mood. New York: Cambridge University Press, 1996.

148. Rush A, Weissenburger J. Melancholic symptom features. DSM-IV Sourcebook, vol 2. Washington, DC: American Psychiatric Association, 1994.

149. Zimmerman M, Spitzer R. Melancholia. From DSM-III to DSM-III-R. Am J Psychiatry 1989; 146: 20–28.

150. Schotte C, Maes M, Cluydis R, Cosyns P. Cluster analytic validation of the DSM melancholic depression. The threshold model: integration of quantitative and qualitative distinctions between unipolar depressive subtypes. Psychiatry Res 1997; 71: 181–195.

151. Rothschild AJ, Schatzberg AF. Diagnosis and treatment of psychotic (delusional) depression. In: Grunhaus L, Greden JF, eds. Severe Depressive Disorders. Washington, DC: American Psychiatric Press, 1994: 195–207.

152. Rush AJ, Weissenberger JE. Melancholic symptom features and DSM-IV. Am J Psychiatry 1994; 151: 489–498.

153. Kendler K. The diagnostic validity of melancholic major depression in a population based sample of female twins. Arch Gen Psychiatry 1997; 54: 299–304.

154. Perry PJ. Pharmacotherapy for major depression with melancholic features: relative efficacy of tricyclic versus selective serotonin reuptake inhibitor antidepressants. J Affect Disord 1996; 39: 1–6.

155. Baca E, González de Chávez M, García Toro M, et al. Sertraline is more effective than imipramine in the treatment of non-melancholic depression: results from a multicentre, randomized study. Prog Neuro-Psychopharmacol Biol Psychiatry 2003; 27: 493–500.

156. Liebowitz MR. Depression with anxiety and atypical depression. J Clin Psychiatry 1993; 54: 10–14.

157. Akiskal HS. The diagnosis of depression. In: Feighner JP, Boyer WF, eds. Perspectives in Psychiatry, vol 2. New York: Wiley, 1991.

158. Parker G, Roy K, Mitchel P, et al. Atypical depression: a reappraisal. Am J Psychiatry 2002; 159: 1470–1479.

159. Davidson JRT. Monoamine oxidase inhibitors. In: Paykel ES, ed. Handbook of Affective Disorders. Edinburgh: Churchill Livingstone, 1992.

160. Coryell W, Winokur G, Shea T, et al. The long-term stability of depressive subtypes. Am J Psychiatry 1994; 151: 199–204.

161. Melartin T, Leskelä U, Rytsälä H, et al. Co-morbidity and stability of melancholic features in DSM-IV major depressive disorder. Psychol Med 2004; 34: 1443–1452.

162. Sullivan PF, Kendler KS. Typology of common psychiatric syndromes. An empirical study. Br J Psychiatry 1998; 173: 312–319.

163. Sullivan PF, Prescott CA, Kendler KS. The subtypes of major depression in a twin registry. J Affect Disord 2002; 68: 273–284.

164. Willeit M, Praschak-Rieder N, Neumeister A, et al. A polymorphism (5-HTTLPR) in

the serotonin transporter promoter gene is associated with DSM-IV depression subtypes in seasonal affective disorder. Mol Psychiatry 2003; 8: 942–946.

165. Fountoulakis KN, Iacovides A, Gerasimouc G, et al. The relationship of regional cerebral blood flow with subtypes of major depression. Prog Neuro-Psychopharmacol Biol Psychiatry 2004; 28: 537–546.

166. Clark LA, Beck AT, Beck JS. Symptom differences in major depression, dysthymia, panic disorder, and generalized anxiety disorder. Am J Psychiatry 1994; 151: 205–209.

167. Watson D, Weber K, Assenheimer JS, et al. Testing a tripartite model: I. evaluating the convergent and discriminant validity of anxiety and depression symptom scales. J Abnorm Psychol 1995; 104: 3–14.

168. Fava M, Alpert JE, Carmin CN, et al. Clinical correlates and symptom patterns of anxious depression among patients with major depressive disorder in STAR*D. Psychol Med 2004; 34: 1299–1308.

169. Jackson A, Cavanagh J, Scott J. A systematic review of manic and depressive prodromes. J Affect Disord 2003; 74: 209–217.

170. Fava GA, Fabbri S, Sonino N. Residual symptoms in depression: an emerging therapeutic target. Prog Neuro-Psychopharmacol Biol Psychiatry 2002; 26: 1019–1027.

171. Kessler RC, Berglund P, Demler O, et al. Epidemiology of major depressive disorder: results from the National Comorbity Survey Replication (NCS-R). JAMA 2003; 289: 3095–3105.

172. Angst J, Gamma A, Sellaro R, et al. Recurrence of bipolar and major depression: a life-long perspective. Eur Arch Psychiatry Clin Neurosci 2003; 253: 236–240.

173. Uebelacker LA, Keitner GI, Ryan CE, Miller IW. Characterizing the long-term course of individuals with major depressive disorder. J Nerv Ment Disord 2004; 192: 65–68.

174. Chen LS, Eaton WW, Gallo JJ, et al. Empirical examination of current depression categories in a population based study: symptoms, course and risk factors. Am J Psychiatry 2000; 157: 573–580.

175. Paykel ES. Life events and affective disorders. Acta Psychiatr Scand 2003; 418: S61–S66.

176. Brown GW, Harris TO, Hepworth C. Life events and endogenous depression: a puzzle re-examined. Arch Gen Psychiatry 1994; 51: 525–534.

177. Mitchell PB, Parker GB, Gladstone GL, et al. Severity of stressful life events in first and subsequent episodes of depression: the relevance of depressive subtype. J Affect Disord 2003; 73: 245–252.

178. Lewinsohn PM, Allen NB, Seeley JR. First onset versus recurrence of depression: differential processes of psychosocial risk. J Abnorm Psychol 1999; 73: 483–489.

179. Caspi A, Sugden K, Moffitt TE, et al. Influence of life stress on depression: moderation by a polymorphism in the S-HTT gene. Science 2003; 301: 386–389.

180. Gillespie NA, Whitfield JB, Williams B, et al. The relationship between stressful life events, the serotonin transporter (5-HTTL-PR) genotype and major depression. Psychol Med 2005; 35: 101–111.

181. Gilman SE, Kawachi I, Fitzmaurice GM, Buka SL. Socio-economic status, family disruption and residential stability in childhood: relation to onset, recurrence and remission of major depression. Psychol Med 2003; 33: 1341–1355.

182. Tennant C. Work-related stress and depressive disorders. J Psychosom Res 2001; 51: 697–704.

183. Parker G, Ritch J. The influence of an uncaring partner on the type and outcome of depression. J Affect Disord 2001; 66: 207–214.

184. Merikangas KR, Zhang H, Avenevoli S, et al. Longitudinal trajectories of depression and anxiety in a prospective community study: the Zurich Cohort Study. Arch Gen Psychiatry 2003; 60: 993–1000.

185. Bagby RM, Ryder AG, Cristi C. Psychosocial and clinical predictors of response to pharmacotherapy for depression. Rev Psychiatry Neurosci 2002; 27: 250–257.

186. Wells JE, Horwood LJ. How accurate is recall of key symptoms of depression? A comparison of recall and longitudinal reports. Psychol Med 2004; 34: 1001–1011.

187. Herrera JM, Lawson WB, Sramek JJ. Cross Cultural Psychiatry. New York: Wiley, 1999.

188. Weissman MM, Bland RC, Canino GJ, et al. Cross-national epidemiology of major depression and bipolar disorder. JAMA 1996; 276: 293–299.

189. Kleinman A. Culture and depression [Editorial]. N Engl J Med 2004; 351: 951–952.

190. Betancourt JR. Cultural competence – marginal or mainstream movement? N Engl J Med 2004; 351: 953–955.

191. Kirmayer LJ. Cultural variations in the clinical presentation of depression and anxiety: implications for diagnosis and treatment. J Clin Psychiatry 2001; 62 (Suppl 13): 22–28.

192. Bhugra D, Mastrogianni A. Globalisation and mental disorders: overview with relation to depression. Br J Psychiatry 2004; 184: 10–20.

193. Patel V, Winston M. The 'universality' of mental disorder revisited: assumptions, artefacts and new directions. Br J Psychiatry 1994; 165: 437–440.

194. Ballenger JC, Davidson JRT, Lecrubier Y, et al. Consensus Statement on Transcultural issues in Depression and Anxiety from the International Consensus Group on Depression and Anxiety. J Clin Pyschiatry 2001; 62 (Suppl 13): 47–55.

195. Katon W, Walker EA. Medically unexplained symptoms in primary care. J Clin Psychiatry 1998; 59 (Suppl 20): 15–21.

196. Goodman A. Organic unity theory: the mind–body problem revisited. Am J Psychiatry 1991; 148: 553–563.

197. Patel V. Cultural factors and international epidemiology. Br Med Bull 2001; 57: 33–45.

198. Escobar JI. Cross-cultural aspects of the somatization trait. Hosp Community Psychiatry 1987; 38: 174–180.

199. Regier DA, Myers JK, Kramer M, et al. The NIMH Epidemiologic Catchment Area Program: historical context, major objectives, and study population characteristics. Arch Gen Psychiatry 1984; 41: 934–941.

200. Robins LN, Regier DA, eds. Psychiatric Disorders in America: The Epidemiologic Catchment Area Study. New York: Free Press, 1991.

201. Guarnaccia PJ, de la Cancela V, Carrillo E. The multiple meanings of ataques de nervios in the Latino community. Med Anthropol Q 1989; 11: 47–62.

202. Parker G, Gladstone G, Chee KT. Depression in the planet's largest ethnic

group: the Chinese. Am J Psychiatry 2001; 158: 857–864.

203. Good B, Good MD. The cultural context of diagnosis and therapy: a view from medical anthropology. In: Miranda MR, Kitano HHL, eds. Mental Health Research in Minority Communities: Development of Culturally Sensitive Training Programs. Rockville, MD: NIMH, 1986.

204. Hoyle RH, Smith GT. Formulating clinical research hypotheses as structural equation models: A conceptual overview. J Consult Clin Psychol 1994; 62: 429–440.

205. Wig NN. DSM-III: a perspective from the third world. In: Spitzer RL, Williams JBW, Skodol AE, eds. International Perspectives on DSM-III. Washington, DC: American Psychiatric Press, 1983.

206. Florsheim P. Cross-cultural views of self in the treatment of mental illness: Disentangling the curative aspects of myth from the mythic aspects of cure. Psychiatry 1990; 53: 304–315.

207. Katon W, Kleinman A, Rosen G. Depression and somatization: a review. Am J Med 1982; 72: 127–135, 241–247.

208. Westermeyer J. Psychiatric diagnosis across cultural boundaries. Am J Psychiatry 1985; 142: 798–805.

209. Griffith EEH, González CA. Fundamentos de la psiquiatría cultural. In: Hales RE, Yudofsky SC, Talbott JA, eds. Tratado de Psiquiatría, 2nd edn. Barcelona: American Psychiatric Press, 1996.

210. Costello EJ, Compton SN, Keeler G, Angold A. Relationships between poverty and psychopathology: a natural experiment. JAMA 2003; 290: 2023–2029.

211. Pérez Urdániz A, Santos García JM, Molina Ramos R, et al. Aspectos socioculturales en la génesis de los trastornos de la personalidad. Actas Españolas Psiquiatría 2001; 29: 47–57.

212. Luthar SS. The culture of affluence: psychological costs of material wealth. Child Dev 2003; 74: 1581–1593.

213. Pérez Urdániz A, Bueno Carrera GM, Santos García JM, Molina Ramos R. Aspectos arquetípicos del envejecimiento: del Eros al Tánatos [Archetypal aspects of ageing: from Eros to Tanathos]. Actas Luso-Españolas Psiquiatría Neurología 1999; 27: 127–132.

214. Pérez Urdániz A, Santos JM. Are personality disorders a syndrome of the postmodern man? Symposium (# S12C) to the VIII International ISSPD Congress, Florence, Italy, 9–12 October, 2003.

215. Eisenbruch M. Cross-cultural aspects of bereavement, II: ethnic and cultural variations in the development of bereavement practices. Culture Med Psychiatry 1984; 8: 315–347.

216. King LM. Suicide from a 'black reality' perspective. In: Bass BA, Wyatt GE, Powell GJ, eds. The Afro-American Family: Assessment, Treatment, and Research Issues. New York: Grune & Stratton, 1982.

217. Shaw CM, Creed F, Tomenson B, et al. Prevalence of anxiety and depressive illness and help seeking behaviour in African Caribbeans and White Europeans: two phase general population survey. BMJ 1999; 318: 302–306.

218. Malgady RG, Rogler LH, Cortes DE. Cultural expression of psychiatric symptoms: idioms of anger among Puerto Ricans. Psychol Assess 1996; 8: 265–268.

219. Jones BE, Gray BA. Problems in diagnosing schizophrenia and affective disorders among

blacks. Hosp Community Psychiatry 1986; 37: 61–65.

220. Bhugra D. Migration and depression. Acta Psychiatr Scand 2003; 418: S67–S72

221. Cheng H, Guarnaccia PJ, Chung H. Self-attention as a mediator of cultural influences on depression. Int J Soc Psychiatry 2003; 49: 192–203.

222. Trimble MR, Krishnamoorthy ES. Neuropsychiatric disorders in epilepsy: Some transcultural issues. Epilepsia 2003; 44 (Suppl 1): 21–24.

223. Gilliam F. Optimizing health outcomes in active epilepsy. Neurology 2002; 58 (Suppl 5): 9–19.

224. Levy JE, Neutra R, Parker D. Life careers of Navajo epileptics and convulsive hysterics. Soc Sci Med 1979; 13: 53–66.

225. Lin KT. Biological differences in depression and anxiety across races and ethnic groups. J Clin Psychiatry 2001; 62 (Suppl 13): 13–19.

226. Giles DE, Perlis ML, Reynolds CF III, Kupfer DJ. EEG sleep in African-American patients with major depression: a historical case control study. Depression Anxiety 1998; 8: 58–64.

227. Sadler JZ, Fulford KWM. Preoccupied with process, annoyed with content. Curr Opin Psychiatry 2003; 16: 659–660.

2

Neurobiology of depression and new opportunities for treatment

Hrvoje Hećimović and Frank G Gilliam

A neurochemical hypothesis of major depressive disorder has been widely accepted. However, multidisciplinary basic and clinical research in mechanisms of mood dysfunction has yielded remarkable progress in understanding the basic concepts of the neurobiology of depression[1–3]. Several new hypotheses of cerebral mechanisms of depression and antidepressant actions have been formulated. It is clear that depression is a complex disorder, and multiple studies are trying to define neurobiological mechanisms, genetic factors for predisposition, and means by which environmental factors influence pathological mood changes. At the same time our knowledge and understanding of the circuits in the brain that are responsible for normal affect have substantially increased[4–6], and several studies have associated depression with cerebral structural and functional abnormalities[7–11]. Important neuroanatomical research in animal models and humans suggests that dysfunction in specific neural networks might be a foundation for mood regulation[6,12,13]. Cellular mechanisms and neurotransmitter pathways that underlie these structural and functional abnormalities are described in this chapter.

ANATOMY OF NEURAL PATHWAYS

Current theories about the neural localization of depression have developed from several complementary approaches. Early observations of mood disturbance following direct stimulation of the frontal regions, such as Brodmann areas (BA) 47 and 11, by Kleist[14] indicated involvement of frontal and limbic brain regions. Later, studies by Broca[15], Papez[16], Yakovlev *et al.*[17–19] and Maclean[20] focused on anatomical details and cytoarchitectonics of frontal cortex and limbic structures that include the hippocampus, amygdala, orbitofrontal cortex, cingulate gyrus and hypothalamus. With these pioneers and others suggesting a neuroanatomical basis of mood dysfunction, the current work, using sophisticated techniques, only confirmed a prominent role of frontal and temporal lobes in the expression and modulation of mood and affect. Of all the structures that are part of this frontolimbic network, the anatomical structure and connections of the hippocampus and amygdala are best examined.

The hippocampus and its projections

Most studies of neuroplasticity in response to stress and to the hormones of the hypothalamus–pituitary–adrenal (HPA) axis have focused on the hippocampus (in Greek meaning 'shaped like a sea horse'). This is a well-defined structure with rich connections that has been used for many years in behavioral and electrophysiological studies of learning and memory (for studying long-term potentiation and long-term depression). In addition, the highest levels of glucocorticoid receptors in the brain are found in the hippocampus, suggesting its prominent role in response to stress.

Hippocampal main afferents arise in the frontal lobe (BA 12, 13 and 25), occipital lobe (BA 19) and temporal lobe (BA 20, 22, 35, 36 and 38). They provide inputs from the sensory systems by direct projections to the entorhinal cortex. This structure then projects further to the Ammon's horn or the hippocampus proper, consisting of the CA1, CA2 and CA3 layers, across the hippocampal fissure. The frontal lobe cortex (BA 9 and 46) and parietal lobe cortex (BA 7 and 23) project directly to the Ammon's horn. In the Ammon's horn information is processed through a local hippocampal network from the dentate gyrus to the Ammon's horn and to the subiculum. Further, visceral fibers from the basal and lateral nuclei of the amygdala project to the hippocampus via area 28. Fibers from the anterior and midline thalamic nuclei pass through the cingulum to the entorhinal cortex, but some of the fibers also project directly to the hippocampus. Projections from the supramamillary region of the hypothalamus pass through the fornix to the entorhinal area and the hippocampus proper[21,22] (Figure 2.1).

Hippocampal efferents mainly originate from the Ammon's horn and the subiculum. The fibers from the subiculum form the fornix, which is divided by the anterior commissure into the precommissural and postcommisural fornix. Axons that are part of the precommissural fornix end in the septal nuclei, nucleus accumbens, preoptic nucleus of the hypothalamus and the anterior olfactory nucleus, and then project to the medial frontal cortex and cortex in the gyrus rectus including BA 11, 12, 13, 25 and 32, respectively. Fibers of the postcommissural fornix project to the interstitial nucleus of the stria terminalis, anterior nucleus of the thalamus, and ventromedial and lateral mamillary nuclei of the hypothalamus. Additional fibers from the subiculum terminate directly in the basal and lateral nuclei of the amygdala, entorhinal cortex and retrosplenial cortex, and some of them pass the cingulum and terminate in the cingulate cortex. Fibers from the Ammon's horn terminate in the septal nuclei via the precommisural fornix (Figure 2.2).

The amygdala and its projections

The amygdala (in Greek 'almond shaped') integrates information from the sensory system and from the visceral afferents. It contains five nuclei organized into three groups: the central nucleus; the cortical and medial nuclei; and the basal and lateral nuclei. There are five main afferent pathways into the amygdala. Projections from the cerebral cortex and temporal lobe terminate in all amygdalar nuclei: the insular fibers and the fibers from BA 20, 21, 22 and 38 project to the central nucleus, the insular fibers and the fibers from BA 20, 21, 22, 35, 36 and 38 and some subiculum efferents project to the basal and lateral nuclei, whereas other subiculum projections terminate in the cortical and medial nuclei of the amygdala[21].

Further, axons from BA 12, 13, 14, 23, 24 and 25 from the frontal lobe connect to the central nucleus, while those from BA 11, 12 and 24 project to the basal and lateral nuclei. Axons from the interstitial nucleus of the stria terminalis join the fibers from the ventromedial nucleus of the hypothalamus and lateral hypothalamic area and terminate in the central, medial and cortical nuclei of the amygdala. Fibers from the olfactory bulb pass the lateral olfactory stria to end in the medial and cortical amygdalar nuclei (Figure 2.3).

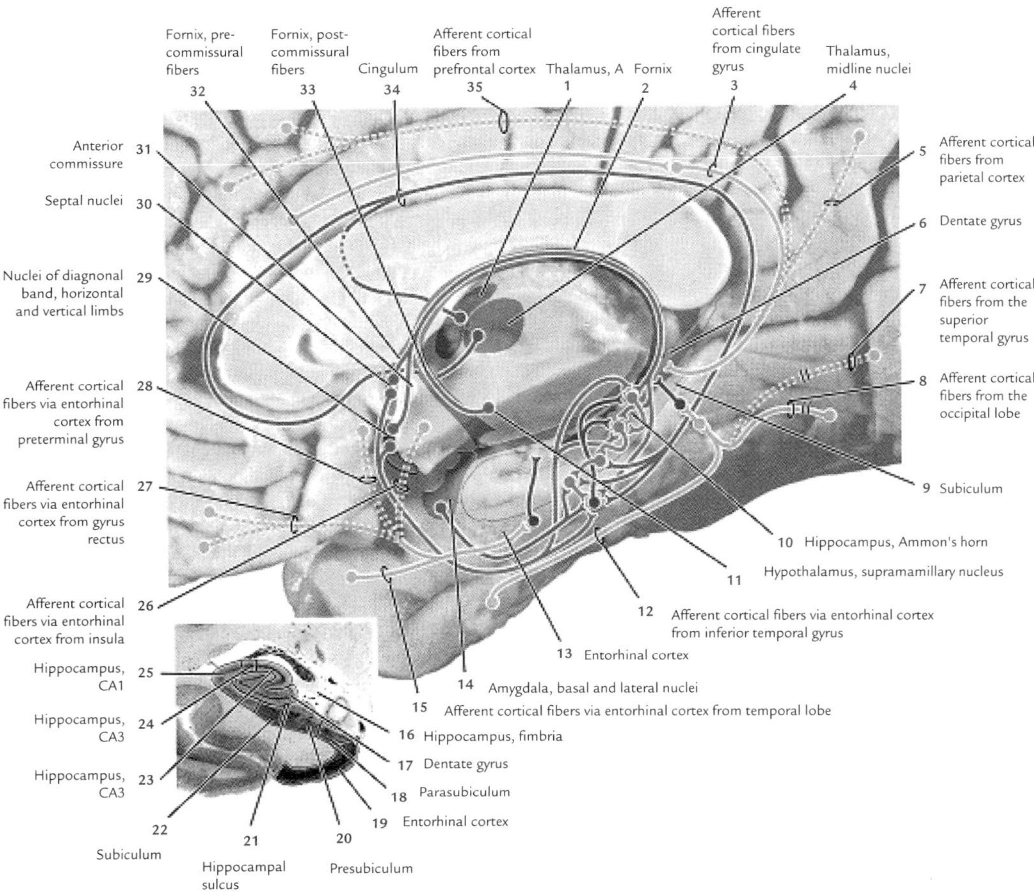

Figure 2.1 Hippocampal afferent pathways. Reproduced from reference 21, with permission. (See also color image on p. xiii)

The five amygdalar nuclei project widely, using four groups of axons throughout the forebrain, midbrain and brain stem. The central nucleus of the amygdala projects to the substantia innominata, nuclei of the diagonal gyrus, the lateral hypothalamic area, the interstitial nucleus of the stria terminalis and the septal nuclei, as shown in Figure 2.4. In addition to these forebrain targets, axons from the central nucleus also course in the medial forebrain bundle to reach their targets in the brain stem. In the midbrain, the fibers terminate in the parafascicular nucleus, the ventral tegmental area, the pars compacta of the substantia nigra, the peripeduncular nucleus, the periaqueductal gray substance and the dorsal raphe nuclei. In the pons and medulla, the projections end in the superior central nucleus, the lateral parabrachial nucleus, the locus coeruleus, the nucleus subcoeruleus, the raphe nuclei magnus,

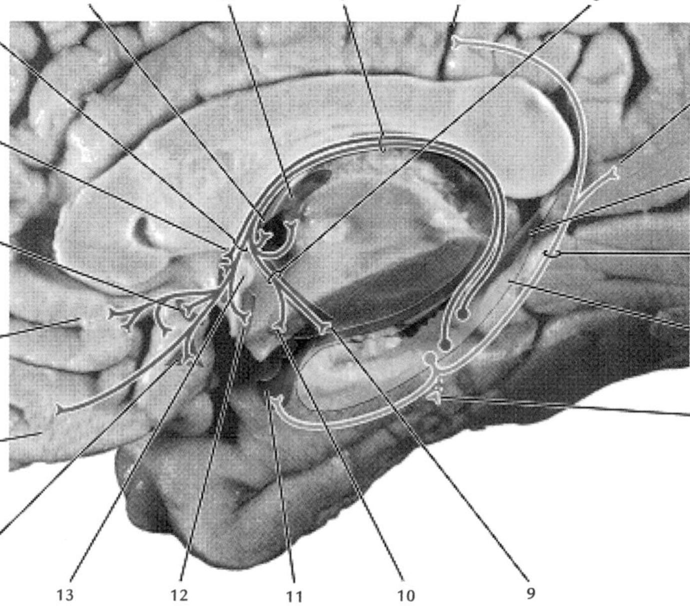

Figure 2.2 Hippocampal efferent pathways. Reproduced from reference 21, with permission. (See also color image on p. xiv)

pallidus and obscurus, the nucleus of the solitary tract and the dorsal nucleus of the vagal nerve.

Amygdalar efferents from the basal, lateral, medial and cortical nuclei travel in the stria terminalis to innervate the interstitial nucleus of the stria terminalis and the paraventricular, anterior, ventromedial and preoptic nuclei of the hypothalamus. Fibers from the basal and lateral nuclei that link with the stria terminalis also innervate the caudate nucleus and putamen, the nucleus accumbens and the olfactory tubercle of the telencephalon. Further, some axons connect directly through the temporal lobe to the septal nuclei and the nucleus of the diagonal gyrus[22].

Prefrontal cortical networks related to mood

Neuroanatomical and pharmacological studies have shown that cortical connections of the amygdala, hippocampus and other limbic structures are primarily directed to the orbital and medial prefrontal cortex in humans[23] and in monkeys[6]. Different staining techniques and

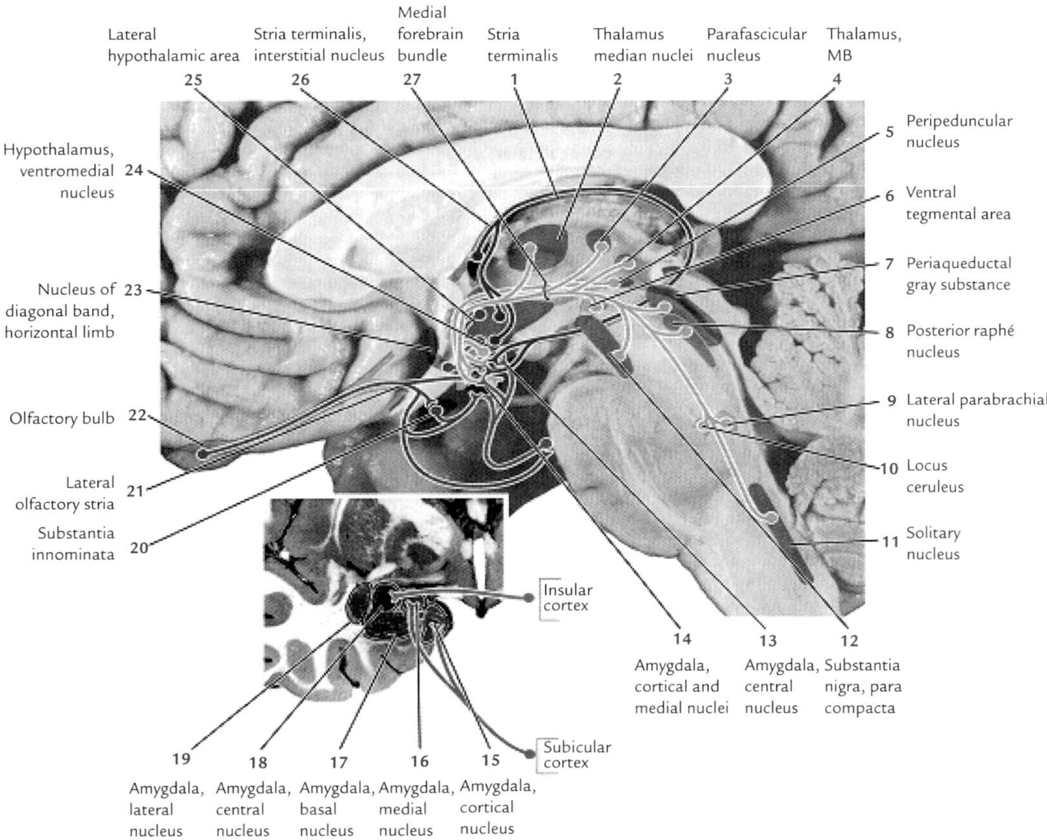

Figure 2.3 Amygdalar afferent pathways. Reproduced from reference 21, with permission. (See also color image on p. xv)

injections of retrograde and anterograde tracers into limited cortical areas suggest that there are at least 22 architectonic areas that can be differentiated within the orbital and medial cortex. Although each of these areas has a distinct structure and connections, they can be grouped into two 'networks' that interconnect and that are defined by corticocortical connections within each network[24] (Figure 2.5). These networks send additional connections to the striatum, medial thalamus and other brain regions[25]. The orbital part of the network consists of the majority of the areas in the orbital prefrontal cortex. Main inputs appear to come from sensory olfactory, gustatory, visceral afferent, visual and somatic sensory fibers. This region also receives many limbic inputs from the amygdala, subiculum, entorhinal cortex and a specific projection from the ventrolateral part of the basal amygdaloid nucleus. The current hypothesis suggests that this network integrates viscerosensory information with affective signals[6].

The medial prefrontal cortical network consists of areas in the medial prefrontal region and some areas in the orbital cortex. Limbic

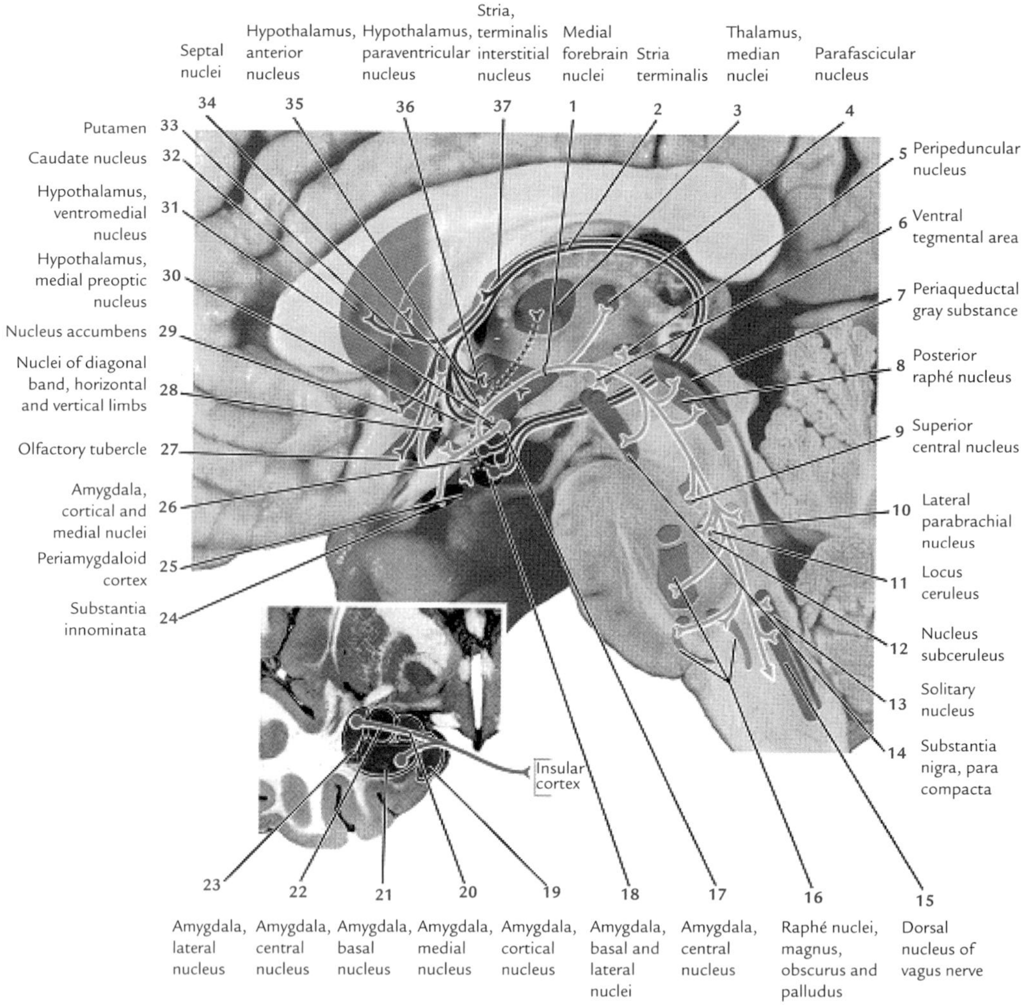

Figure 2.4 Amygdalar efferent pathways. Reproduced from reference 21, with permission. (See also color image on p. xvi)

inputs to these areas are somewhat different and consist of fibers from the ventromedial part of the basal amygdaloid nucleus, and there are less direct sensory inputs from other areas. The main function of this network is to provide major output from the prefrontal cortex to the hypothalamus and to the periaqueductal gray in the brain stem[6,26–30]. Thus, it has been suggested that the medial prefrontal network integrates a visceromotor system, by which the prefrontal cortex controls autonomic and endocrine functions.

It is highly likely that, because of numerous connections between the two networks, they

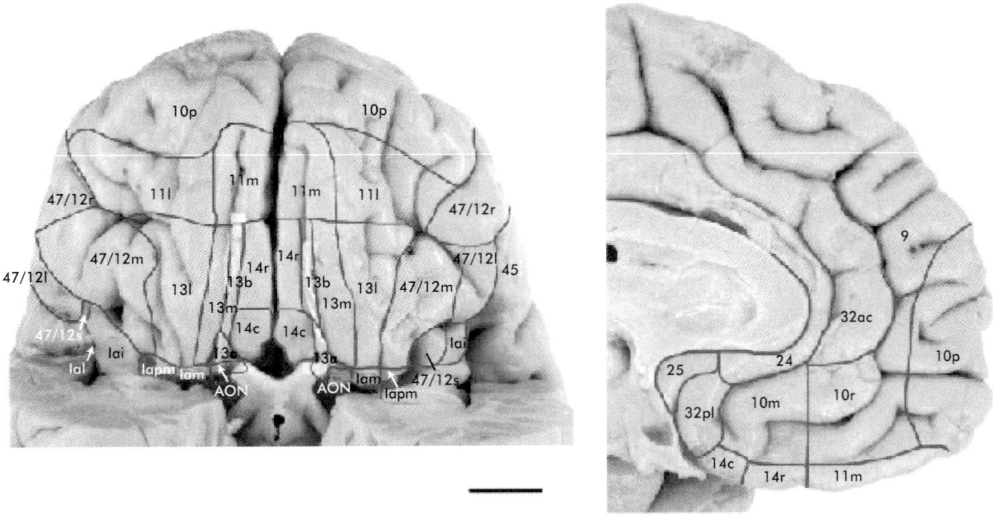

Figure 2.5 Architectonic subdivisions of orbital (left) and medial (right) prefrontal regions. Scale bar = 10mm. Reproduced from reference 23, with permission

also mediate information exchange from viscerosensory to visceromotor systems, and this might be an important feature in mood regulation (Figure 2.6). Recent neuroimaging and quantitative analyses of these two networks showed that many of these areas have significant volume changes and abnormalities in glucose metabolism related to affective symptoms. Quantitative studies of the subgenual prefrontal cortex and the amygdala in patients with depressive symptoms showed volumetric abnormalities that might be due either to glial[31] or neuronal[32] cell loss. Some groups[31,33], but not others[32,34], suggested oligodendroglial cell loss in familial major depression. Functional studies also indicate that patients with unipolar and bipolar depression have consistent abnormalities in medial and/or orbital prefrontal regions[9,11,35,36].

CURRENT MODELS OF DEPRESSION

During the past three to four decades, significant discoveries have been made in understanding the cellular basis of depression[37]. A majority of this work investigated the neurochemical imbalance of two monoaminergic, noradrenergic and serotonergic neurotransmitter pathways that lead to mood changes. This approach was supported by observations that patients with a history of depression who were exposed to high doses of reserpine, which is a drug that depletes monoamine stores, had a recurrence of symptoms of depression[38]. Dietary restrictions of tryptophan, which is a precursor of serotonin synthesis, caused a transient return of depression in patients who were in remission (so-called 'tryptophan depletion challenge')[39]. Following an exposure to

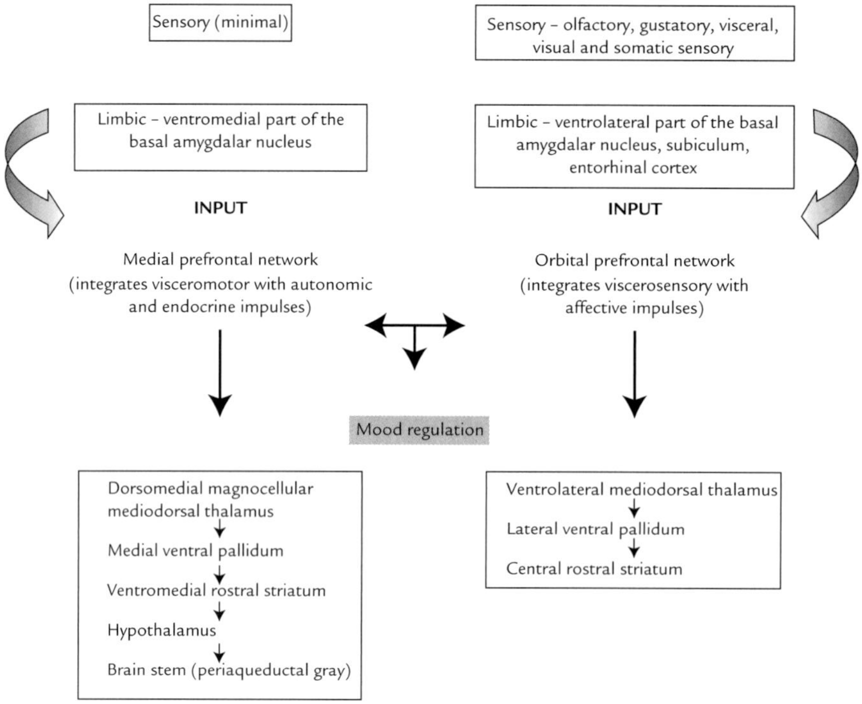

Figure 2.6 Prefrontal cortical networks related to mood. Modified from reference 6

α-methyl-paratyrosine (AMPT), an inhibitor of tyrosine hydroxylase which is the enzyme responsible for the synthesis of catecholamines (the 'AMPT challenge'), an impairment in norepinephrine (noradrenaline) and dopamine synthesis was demonstrated with clinical recurrence of symptoms of depression[40]. A dysfunction of the α_2- and β-adrenergic receptors was observed by their increased density in the brains at autopsy of the patients who committed suicide[41]. These findings indicated a decrease in the activity of noradrenergic neurons in depressed patients.

This research guided the development of most antidepressants that act at the monoaminergic receptors and modulate synaptic concentrations of these neurotransmitters. Subsequent studies showed that long-term antidepressant treatment induced a down-regulation of limbic β-receptors[42], and some but not all antidepressants down-regulated 5-hydroxytryptamine $(5\text{-HT})_2$ receptors[43], whereas tricyclic antidepressants up-regulated 5-HT_{1A} receptors[44]. These observations suggested a monoamine hypothesis of depression, which postulated that depression results from insufficient monoaminergic transmission in the brain, and that a reconstitution of normal synaptic serotonin (5-HT) and norepinephrine levels provides alleviation of depressive symptoms[45,46]. However, it is unclear whether these neurotransmitters provide a final common pathway for the therapeutic effect of antidepressant drugs, and how their interaction contributes to their actions.

Activation of the HPA axis is important in understanding the etiology of depression. Several brain structures control the activity of the HPA axis, including the hippocampus (with an inhibitory effect on hypothalamic neurons) and the amygdala (with a direct excitatory influence)[47]. Glucocorticoids from the circulation exert powerful feedback on the HPA axis, and under physiological conditions appear to enhance hippocampal inhibition of HPA activity, and probably hippocampal function in general[48]. Dysfunction of this axis also presents a challenging hypothesis in the neurobiology of depression.

The neurotrophic hypothesis of depression focuses on neurotrophic factors and their role in the etiology of depression and its treatment. Recent studies have described the role of neurotrophic factors in regulation of neuronal growth and differentiation during development, and suggested their importance in the plasticity and survival of adult neurons and glia[49].

With increasing knowledge of this complex disorder, several promising hypotheses of the neurobiology of depression and antidepressant action have been proposed[12,48]. These include the monoamine hypothesis of depression, dysfunction of the hippocampus and the HPA axis, and the neurotrophic hypothesis of depression.

The monoamine hypothesis of depression

Detailed investigation of the monoaminergic neurotransmitter actions started with discoveries that two drug classes can elevate mood: the tricyclic antidepressants and the monoamine oxidase (MAO) inhibitors. The first tricyclic agents (imipramine) were used in studies with antihistaminic drugs, and the effects of the MAO inhibitors (iproniazid) were observed during antitubercular treatment. This is also historically important, as it provided the first indication that chemical changes in the brain modulate depressive symptoms. Investigation of their mechanisms of action showed that MAO inhibitors and tricyclic antidepressants elevate synaptic levels of norepinephrine and 5-HT. Inhibition of serotonin or norepinephrine reuptake transporters by tricyclic antidepressants, and inhibition of MAO by MAO inhibitors led to the development of second-generation medications, selective serotonin reuptake inhibitors (SSRIs) and selective norepinephrine reuptake inhibitors that are both widely used in clinical practice[50,51] (Table 2.1).

In addition to investigations in the role of serotonin and norepinephrine in mood dysfunction, a role of a dopaminergic mechanism in depression has also been considered, and is supported by several lines of evidence[51–53]. L-Dopa has been shown to relieve mood symptoms in patients with Parkinson's disease, and amphetamines elevate mood in depressed and non-depressed individuals[54,55]. Dopamine reuptake inhibitors, such as bupropion, have been studied as antidepressants[56]. The potential role of dopamine in the neurobiology of depression is further supported from the regional specificity of dopaminergic projections from the ventral tegmental area to the orbital prefrontal cortex, striatum and anterior cingulate area that are typically identified in functional neuroimaging studies in patients with depression. However, none of the studies clearly showed whether dopaminergic stimulation alone can alleviate depressive symptoms.

A GABAergic deficit in central neurons has also been suggested to result in depressive symptoms, and low levels of γ-aminobutyric acid (GABA) in the plasma and a decreased

Table 2.1 Current evidence for neurotransmitter systems involvement in mood changes. Table organized from references 12, 42–46, 56, 57 and 138

Serotonergic system	*GABAergic system*
Decrease in CSF 5-HIAA	Reduced CSF and plasma GABA
Reduced 5-HT_{1A} receptor binding	Reduced $GABA_B$ density
Antidepressant efficacy of agents that increase synaptic 5-HT	Lithium, VPA and some other AED increase GABA signaling
Relapse of depressive symptoms after tryptophan depletion in treated patients in remission	*Cholinergic system*
Some antidepressants decrease 5-HT_2 density	Some cholinomimetics cause depressive symptoms
Noradrenergic system	Enhanced cholinergic sensitivity
Increase in plasma NE concentration	*Glutamatergic system*
Reduced CSF and urinary MHPG	Stress potentiates Glu signaling
Changes in α_2AR and β-AR density in the hippocampus and amygdala	Lamotrigine probably decreases Glu release
Some antidepressants increase NE concentration	Ketamine may have antidepressant effects
Decreased density of NE transporter	*Corticotropin releasing factor (CRF)*
Dopaminergic system	Prolonged hypercortisolemia and resistance to feedback inhibition, but sensitive to chronic antidepressant treatment
Reduced CSF HVA	Adrenal and pituitary hypertrophy
Some antidepressants increase DA levels	Increased CSF CRF, and reduced CRF receptors in postmortem brain
Some DA reuptake inhibitors are potential antidepressants	Indication for anxiolytic effects of some CRF agonists
AMPT and reserpine cause depressive symptoms in susceptible individuals	
Increased prevalence of depression in PD	
L-Dopa improves mood in some PD	

5-HIAA, 5-hydroxyindole acetic acid; AED, antiepileptic drug, AMPT, methyl-*para*-tyrosine; CSF, cerebrospinal fluid; HVA, homovanillic acid; MHPG, 3-methoxy 4-hydroxyphenylglycol; VPA, valproic acid; NE, norepinephrine; AR, adrenoreceptor; DA, dopaminergic receptor; PD, Parkinson's disease

quantity of $GABA_B$-binding sites were found in the brains of patients with and without depression who committed suicide[57]. Some investigators suggested that GABA has an inhibitory effect on the biogenic amine neurotransmitters, and that this inhibition may be implicated in local networks and interneurons. It has further been suggested that the hypothesis of a GABA deficit in mood disorders does not compete with, but complements, the hypothesis of alterations in norepinephrine and 5-HT function in mood disorders. Also, studies that will evaluate antiepileptic drugs, such as phenobarbital and vigabatrin that are known to enhance a GABAergic response, should also examine their antidepressant properties.

Serotonergic mechanism of action

The serotonin system is described in more detail, because it has been most extensively studied. The vast majority of serotonergic nerve terminals in the brain originate in the cell bodies of the raphe nuclei in the brain stem. The raphe nuclei are localized along the midline of the brain-stem tegmentum and only the dorsal raphe nucleus lies in the ventral quadrant of the periaqueductal gray. Approximately 30–80% of the neurons of the raphe nuclei synthesizes serotonin. The raphe nuclei are subdivided into two groups: the rostral group, that includes caudal linear, the median raphe and the dorsal raphe nucleus with efferent projections directed rostrally; and the caudal group, which includes the raphe magnus, the raphe obscurus, the raphe pallidus nuclei and a cluster of neurons in the lateral reticular formation of the myelencephalon, project mainly to the caudal brain stem. The observation that the axonal arborization of a single typical serotonergic neuron could reach large brain areas, with different functions, introduced a concept of a diffuse system operating jointly with other aminergic projections in the brain.

5-HT receptors are the most numerous and diverse of the monoamine transmitter system receptors. The central 5-HT system is thought to function as a behavioral inhibitory system and is involved in the regulation of food intake, circadian rhythms, mood, anxiety, aggression and impulsivity. It exerts its action by modulating various other transmitter systems, by integrating and connecting other spatially distinct systems. Serotonergic dysfunction has been implicated in a number of neuropsychiatric disorders.

At the cellular level, 14 types of 5-HT receptor have been identified and classified on the basis of structural similarity, and have been subdivided into seven families[58], as shown in Table 2.2. Each of them has a specific localization and distinct effector pathways. The largest family of 5-HT receptors is the 5-HT_1 family with five 5-HT receptor subtypes. 5-HT_{1A} receptors act as autoreceptors in the raphe nuclei and postsynaptically in the hippocampus, septum and amygdala, where they initiate neuronal inhibition. A knockout transgenic mouse that lacks the 5-HT_{1A} receptors displays a spontaneous phenotype with increased sensitivity to anxiety that is caused by an increased serotonergic tone due to failed negative feedback of the 5-HT_{1A} autoreceptor. This phenotype persists if the 5-HT_{1A} receptor is restored[59].

The 5-$HT_{1,2,4–7}$ receptor families are members of the G-protein-coupled receptor superfamily, whereas the 5-HT_3 receptor differs and belongs to the gated ionophore receptor superfamily. In the central nervous system (CNS), 5-HT is believed to act predominantly as an inhibitory neurotransmitter, although there are some variations of the intracellular effects on the second messenger systems among the 5-HT receptor subtypes. Following its release, 5-HT that is present in the synaptic cleft is inactivated and transported back into the presynaptic terminal by the serotonin transporter (SERT). It is then either degraded by MAO to 5-hydroxyindoleacetic acid (5-HIAA), or repackaged into secretory vesicles via the serotonin transporter[60].

Following cloning of the human SERT[61], a number of biochemical and pharmacological studies showed that this transport process may alter the excitability of the neurons, and that the transporter exhibits voltage dependence that affects transport efficiency[62]. Further research showed that decreased brain and platelet SERT binding was a remarkably consistent observation in depression[60]. Therefore, some studies tried to determine whether SERT dysregulation had a role in the pathophysiology of depression. It appears that both the tricyclic antidepressants (TCAs) and the newer SSRIs bind to the serotonin transporter and inhibit the uptake of 5-HT, so that many, if not most, of the antidepressants used in clinical practice are potent SERT antagonists.

Table 2.2 5-hydroxytryptamine (HT) receptor classification and mechanism of action. Table derived from references 58–60

Receptor class	*Mechanism of action*	*Clinical expression*	*Pharmacotherapy*
5-HT_{1A}	(-)AC	Knockout transgenic mouse anxious, benzodiazepine-resistant	Buspirone, Pindolol, Gepirone
5-HT_{1B}	(-)AC	=5-$HT_{1D\beta}$; knockout transgenic mouse aggressive	Sumatriptan
5-HT_{1D}	(-)AC	=5-$HT_{1D\alpha}$	Sumatriptan
5-HT_{1E}	(-)AC	=S31	
5-HT_{1F}	(-)AC	=5-HT_{1E}	
5-HT_{2A}	PLC	=5-HT_2	Clozapine, LSD
5-HT_{2B}	PLC	=5-HT_{2F}	
5-HT_{2C}	PLC, GC	=5-HT_{1C}; knockout transgenic mouse obese, promotes epileptogenesis	Clozapine, LSD
5-HT_{3A}	Na^+/Ca^{2+} channel	Forms heteromer with 5-HT_{3B}, gut motility	Ondansetron
5-HT_{3B}	Na^+/Ca^{2+} channel	Forms heteromer with 5-HT_{3A}	Ondansetron
5-HT_4	(+)AC	2 splice variants: 5-HT_{4L}, 5-HT_{4S}; atrial arrhythmias, gut motility	Zacopride
5-HT_{5A}		Knockout transgenic mouse more active	LSD
5-HT_6	(+)AC	=St-B17	LSD
5-HT_7	(+)AC	80–90% homologous pseudogene; gut motility	Clozapine, LSD

AC, adenylyl cyclase; GC, guanylyl cylase; LSD, lysergic acid diethylamide; PLC, phospholipase C

Noradrenergic mechanism of action

Norepinephrine is the principal sympathetic neurotransmitter in the periphery and is prevalent throughout the brain. Its terminals are also diffusely distributed throughout the CNS. Their cell bodies reside predominantly in the locus coeruleus of the dorsal pons. The norepinephrine system is similar to the 5-HT system, in that it has abundant projections throughout the entire cortex. Many areas that are known to participate in mediation of stress responses and autonomic functions, such as the paraventricular nucleus, the lateral hypothalamus and the preoptic area, project their fibers towards the locus coeruleus, adding further support for a role of the norepinephrine system in stress and arousal; main locus coeruleus efferents are cortical terminal fibers primarily ending in the layers I, IV and V, suggesting a role of norepinephrine in modulating intracortical and thalamocortical gating. These findings indicate that projections to the thalamus might have a role in co-ordinating activity related to levels of arousal, along with a role in sensory and motor gating. Other studies have indicated that its hippocampal projections are important in learning and memory, and hypothalamic projections have a role in neuroendocrine effects related to the state of arousal[63].

Molecular studies showed that there are two subclasses of the α receptor, the α_1 and α_2 receptors, which are as distinct as the α and β classes. Several investigations using α_{2a}- and α_{2c}-adrenergic receptor knockout mice have suggested that the α_{2a}-adrenergic receptor is 'stress-protective', and the α_{2c}-adrenergic receptor mediates stress susceptibility. All of the norepinephrine receptors belong to the G-protein-linked receptor superfamily and thus mediate slower neuromodulatory postsynaptic responses.

Major observations supporting norepinephrine system dysfunction include both decreased and increased norepinephrine metabolism in depressed patients[64,65]. Prolonged exposure to stress decreases α_{2a}-receptor density in the amygdala and the hippocampus. Further, the locus coeruleus, which is the main norepinephrine-containing nucleus in the brain, showed increased density of the α_2 adrenergic receptors in suicide victims with a history of major depression. The results also showed increased activity of tyrosine hydroxylase in the locus coeruleus in addition to decreased density of the norepinephrine transporter. The majority of these effects are reproduced by norepinephrine depletion from the brain, and are significantly reversed by antidepressant administration[66].

Dopaminergic mechanism of action

Dopamine is transported into storage vesicles in both dopamine- and norepinephrine-producing neurons via amine-specific transporters. Norepinephrine-specific neurons, however, contain dopamine-β-hydroxylase within the vesicles that rapidly transform dopamine to norepinephrine via hydroxylation of the β-carbon. After norepinephrine is released into the synaptic cleft following vesicle fusion, it is rapidly returned to the synaptic terminals via the norepinephrine transporter. These specific carriers are located on the outer membrane of the synaptic terminals. Inside the terminal, the neurotransmitter that is not rapidly returned to the storage vesicles is biochemically degraded. The norepinephrine transporter is inhibited by many classes of antidepressants, which acutely increase the synaptic norepinephrine concentration[63].

Intracellular signaling pathways

The mechanism by which antidepressants regulate mood functioning appears to be highly complex. One would presume that these drugs simply enhance the acute actions of the neurotransmitters. However, multiple observations have suggested that the mood elevation occurs only after a prolonged administration, usually after several weeks of continuous treatment. This indicates that enhanced neurotransmission *per se* is not the major determining factor of the clinical actions of these drugs, but that they generate a gradually evolving mechanism that mediates their antidepressant actions.

During the past decade, research on the pathophysiology and treatment of depression has focused investigation on the intracellular signaling pathways[67]. Multiple levels of cellular signaling pathways form complex neural networks that modulate signals generated by multiple neurotransmitter and neuropeptide systems. These signaling pathways are involved in neuroplasticity and regulate complex psychological and cognitive processes, as well as autonomic and motor functions.

Amongst the better described intracellular signaling pathway that mediates effects of antidepressant treatment is the cAMP cascade, which appears to be regulated by both 5-HT and norepinephrine[47] (Figure 2.7). This pathway also represents a common target for several classes of antidepressant. Their proposed mechanism of action suggests that the receptor activation leads to the generation of cAMP via the stimulation of adenyl cyclase by the G-protein subtype Gs_{α}. Intracellular Ca^{2+} levels also regulate certain types of adenyl cyclase. The generation of the cAMP then results in activation of cAMP-dependent protein kinase (PKA)[67], and one of its substrates is the transcription factor cAMP response element binding protein (CREB). This protein in its phosphorylated form regulates transcriptional activity[68]. Overexpression of CREB in the hippocampus results in an antidepressant effect, and suggests that CREB may serve as a potential molecular target for novel therapeutic agents.

This evidence suggests that the cause of depression is far more complex than just alteration of synaptic levels of norepinephrine and 5-HT, as suggested by the monoamine hypothesis of depression. Furthermore, there is still no direct evidence that depression is caused by disturbances in monoamine circuits, but rather in dysfunction of the higher neural systems network.

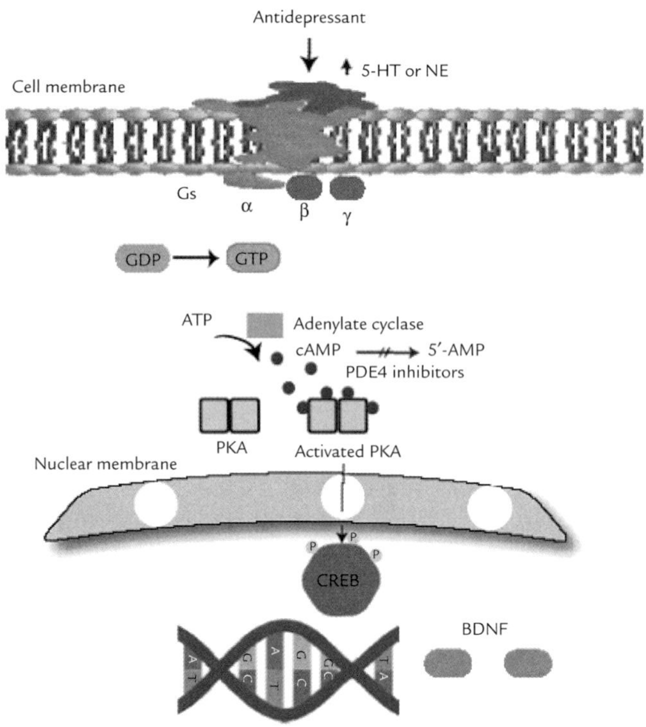

Figure 2.7 Intracellular cAMP signaling pathway. NE, norepinephrine; PKA, protein kinase A; CREB, cAMP response element binding protein, PDE4, phosphodiesterase 4; BDNF, brain-derived neurotrophic factor. (See also color image on p. xvii)

Potential role of antidepressants and neurotransmitters in treatment of epilepsy

Early studies showed that depletion of monoamines with reserpine was associated with an increase in frequency and severity of seizures in patients with epilepsy[69]. Although the data for antidepressant drug effects in human epilepsy are not as systematic as those reported in the animal models, some observations appear promising. Imipramine, a tricyclic antidepressant that blocks reuptake of norepinephrine and 5-HT, was reported to suppress absence and myoclonic seizures[70]. Other groups[71] reported a decrease in seizure frequency in patients treated with doxepin, trazodone and desipramine. Some studies[72] found a seizure frequency reduction with introduction of fluoxetine, but this effect appears to be more complex[73,74].

Recently, it has been suggested that the anticonvulsant effect of vagal nerve stimulation (VNS) could be mediated by the activation of monoaminergic transmission. It has been shown that VNS activates the locus coeruleus neurons in the rat[75], and that this effect may be responsible for its antidepressant action[76,77]. Additional studies with inactivation of norepinephrine and 5-HT neurons in the rat brain prevented or significantly reduced the anticonvulsant effect of VNS against electroshock- or pentylenetetrazol-induced seizures[78].

These studies provided data supporting a speculation that depression and epilepsy may share common etiological mechanisms, and indicate that the occurrence of one of them may facilitate the development of the other.

Genetic and environmental studies

The majority of genetic studies have been carried out using animal models. In recent years we have started to gain better understanding of the interaction between genes and the environment in humans, particularly in early life, with further efforts to determine the genetic components of the clinical expression of the diverse symptoms of depression.

One of the most important findings in understanding the genetic background of depression was achieved in analysis of the 5-HT transporter (5-HTT) gene *SLC6A4*, located on chromosome 17q11[79]. This gene exhibits a functional polymorphism in the promotor region and consists of either long (l) or short (s) alleles. Animal studies found that animals with two l alleles exhibited less behavioral deterioration following stress. In contrast, mice with one or two copies of the s allele showed more fearful reaction to stress. Monkeys with the double s allele that were raised in stressful conditions had impaired serotonergic transmission[80]. However, in the absence of stress, there was no difference in the expression of genotype. 5-HTT-deficient mice display increased anxiety, apparently due to increased serotonergic neurotransmission resulting in the down-regulation of 5-HT_{1A} receptors[81,82].

Research in human subjects with two s alleles showed that the probability of a major depressive episode was 43% among those who had four or more stressful experiences in their lifetime[83]. That was more than double the risk for the subjects with two l alleles who were similarly exposed to stress. The average score on a depression symptom inventory was likewise more than twice as high for stressed people with two s alleles as for those homozygous for the l alleles. Further, abuse during childhood predicted depression after the age of 18 only in people carrying at least one s allele. Among the 11% who experienced severe maltreatment, the subjects with the double s allele ran a 63% risk of a major depressive episode. The l homozygous participants averaged a 30% risk, regardless of whether they were abused in childhood. These were the first studies to combine the genetic background of the serotonin transporter gene with environmental influences. These need further intra- and intercultural validation.

Neuroimaging data in humans also support the role of 5-HT in the stress response. One smaller study indicated that humans with one or two copies of the s allele exhibited greater amygdala neuronal activity to fearful stimuli compared to individuals homozygous for the l allele[84,85].

There are a number of other genes currently investigated for their role in major depression. 5-HT is synthesized from tryptophan, and the rate-limiting enzyme is tryptophan hydroxylase (TPH), which is restricted to the raphe neurons. The *TPH* gene has been assigned to chromosome 11p15.3-p14[86], and recently a link between *TPH* gene polymorphism and a correlation with aggression and impulsivity has been proposed[87]. In one study[88], it was shown that the presence of the L variant of the *TPH* gene was associated with decreased cerebrospinal fluid (CSF) 5-HIAA concentrations, and that this polymorphism might predict suicidal behavior. In another study, a similar association was found in schizophrenic patients between the L allele in male subjects and a history of violent behavior[89]. Further, in suicidal and depressed patients, a less common TPH U allele occurred with increased frequency in a Caucasian population[90]. This supports a role for the serotonin system-related genes in depression[91]. Nevertheless, these studies lack intercultural validity and need to be repeated in larger populations.

Although genomic variations have been identified in all five dopamine receptor genes, only the role of the D_2, D_3 and D_4 receptor genes (*DRD2, DRD3* and *DRD4*, respectively) have been studied in personality disorders and patients with Parkinson's disease. *DRD4*, which is a molecular target for atypical antipsychotics, has been mapped to human chromosome 11p15.5[92]. The potential role of *DRD4* in personality traits and affective behavior has attracted considerable attention.

Dysfunction of the hippocampus and hypothalamic–pituitary–adrenal axis

The HPA axis is important in understanding the etiology of depression. Neurons in the paraventricular nucleus (PVN) of the hypothalamus secrete corticotropin releasing factor (CRF), which stimulates the synthesis and release of adrenocorticotropic hormone (ACTH) from the anterior part of the pituitary gland. ACTH stimulates the synthesis and release of glucocorticoids from the adrenal cortex. Glucocorticoids then exert their effect on various brain regions. Diverse brain structures control the activity of the HPA axis, including the hippocampus and amygdala[47]. Further, glucocorticoids, once in the circulation, exert powerful feedback on the HPA axis, and under normal physiological condition appear to enhance hippocampal inhibition of HPA activity, and probably hippocampal function in general[48] (Figure 2.8). On the other hand, prolonged increased concentration of glucocorticoids, for example during severe stress, can be harmful and damage the hippocampal neurons, particularly CA3 pyramidal neurons, possibly by reduction of dendritic branching and loss of dendritic spines that are included in glutamatergic synaptic inputs[93]. In addition, it was also proposed that the resulting

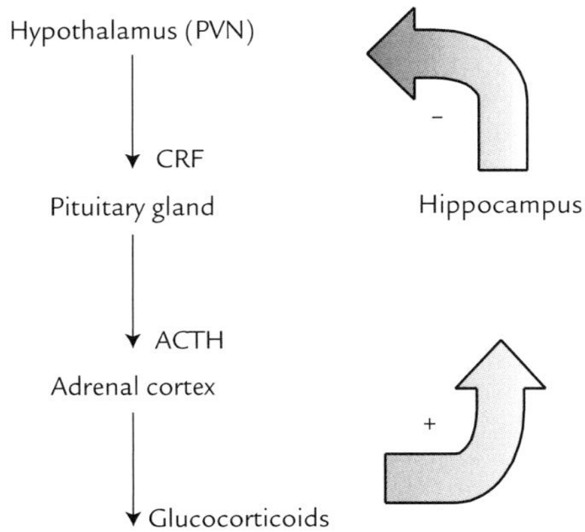

Figure 2.8 The hypothalamic–pituitary–adrenal axis in normal mood regulation. CRF, corticotropin releasing factor; ACTH, adrenocorticotropic hormone

hypercortisolemia interferes with development of new granular cell neurons in the adult hippocampal dentate gyrus[94]. A significant percentage of patients with Cushing's disease, in which pituitary adenomas result in cortisol hypersecretion, exhibit depressive symptoms, as well as hippocampal atrophy[95]. Some authors have suggested an inverse relationship between hippocampal atrophy and plasma cortisol concentrations; following a surgical procedure, the hippocampal volume increases in proportion to the decrement in urinary free cortisol concentrations[96]. In addition, recurrent stress, associated with relapses of major depression episodes and hypercortisolemia, might lower the threshold for cell death in response to a variety of physiological (such as aging) and pathological events.

Clinical studies support this hypothesis and show that abnormal, excessive activation of the HPA axis is observed in almost half of the individuals with depression, and that this effect is reversed by antidepressant treatment[97]. The exact mechanism by which glucocorticoids affect the hippocampus is unclear; some groups have proposed the facilitation of glutamatergic signaling and inhibition of glucose transport. It has been shown that the stress-induced secretion of glutamate in the hippocampus resulted in atrophy of CA3 pyramidal neurons, and that the effect was attenuated after application of an NMDA antagonist[93]. This increase in extracellular levels of glutamate resulted in high intracellular levels of calcium, which can cause neurotoxicity and contribute to the pathogenesis of seizures and stroke, by increasing the vulnerability of neural tissue.

Other groups have suggested that the decrease in hippocampal volume is due, at least partially, to impaired neurogenesis that occurs in response to both acute and chronic stress[94]. For example, adrenalectomy increases neurogenesis, and treatment with high levels of glucocorticoids results in down-regulation of neurogenesis in response to stress. Aging also influences the rate of neurogenesis. Although neurogenesis continues into late life, the process becomes significantly reduced[98]. This decreased rate of cell generation could occur for two reasons: as a result of the up-regulation of the HPA axis, or from higher basal levels of adrenal steroids that occur later in life. A support for this comes from studies in which lowering glucocorticoid levels in aged animals restored neurogenesis to levels observed in younger animals[99]. Interesting experiments have provided us with additional information. Studies in glucocorticoid receptor knockout mice showed significant alterations in hippocampal neurogenesis[100]. A reduction of granular cell neurogenesis (up to 65% of control levels) was found in the mineralocorticoid receptor $(MR)^{-/-}$ mice, whereas glucocorticoid receptor $(GR)^{-/-}$ mice did not show neurogenic disruption, thus indicating a role for the MR receptor in the pathogenesis of the hippocampal changes observed in chronic stress. These preclinical observations raise an interesting possibility that CRF and GR antagonists, currently being developed for the treatment of mood and anxiety disorders, might be more beneficial in the treatment of elderly depressed patients.

Further, ovariectomy decreases the proliferation of new cells in the hippocampus, an effect that is reversed by estrogen replacement. The rate of neurogenesis fluctuates over the course of the estrus cycle in rodents, and the total rate of cell birth is higher in female rodents relative to males. This suggests its potential role in the beneficial cognitive effects of estrogen. Additionally, the regulation of neurogenesis by this gonadal steroid may also provide important evidence about certain sexually dimorphic characteristics of mood disorders.

Most patients exhibit increased cortisol production, which can be measured by elevation of urinary free cortisol, and the decreased ability of dexamethasone to suppress plasma levels of cortisol, ACTH and β-endorphin. Other reports[101,102] suggested hypersecretion of CRF in some depressed patients, which was also corrected by antidepressant treatment. CRF is a major neuropeptide mediator of the stress responses in the CNS and is expressed in the paraventricular nucleus of the hypothalamus to co-ordinate the release of ACTH from the anterior pituitary[97]. It has been observed that levels of CRF are increased in the CSF in suicide victims and in persons with depression[103]. So far, two CRF receptors have been identified in the brain – CRF_1 and CRF_2 – both widely distributed in the brain, and there is some evidence that both are involved in the response to stress[104,105].

Recent data have suggested that CRF is important in long-term effects on early-life stress and hippocampal function[105]. Administration of this substance in the brains of immature rats reduced their memory function throughout life. Anatomically, there is a progressive loss of hippocampal CA3 neurons and chronic up-regulation of hippocampal CRF expression. Central administration of the CRF1 antisense oligodeoxynucleotides has anxiolytic effects against both CRF concentration and psychological stressors. There were also observations[60] that there is a down-regulation in the CRF receptor number in the frontal cortex of depressed patients compared to controls. Following administration directly into the CNS of laboratory animals, CRF produced many effects that are similar to the symptoms observed in depressed patients[106]. These included decreased appetite and weight loss, decreased sexual activity, decreased sleep and psychomotor alterations. Further, in another study, chronic administration of CRF to normal volunteers resulted in HPA axis alterations indistinguishable from those of patients with major depression[101]. In recent years some studies have evaluated a family of CRF-related peptides, such as urocortin[102], which is also regulated by stress and provides more insight into regulation of the HPA axis. It is expected that focusing on the CNS-permeable CRF antagonists will allow further investigation of their clinical use in the treatment of depression and anxiety.

From this extensive work it appears that stress and release of glucocorticoids can directly cause neuronal atrophy and reduce cellular resilience. As a consequence, affected neurons become vulnerable to insults, such as ischemia, hypoglycemia and excitatory amino acid toxicity. Therefore, the proposed mechanisms by which glucocorticoids reduce cellular resilience possibly involve an inhibition of glucose transport, diminishing glucose metabolism and brain energy production, and also by the excessive facilitation of glutamatergic signaling.

Nonetheless, it is still unclear whether these pathophysiological abnormalities are the primary cause of depression or part of some secondary pathways in the neural network. Future research should examine the function of these substances and also of CRF in extrahypothalamic areas, particularly in the amygdala, prefrontal cerebral cortex, locus coeruleus, raphe nuclei and other parts of the frontolimbic network.

Neurotrophic hypothesis of depression

There has been remarkable progress in our understanding of the actions of antidepressant action on intracellular signaling pathways. Nevertheless, the current data represent a tip of the iceberg only to provide a hypothetical framework for identifying the critical signaling pathways. The neurotrophic hypothesis of depression focuses on neurotrophic factors and their role in the etiology of depression and its treatment. Earlier studies described the role of neurotrophic factors in the regulation of neuronal growth and differentiation during development. However, more recent studies have suggested their importance in the plasticity and survival of adult neurons and glia[49]. Brain-derived neurotrophic factor (BDNF) is a member of the nerve growth factor family, and it has been shown to increase synaptic strength, survival and growth of mature neurons through activation of a transmembrane receptor that contains intrinsic tyrosine kinase B (TrkB) activity.

It has been shown in animals that acute and chronic stress decreased the levels of BDNF expression in the dentate gyrus and pyramidal cell layer of the hippocampus[107]. This reduction appeared to be mediated by glucocorticoid action and by an increase in serotonergic transmission. Chronic, but not an acute, administration of antidepressant drugs or chronic electroconvulsive seizure increased BDNF expression and also prevented a stress-induced decrease in BDNF levels[108] (Table 2.3). Studies further showed that induction of BDNF by antidepressants is at least partly mediated via the transcription factor cAMP response element binding protein (CREB). Levels of CREB are reduced in the temporal cortex in depressed subjects[109], but it has been shown that virtually all major classes of antidepressants increased levels of CREB expression in several brain regions, including the hippocampus[110].

Some studies have shown that BDNF can promote epileptogenesis. Data from transgenic mice with increased BDNF signaling due to overexpression of its receptor TrkB have suggested a role of BDNF and TrkB in seizure generation and acute cellular damage after kainate-induced status epilepticus[111]. Other groups have found that BDNF action in the

Table 2.3 Brain-derived neurotrophic factor (BDNF) and mood regulation. Current concept of action

Increases synaptic strength, survival and growth of neurons
Acute and chronic stress decrease levels of BDNF in the hippocampus
Chronic antidepressant administration increases BDNF expression, partly mediated by cAMP response element binding protein
Promotes epileptogenesis

ventral tegmental area–nucleus accumbens (VTA-NAc) pathway might be related to development of a depression-like phenotype[112]. This interpretation is intriguing in that it suggests a role for BDNF in the VTA-NAc that is opposite to the proposed role for BDNF in the hippocampus.

Thus, most antidepressants appear to up-regulate the expression of BDNF and TrkB by increasing CREB and cAMP response element-mediated gene expression. In line with this, some groups[113] have found, in postmortem studies, increased BDNF expression in dentate gyrus, hilus and supragranular regions in subjects treated with antidepressant medications, compared with those not treated with antidepressants. BDNF and related neurotrophic factors appear to be necessary for the survival and function of neurons, suggesting that a sustained reduction of these factors can affect neuronal viability. Despite early belief that endogenous neurotrophic factors increase cell survival by providing optimal trophic support, more detailed studies from sympathetic neurons suggest that this effect is mediated by an inhibition of cell death cascades[114]. There is increasing evidence that neurotrophic factors inhibit cell death cascades by activating the mitogen-activated protein (MAP) kinase signaling pathway and that phosphatidylinositol-3 kinase inhibits apoptosis by increasing expression of Bcl-2 protein[115]. Thus, it appears that the cascade of the neurotrophic factor/MAP kinase/Bcl-2 components have a critical role in cell survival. If this is true, then minor oscillations in this signaling cascade could profoundly affect cellular viability.

Another confirmation for the role of BDNF has been suggested using animal models and the learned helplessness paradigm. In this paradigm rats are subjected to repeated stress that results eventually in failure to escape during subsequent testing. This behavior was immediately reversed by a single, bilateral hippocampal infusion of BDNF, or by repeated systemic administration of chemical antidepressants[49]. However, inhibitors of TrkB or of MAP kinase suppressed the effect of BDNF, suggesting a role for the TrkB–MAP kinase cascade in the mechanism of action of antidepressive drugs[49].

This indicates that antidepressants possibly increase the synaptic concentration of monoamine neurotransmitters such as 5-HT or norepinephrine and activate the extracellular signal-regulated kinase (ERK)/MAP kinase cascade by increased stimulation of G-protein-coupled receptor pathways. Subsequently, these receptors activate the MAP kinase cascade via stimulation of second messengers, such as cAMP, inositol triphosphate (IP_3) and Ca^{2+}, and their protein kinases or some other signaling pathways upstream of MAP kinase. The activation of CREB increases BDNF and TrkB expression in specific regions, resulting in neuronal sprouting, increased neurogenesis and synaptic plasticity, and contributes to their behavioral aspects (see Figure 2.7).

Animal models in the future will demonstrate whether alterations of these signal transduction proteins are sufficient to produce cellular and behavioral effects similar to those of drug treatments. Current data indicate that antidepressant-induced up-regulation of BDNF can hypothetically repair damage to hippocampal neurons and protect vulnerable neurons from additional damage. One group suggested that valproate has neurotrophic properties, as it activates the MAP kinase cascade to promote neurite outgrowth and enhances neuronal survival in cell culture[116]. The precise mechanism of this process has not been fully investigated.

It has been suggested that chronic antidepressant treatment upregulates the cAMP cascade in the hippocampus and cerebral cortex, increases levels of cAMP-dependent protein kinase (PKA), and enhances expression of CREB[67,68]. These observations indicate that agents that activate this pathway could potentially be useful in the treatment of depression. cAMP-specific phosphodiesterase (PDE4) is responsible for the breakdown of cAMP, and phosphodiesterase inhibitors were shown to prevent cAMP degradation[117]. A proposal that phosphodiesterase inhibitors might reduce the symptoms of depression generated a novel strategy for antidepressant drug development. This is additionally supported by preliminary clinical trials that showed that some selective type-4 phosphodiesterase inhibitors could reduce symptoms of depression, but also had major side-effects[117]. Further research showed that there are currently four subtypes of PDE4 genes[118], and demonstration of their regional specificity in the brain will potentially guide the development of more selective agents.

Interestingly, evaluation of the additional role of neurotrophic factors outside the hippocampus revealed that BDNF expression in the neocortex and amygdala was decreased by stress, similar to hippocampal neurons, and that it was increased in the hypothalamus[107]. Infusion of BDNF into the midbrain caused antidepressant-like effects similar to those following intrahippocampal administration[119], thus providing another avenue for development of new compounds.

ANIMAL MODELS OF DEPRESSION

Some reports have suggested that approximately 40–50% of the risk for depression is inherited[37,120]. However, the search for specific genes that are responsible for depressive symptoms is still at an early stage. In addition, nongenetic factors such as stress and emotional disturbance, viral infections and other environmental factors also contribute to the etiology of depression[37,83].

One of the major problems in the research on the neurobiology of depression is the lack of validated animal models. In laboratory animals we try to generate specific depressive symptoms to study neurobiological mechanisms that could lead to new treatment modalities. However, such animal models do not really reflect the identical mechanisms involved in humans. A different approach would be to model a disease mechanism in a laboratory animal with a variety of etiological factors that we suppose cause depression in humans. Such models are generally reliable and have already been used in creating animal models for several neurological conditions, such as familial Alzheimer's dementia and Huntington's disease. It is more reliable to use such models for new drug applications and try to reverse the natural mechanisms of the disease. In models of depression we still lack clear knowledge of the etiology of the disease and its pathophysiology. Also, the most common symptoms of depression, such as mood changes and suicidality, involve higher cortical functions that cannot easily be reproduced in animal models. Most of the current animal models are used to test a particular aspect of depression, such as cognitive function, sleeping, appetite, etc.[110,121,122]. Nevertheless, some of the tests, such as the forced swim test, have been very useful in testing antidepressive drugs[122], and most of them might prove useful in future studies of the neurobiological and genetic mechanisms underlying mood changes.

Forced swim test (Porsolt test)

This is the most widely used animal model in research of depression, and more reproducible in testing antidepressants[122]. In this easily applicable test a rat or a mouse is placed in a tank filled with water and a researcher measures the duration while the animal is immobile (when it begins to float) or the latency when it swims or struggles to become immobile. Most antidepressants increase the latency to immobility and decrease the amount of immobility time. This test might present a relatively rapid method for screening agents with antidepressant-like activity. A potential problem with the test is that psychostimulants can also decrease immobility and produce a response similar to that of antidepressant drugs.

Learned helplessness

Learned helplessness is another widely used test that examines cognitive functioning. The main intention of this test is to study animals that are exposed to inescapable shock and subsequently fail to escape from a situation in which the escape is possible. They learn that voluntary actions are no longer effective. This test can produce several neurovegetative changes that are reminiscent of depression, such as sleep alteration, rapid eye movement, weight loss, decreased sexual behavior and elevated CRF and corticosterone levels[123]. Repeated administration of antidepressants or chronic electroconvulsive seizures reduce the latency to escape and decrease numbers of animals that exhibit learned helplessness.

Multiple studies have shown that this test provides a comprehensive assessment of depressive symptomatology that is seen in humans. However, it generates other ethical questions and whether the effects are strictly cognitive or induced by stress. Despite a number of problems with application of the test, it is commonly used in studies of post-traumatic stress behavior[124].

Water maze (Morris water maze)

This model is used for spatial localization and working memory testing. Rats are placed in a large circular pool of opaque water, and they quickly learn to escape by finding and climbing onto a small platform hidden in the water. Apparently the animals require only a few trials in order to learn to swim directly toward it, using the shortest route, despite the absence of cues from the platform itself or from a novel starting point. A number of groups reported that total hippocampal lesions cause a profound and lasting place–navigational impairment that could be dissociated from correlated motor, motivational and reinforcement aspects of the procedure[125].

Chronic stress

This test is related to the learned helplessness test. Study animals are exposed to mild chronic stress (changes in light–dark cycle, intermittent food and water deprivation) and most of the studies utilize rats. Similar, but more robust, experiments are performed by putting the tested animal in a cage with a dominant and aggressive animal to produce changes in cardiovascular and neuroendocrine systems of the test animal. Such experiments suffer from the same drawbacks as the learned helplessness test, and their reproducibility is variable.

Early life stress

This model has been successfully used in rodents and in primates. Typically, the model includes manipulation of the early life environment, including prenatal stress, early postnatal handling and maternal separation. This type of stress test produces cardiovascular and neuroendocrine changes in rodents and primates that persist into adulthood. For example, these animals that were subjected to early stress show a hyperactive HPA axis, and elevated CRF and glucocorticoid levels in response to stress[126,127]. Importantly, some groups, but not all, showed that many of the resulting abnormalities could be reversed by antidepressant treatment. Other limitations of the test include lower reliability of the measures used and complex data analysis.

Non-human primates

In recent years models of depression in primates have expanded and provided additional information about the cognitive and emotional component of depression in animals[128,129]. Well-designed protocols can test several paradigms, types of stress and multiple outcome measures. In general, primates exposed to early life stress in the form of maternal separation showed some signs similar to humans with depression[130], and mothers exposed to unpredictable stress were likely to raise offspring with altered HPA axis and depression-like behavior as in the adults. A major limitation is that it takes 4–5 years for primates to reach sexual maturity, and studies are financially demanding. The primate models should also consider special behaviors and environmental requirements of the species tested and examine animals living in more naturalistic settings.

It is obviously important to develop further animal models of depression. Such animal models will help us to identify and test neural networks in the brain related to depressive symptomatology and possibly other disorders.

NEURAL NETWORKS OF DEPRESSION

In summary, there is growing evidence that recurrent mood disorders are associated with significant reductions in the volume of specific regions of the brain, as well as in the numbers or shapes of neurons and glia in those regions[33,34]. Recent data in both rodents and primates showed stress-induced atrophy and reductions in neurogenesis in the adult hippocampus that were reversed by chronic antidepressant treatment. Apoptosis is also regulated by BDNF and other neurotrophic factors and the MAP kinase cascade. This has focused on a possible role of apoptosis in mood disorders. Further, it has been shown that stress decreases neurogenesis in the hippocampus and specifically in the dentate gyrus, where the most rapid neurogenesis in the brain occurs. This led to speculations that such a process inevitably causes atrophy of the hippocampus that is seen in affective disorders and in some other disorders, such as post-traumatic stress syndrome. Hypothetically, a reduction in the number of hippocampal neurons could also contribute to the cognitive and neurovegetative impairments in depression. However, we have an indication that, in some animal models, chronic antidepressant administration increased neurogenesis in the adult rat hippocampus, and it will be interesting to see the results of hippocampal volume change in humans.

There is still little knowledge about neural circuitry that underlies normal mood. Hence, the understanding of neural networks in mood pathology is inevitably more complex, owing to the possibility that multiple brain regions mediate the diverse symptoms of depression. Despite the fact that most preclinical studies have focused on the hippocampus as the site involved in the generation and treatment of depression, it is unlikely that the hippocampus alone regulates all these phenomena[47]. Some groups[11,131] recently proposed a model of a network of anatomical regions that subserve specific but related functions, such as mood, cognition and circadian rhythm. This hypothesis has been supported by human brain imaging studies that demonstrated changes in cerebral blood flow in several brain areas, including the hippocampus, amygdala, nucleus accumbens, striatum, thalamus and regions of the prefrontal and cingulate cortex[132,133]. In accord with this, a number of postmortem studies found abnormalities in many of the same brain regions[47].

Thus, understanding of the normal function of these brain areas will facilitate understanding of various aspects of depression to which they may contribute. For example, the neocortex and hippocampus may mediate cognitive aspects of depression, such as memory impairments and thoughts of hopelessness, guilt and suicidality. Dysfunction of the nucleus accumbens, a part of the mesolimbic dopamine system that is connected with other limbic structures through the ventral tegmental area (VTA) of the midbrain, is also implicated in depression. The connected structures including the amygdala and related brain areas are important in emotional states, and they have a role in mediating sadness, anxiety, anhedonia and reduced motivation. The amygdala has been best studied for its role in conditioned fear[134]. It has been proposed that the circuits that comprise the 'extended amygdala' including the nucleus accumbens, bed nucleus of the stria terminalis and other brain regions, are critical for emotional memory[135,136]. The molecular basis of the plasticity that occurs in the amygdala and its importance in emotional memory has not been as thoroughly studied as that in the hippocampus; however, it appears that similar molecular mediators are implicated. Preliminary studies showed that the cAMP pathway and CREB in the amygdala appear to promote the formation of both aversive and rewarding associations[137]. Stress decreases the expression of BDNF in the amygdala, similar to the hippocampus, although the mechanism remains poorly understood[107,138]. Current research, supported by neuroimaging studies of the adult brain, emphasizes the importance of the amygdala. Symptoms such as anxiety and fear, and abnormal responses to pleasurable stimuli, are prominent in the majority of depressed patients.

Further, the hypothalamus has long been known to mediate many neuroendocrine and neurovegetative functions. It has largely been studied as part of the HPA axis; however, other hypothalamic functions remained unexplored in research of depression, despite the fact that a number of hypothalamic nuclei and the peptide transmitters they secrete are crucial for regulating appetite, sleep and circadian rhythms, which are all abnormal in many depressed patients[97]. In addition, hypothalamic mechanisms can also contribute to the greatly increased risk of depression among females.

Although remarkable progress has been made in our understanding of the underlying mechanisms of antidepressant medication on intracellular signaling pathways, it should be clearly stated that this work has just begun. Future studies will focus on research that combines genetic studies, and potential endocrine, metabolic and environmental factors, to identify the site of neural network disruption and clarify its clinical expression.

New compounds that are capable of reversing the hypothesized impairments in cellular resilience, reductions in volume of specific structures, and cell death in depression have a potential for becoming new classes of antidepressant drug (Table 2.4). Currently proposed primary molecular targets will include

Table 2.4 Potential substrates for future mood-stabilizing drugs. From reference 48, with permission

Molecule	*Hypothesized involvement in mood disorders or treatment*	*Function plausibly relevant to mood disorders*
5-HT_{1A}/5-HT_{1B} antagonists	5-HT_{1A}/5-HT_{1B} antagonists may augment AD response	Somatodendritic 5-HT_{1A} receptors regulate 5-HT neuron firing; nerve terminal 5-HT_{1B} receptors regulate 5-HT release. Receptor blockade increases 5-HT transmission
5-HT_2 antagonists	5-HT_2 antagonists may possess AD effects	CNS distribution; may regulate AD throughput; important roles in regulating sleep and appetite
5-HT_7 antagonists	5-HT_7 antagonists may have AD effects or ameliorate sleep/circadian disturbances	CNS distribution; may regulate circadian rhythms; positively linked to cAMP generation
α_2 adrenergic antagonists	Such antagonists would increase firing of noradrenergic neurons and may regulate 5-HT neurons	α_2 receptors serve as autoreceptors on noradrenergic neurons and heteroreceptors on 5-HT neurons
CRF antagonists	Enhanced throughput of CRF receptors may mediate some of the signs and symptoms of depression and anxiety; CRF antagonists may be effective AD or anxiolytics	Regulates NE LC firing; activated by stress; CRF receptors are well placed to regulate many of the neurovegetative symptoms of depression
NK_1 antagonists	Enhanced NK_1 function in depression? antagonists may be affective ADs	Play important roles in mediating pain; NK_1 receptors reduce 5-HT neurotransmission
NPY receptor agonists	NPY may serve as an endogenous anti-stress, anxiolytic agent. NPY agonists may be efficacious for certain symptoms of depression and anxiety	May counter many of the deleterious effects of CRF and stress
NMDA antagonists	Enhanced transmission at NMDA receptors may contribute to brain regional volumetric changes observed in depression; NMDA antagonists may possess AD efficacy	Key regulator of many forms of synaptic plasticity; plays important role in stress-induced hippocampal atrophy and reduction of neurogenesis; implicated in many forms of cell atrophy and death

continued

Table 2.4 *continued*

Molecule	*Hypothesized involvement in mood disorders or treatment*	*Function plausibly relevant to mood disorders*
PDE4 inhibitors	Reduced throughput of the cAMP signaling cascade may be involved in depression; enhancement of cAMP signaling may be AD	Enhanced cAMP signaling and downstream gene expression promoted synaptic plasticity and cell survival in several cell systems
MAP kinase phosphatase inhibitors	Enhancing neurotrophic factor signaling by inhibiting turn-off reaction may be efficacious in the treatment of depression	MAP kinase signaling cascades are critical mediators of the effects of neurotrophic factors (e.g. BDHF), and play important roles in synaptic and structural plasticity
Isozyne selective PKC inhibitors	Enhancement of PKC activity may play a role in the symptomatology of mania; PKC inhibitors may be antimanic	PKC plays a major role in regulating neuronal firing and neurotransmitter release; may modulate psychostimulant-mediated catecholamine release
GSK-3 inhibitors, β-catenin upregulators	GSK-3 inhibitors, β-catenin upregulators may have mood-stabilizing effects	Play important roles in structural plasticity and regulate cell death pathways
Bcl-12	Up-regulating Bcl-12 will exert trophic effects and enhance cellular resilience in the treatment of mood disorders	One of the major cell survival signals; and a major downstream effector of neurotrophic factors. Likely to play an important role in neurite outgrowth, neurogenesis and other forms of neural plasticity

AD, antidepressive; CRF, corticotropin releasing factor; NPY, neuropeptide Y; NMDA, *N*-methyl-D-aspartate; PDE4, phosphodiesterase-4; MAP, mitogen-activated protein; PKC, protein kinase C; GSK, glycogen synthase kinase; NE, norepinephrine; LC, locus coeruleus phosphodiesterase inhibitors that increase CREB phosphorylation; MAP kinase phosphatase inhibitors that increase expression of the antiapoptotic protein Bcl-2; presynaptic glutamate receptor subtypes that attenuate glutamate release; drugs that increase BDNF expression and NMDA antagonists that enhance plasticity and cell survival. Research that determines relevant genotypes and defines neurobiological markers will clarify the relation between environmental and genetic risk factors and possibly quantify the risk for developing depression more precisely. This will substantially change current diagnostic tools and generate new approaches to the prevention and treatment of depression, one of the most serious of all medical conditions.

REFERENCES

1. Murray CJ, Lopez AD. Alternative projections of mortality and disability by cause 1990–2020: Global Burden of Disease Study. Lancet 1997; 349: 1498–1504.
2. Schmitz EB, Moriarty J, Costa DC, et al. Psychiatric profiles and patterns of cerebral blood flow in focal epilepsy: interactions between depression, obsessionality, and perfusion related to the laterality of the epilepsy. J Neurol Neurosurg Psychiatry 1997; 62: 458–463.
3. Watson C, Andermann F, Gloor P, et al. Anatomic basis of amygdaloid and hippocampal volume measurement by magnetic resonance imaging. Neurology 1992; 42: 1743–1750.
4. Kondo H, Saleem KS, Price JL. Differential connections of the temporal pole with the orbital and medial prefrontal networks in macaque monkeys. J Comp Neurol 2003; 465: 499–523.
5. Anderson SW, Bechara A, Damasio H, et al. Impairment of social and moral behavior related to early damage in human prefrontal cortex. Nat Neurosci 1997; 2: 1032–1037.
6. Price JL. Prefrontal cortical networks related to visceral function and mood. Ann NY Acad Sci 1999; 877: 383–396.
7. Hecimovic H, Goldstein JD, Sheline YI, Gilliam FG. Mechanisms of depression in epilepsy from a clinical perspective. Epilepsy Behav 2003; 4 (Suppl 3): S25–S30.
8. Sheline YI, Sanghavi M, Mintun MA, Gado MH. Depression duration but not age predicts hippocampal volume loss in medically healthy women with recurrent major depression. J Neurosci 1999; 19: 5034–5043.
9. Drevets WC. Neuroimaging studies of mood disorders. Biol Psychiatry 2000; 48: 813–829.
10. Drevets WC. Prefrontal cortical–amygdalar metabolism in major depression. Ann NY Acad Sci 1999; 877: 614–637.
11. Seminowicz DA, Mayberg HS, McIntosh AR, et al. Limbic–frontal circuitry in major depression: a path modeling metanalysis. Neuroimage 2004; 22: 409–418.
12. Charney DS, Manji HK. Life stress, genes, and depression: multiple pathways lead to increased risk and new opportunities for intervention. SciSTKE 2004; re5.
13. Bromfield EB, Altshuler L, Leiderman DB, et al. Cerebral metabolism and depression in patients with complex partial seizures. Arch Neurol 1992; 49: 617–623.
14. Kleist K. Bericht über die Gehirnpathologie in ihrer Bedeutung für Neurologie und Psychiatrie. Z Gesamte Neurol Psychiatrie 1937; 158: 159–193.

15. Broca P. Anatomie comparée des circonvolutions cérébrales. Le grant lobe limbique et la scissure limbique dans la série des mamméres. Rev Anthropol 1878; 1: 385–498.

16. Papez JW. A proposed mechanism of emotion. 1937. J Neuropsychiatry Clin Neurosci 1995; 7: 103–112.

17. Chapman WP, Schroeder HR, Geyer G, et al. Physiological evidence concerning importance of the amygdaloid nuclear region in the integration of circulatory function and emotion in man. Science 1954; 120: 949–950.

18. Van Buren JM, Yakovlev PI. Connections of the temporal lobe in man. Acta Anat (Basel) 1959; 39: 1–50.

19. Yakovlev PI, Locke S, Koskoff DY, Patton RA. Limbic nuclei of thalamus and connections of limbic cortex. Arch Neurol 1960; 3: 620–641.

20. Maclean PD. Some psychiatric implications of physiological studies on frontotemporal portion of limbic system (visceral brain). Electroencephalogr Clin Neurophysiol 1952; (Suppl 4): 407–418.

21. Woolsey T, Hanaway J, Gado M. The Brain Atlas: A Visual Guide to the Human Central Nervous System. Hoboken, NJ: John Wiley & Sons, 2003.

22. Duvernoy HM. The Human Brain: Surface, Three-dimensional Sectional Anatomy with MRI, and Blood Supply. New York: Springer-Verlag, 1999.

23. Ongur D, Ferry AT, Price JL. Architectonic subdivision of the human orbital and medial prefrontal cortex. J Comp Neurol 2003; 460: 425–449.

24. Carmichael ST, Price JL. Connectional networks within the orbital and medial prefrontal cortex of macaque monkeys. J Comp Neurol 1996; 371: 179–207.

25. Carmichael ST, Price JL. Limbic connections of the orbital and medial prefrontal cortex in macaque monkeys. J Comp Neurol 1995; 363: 615–641.

26. Rempel-Clower NL, Barbas H. Topographic organization of connections between the hypothalamus and prefrontal cortex in the rhesus monkey. J Comp Neurol 1998; 398: 393–419.

27. Bandler R, Shipley MT. Columnar organization in the midbrain periaqueductal gray: modules for emotional expression? Trends Neurosci 1994; 17: 379–389.

28. Floyd NS, Price JL, Ferry AT, et al. Orbitomedial prefrontal cortical projections to distinct longitudinal columns of the periaqueductal gray in the rat. J Comp Neurol 2000; 422: 556–578.

29. Floyd NS, Price JL, Ferry AT, et al. Orbitomedial prefrontal cortical projections to hypothalamus in the rat. J Comp Neurol 2001; 432: 307–328.

30. An X, Bandler R, Ongur D, Price JL. Prefrontal cortical projections to longitudinal columns in the midbrain periaqueductal gray in macaque monkeys. J Comp Neurol 1998; 401: 455–479.

31. Ongur D, Drevets WC, Price JL. Glial reduction in the subgenual prefrontal cortex in mood disorders. Proc Natl Acad Sci USA 1998; 95: 13290–13295.

32. Rajkowska G. Histopathology of the prefrontal cortex in major depression: what does it tell us about dysfunctional monoaminergic circuits? Prog Brain Res 2000; 126: 397–412.

33. Hamidi M, Drevets WC, Price JL. Glial reduction in amygdala in major depressive

disorder is due to oligodendrocytes. Biol Psychiatry 2004; 55: 563–569.

34. Rajkowska G, Miguel-Hidalgo JJ, Wei J, et al. Morphometric evidence for neuronal and glial prefrontal cell pathology in major depression. Biol Psychiatry 1999; 45: 1085–1098.

35. Mayberg HS, Liotti M, Brannan SK, et al. Reciprocal limbic–cortical function and negative mood: converging PET findings in depression and normal sadness. Am J Psychiatry 1999; 156: 675–682.

36. Davidson RJ, Lewis DA, Alloy LB, et al. Neural and behavioral substrates of mood and mood regulation. Biol Psychiatry 2002; 52: 478–502.

37. Fava M, Kendler KS. Major depressive disorder. Neuron 2000; 28: 335–341.

38. Schildkraut JJ. Norepinephrine metabolites as biochemical criteria for classifying depressive disorders and predicting responses to treatment: preliminary findings. Am J Psychiatry 1973; 130: 695–699.

39. Delgado PL, Price LH, Miller HL, et al. Serotonin and the neurobiology of depression. Effects of tryptophan depletion in drug-free depressed patients. Arch Gen Psychiatry 1994; 51: 865–874.

40. Miller HL, Delgado PL, Salomon RM, et al. Effects of alpha-methyl-para-tyrosine (AMPT) in drug-free depressed patients. Neuropsychopharmacology 1996; 14: 151–157.

41. Crow TJ, Cross AJ, Cooper SJ, et al. Neurotransmitter receptors and monoamine metabolites in the brains of patients with Alzheimer-type dementia and depression, and suicides. Neuropharmacology 1984; 23: 1561–1569.

42. Vetulani J, Sulser F. Action of various antidepressant treatments reduces reactivity of noradrenergic cAMP-generating system in limbic forebrain. Nature 1975; 25: 7495–7496.

43. Peroutka SL, Snyder SH. Long-term antidepressant treatment decreases spiroperidol-labeled serotonin receptor binding. Science 1980; 210: 88–90.

44. Blier P, de Montigny C. Current advances and trends in the treatment of depression. Trends Pharmacol Sci 1994; 15: 220–226.

45. Laifenfeld D, Klein E, Ben Shachar D. Norepinephrine alters the expression of genes involved in neuronal sprouting and differentiation: relevance for major depression and antidepressant mechanisms. J Neurochem 2002; 83: 1054–1064.

46. Schildkraut JJ, Schanberg SM, Breese GR, Kopin IJ. Norepinephrine metabolism and drugs used in the affective disorders: a possible mechanism of action. Am J Psychiatry 1967; 124: 600–608.

47. Nestler EJ, Barrot M, DiLeone RJ, et al. Neurobiology of depression. Neuron 2002; 34: 13–25.

48. Nestler EJ, Gould E, Manji H, et al. Preclinical models: status of basic research in depression. Biol Psychiatry 2002; 52: 503–528.

49. Shirayama Y, Chen AC, Nakagawa S, et al. Brain-derived neurotrophic factor produces antidepressant effects in behavioral models of depression. J Neurosci 2002; 22: 3251–3261.

50. Mann JJ, Aarons SF, Wilner PJ, et al. A controlled study of the antidepressant efficacy and side effects of (-)-deprenyl. A selective monoamine oxidase inhibitor. Arch Gen Psychiatry 1989; 46: 45–50.

51. Stahl SM, Grady MM. Differences in mechanism of action between current and future antidepressants. J Clin Psychiatry 2003; 64 (Suppl 13): 13–17.

52. Booij L, Van der Does AJ, Riedel WJ. Monoamine depletion in psychiatric and healthy populations: review. Mol Psychiatry 2003; 8: 951–973.

53. Richelson E. Interactions of antidepressants with neurotransmitter transporters and receptors and their clinical relevance. J Clin Psychiatry 2003; 64 (Suppl 13): 5–12.

54. Murphy DL, Brodie HK, Goodwin FK, Bunney WE Jr. Regular induction of hypomania by L-dopa in 'bipolar' manic-depressive patients. Nature 1971; 229: 135–136.

55. Meier MJ, Martin WE. Intellectual changes associated with levodopa therapy. JAMA 1970; 213: 465–466.

56. Thase ME, Sachs GS. Bipolar depression: pharmacotherapy and related therapeutic strategies. Biol Psychiatry 2000; 48: 558–572.

57. Shiah IS, Yatham LN. GABA function in mood disorders: an update and critical review. Life Sci 1998; 63: 1289–1303.

58. Hoyer D, Martin G. 5-HT receptor classification and nomenclature: towards a harmonization with the human genome. Neuropharmacology 1997; 36: 419–428.

59. Gross C, Zhuang X, Stark K, et al. Serotonin1A receptor acts during development to establish normal anxiety-like behaviour in the adult. Nature 2002; 416: 396–400.

60. Nemeroff CB, Owens MJ. Treatment of mood disorders. Nat Neurosci 2002; 5 (Suppl): 1068–1070.

61. Ramamoorthy S, Bauman AL, Moore KR, et al. Antidepressant- and cocaine-sensitive human serotonin transporter: molecular cloning, expression, and chromosomal localization. Proc Natl Acad Sci USA 1993; 90: 2542–2546.

62. Purselle DC, Nemeroff CB. Serotonin transporter: a potential substrate in the biology of suicide. Neuropsychopharmacology 2003; 28: 613–619.

63. Ressler KJ, Nemeroff CB. Role of serotonergic and noradrenergic systems in the pathophysiology of depression and anxiety disorders. Depress Anxiety 2000; 12 (Suppl 1): 2–19.

64. Wong ML, Kling MA, Munson PJ, et al. Pronounced and sustained central hypernoradrenergic function in major depression with melancholic features: relation to hypercortisolism and corticotropin-releasing hormone. Proc Natl Acad Sci USA 2000; 97: 325–330.

65. Lambert G, Johansson M, Agren H, Friberg P. Reduced brain norepinephrine and dopamine release in treatment-refractory depressive illness: evidence in support of the catecholamine hypothesis of mood disorders. Arch Gen Psychiatry 2000; 57: 787–793.

66. Ordway GA. Pathophysiology of the locus coeruleus in suicide. Ann NY Acad Sci 1997; 836: 233–252.

67. Coyle JT, Duman RS. Finding the intracellular signaling pathways affected by mood disorder treatments. Neuron 2003; 38: 157–160.

68. Chen AC, Shirayama Y, Shin KH, et al. Expression of the cAMP response element binding protein (CREB) in hippocampus produces an antidepressant effect. Biol Psychiatry 2001; 49: 753–762.

69. Gross RA, Ferrendelli JA. Effects of reserpine, propranolol, and aminophylline on seizure activity and CNS cyclic nucleotides. Ann Neurol 1979; 6: 296–301.

70. Fromm GH, Wessel HB, Glass JD, et al. Imipramine in absence and myoclonic–astatic seizures. Neurology 1978; 28: 953–957.

71. Ojemann LM, Friel PN, Trejo WJ, Dudley DL. Effect of doxepin on seizure frequency in depressed epileptic patients. Neurology 1983; 33: 646–648.

72. Favale E, Rubino V, Mainardi P, et al. Anticonvulsant effect of fluoxetine in humans. Neurology 1995; 45: 1926–1927.

73. Walker MC, Sander JW. Fluoxetine and seizures. Neurology 1996; 47: 850.

74. Pisani F, Spina E, Oteri G. Antidepressant drugs and seizure susceptibility: from in vitro data to clinical practice. Epilepsia 1999; 40 (Suppl 10): S48–S56.

75. Naritoku DK, Terry WJ, Helfert RH. Regional induction of fos immunoreactivity in the brain by anticonvulsant stimulation of the vagus nerve. Epilepsy Res 1995; 22: 53–62.

76. Krahl SE, Clark KB, Smith DC, Browning RA. Locus coeruleus lesions suppress the seizure-attenuating effects of vagus nerve stimulation. Epilepsia 1998; 39: 709–714.

77. Jobe PC, Dailey JW, Wernicke JF. A noradrenergic and serotonergic hypothesis of the linkage between epilepsy and affective disorders. Crit Rev Neurobiol 1999; 13: 317–356.

78. Browning KN, Clark KB, Naritoku DK, et al. Loss of anticonvulsant effect of vagus nerve stimulation in the pentylenetetrazol seizure model following treatment with 6-hydroxydopamine or 5,7-dihydroxytryptamine. Proc Soc Neurosci 1997; 23: 2424.

79. Shen S, Battersby S, Weaver M, et al. Refined mapping of the human serotonin transporter (SLC6A4) gene within 17q11 adjacent to the CPD and NF1 genes. Eur J Hum Genet 2000; 8: 75–78.

80. Bennett AJ, Lesch KP, Heils A, et al. Early experience and serotonin transporter gene variation interact to influence primate CNS function. Mol Psychiatry 2002; 7: 118–122.

81. Li Q, Wichems C, Heils A, et al. Reduction in the density and expression, but not G-protein coupling, of serotonin receptors (5-HT1A) in 5-HT transporter knock-out mice: gender and brain region differences. J Neurosci 2000; 20: 7888–7895.

82. Li Q, Wichems C, Heils A, et al. Reduction of 5-hydroxytryptamine (5-HT)(1A)-mediated temperature and neuroendocrine responses and 5-HT(1A) binding sites in 5-HT transporter knockout mice. J Pharmacol Exp Ther 1999; 291: 999–1007.

83. Caspi A, Sugden K, Moffitt TE, et al. Influence of life stress on depression: moderation by a polymorphism in the 5-HTT gene. Science 2003; 301: 386–389.

84. Hariri AR, Weinberger DR. Functional neuroimaging of genetic variation in serotonergic neurotransmission. Genes Brain Behav 2003; 2: 341–349.

85. Hariri AR, Mattay VS, Tessitore A, et al. Serotonin transporter genetic variation and the response of the human amygdala. Science 2002; 19: 400–403.

86. Craig SP, Boularand S, Darmon MC, et al. Localization of human tryptophan hydroxylase (TPH) to chromosome 11p15.3–p14 by in situ hybridization. Cytogenet Cell Genet 1991; 56: 157–159.

87. Reif A, Lesch KP. Toward a molecular architecture of personality. Behav Brain Res 2003; 139: 1–20.

88. Nielsen DA, Goldman D, Virkkunen M, et al. Suicidality and 5-hydroxyindoleacetic acid concentration associated with a tryptophan hydroxylase polymorphism. Arch Gen Psychiatry 1994; 51: 34–38.

89. Nolan KA, Volavka J, Lachman HM, Saito T. An association between a polymorphism of the tryptophan hydroxylase gene and aggression in schizophrenia and schizoaffective disorder. Psychiatr Genet 2000; 10: 109–115.

90. Mann JJ, Malone KM, Nielsen DA, et al. Possible association of a polymorphism of the tryptophan hydroxylase gene with suicidal behavior in depressed patients. Am J Psychiatry 1997; 154: 1451–1453.

91. Zill P, Baghai TC, Zwanger P, et al. SNP and haplotype analysis of a novel tryptophan hydroxylase isoform (TPH2) gene provide evidence for association with major depression. Mol Psychiatry 2004; 9: 1030–1036

92. Gelernter J, Kennedy JL, van Tol HH, et al. The D4 dopamine receptor (DRD4) maps to distal 11p close to HRAS. Genomics 1992; 13: 208–210.

93. Sapolsky RM. Glucocorticoids and hippocampal atrophy in neuropsychiatric disorders. Arch Gen Psychiatry 2000; 57: 925–935.

94. Fuchs E, Gould E. Mini-review: in vivo neurogenesis in the adult brain: regulation and functional implications. Eur J Neurosci 2000; 12: 2211–2214.

95. Starkman MN, Giordani B, Gebarski SS, et al. Decrease in cortisol reverses human hippocampal atrophy following treatment of Cushing's disease. Biol Psychiatry 1999; 46: 1595–1602.

96. Starkman MN, Gebarski SS, Berent S, Schteingart DE. Hippocampal formation volume, memory dysfunction, and cortisol levels in patients with Cushing's syndrome. Biol Psychiatry 1992; 32: 756–765.

97. Holsboer F. Stress, hypercortisolism and corticosteroid receptors in depression: implications for therapy. J Affect Disord 2001; 62: 77–91.

98. Cameron HA, McKay RD. Restoring production of hippocampal neurons in old age. Nat Neurosci 1999; 2: 894–897.

99. Cameron HA, Gould E. Adult neurogenesis is regulated by adrenal steroids in the dentate gyrus. Neuroscience 1994; 61: 203–209.

100. Gass P, Kretz O, Wolfer DP, et al. Genetic disruption of mineralocorticoid receptor leads to impaired neurogenesis and granule cell degeneration in the hippocampus of adult mice. EMBO Rep 2000; 1: 447–451.

101. Holsboer F. Corticotropin-releasing hormone modulators and depression. Curr Opin Investig Drugs 2003; 4: 46–50.

102. Reul JM, Holsboer F. Corticotropin-releasing factor receptors 1 and 2 in anxiety and depression. Curr Opin Pharmacol 2002; 2: 23–33.

103. Nemeroff CB. The neurobiology of depression. Sci Am 1998; 278: 42–49.

104. Brunson KL, Eghbal-Ahmadi M, Bender R, et al. Long-term, progressive hippocampal cell loss and dysfunction induced by early-life administration of corticotropin-releasing hormone reproduce the effects of early-life stress. Proc Natl Acad Sci USA 2001; 98: 8856–8861.

105. Brunson KL, Khan N, Eghbal-Ahmadi M, Baram TZ. Corticotropin (ACTH) acts directly on amygdala neurons to down-reg-

ulate corticotropin-releasing hormone gene expression. Ann Neurol 2001; 49: 304–312.

106. Nemeroff CB. Recent advances in the neurobiology of depression. Psychopharmacol Bull 2002; 36 (Suppl 2): 6–23.

107. Smith MA, Makino S, Kvetnansky R, Post RM. Effects of stress on neurotrophic factor expression in the rat brain. Ann NY Acad Sci 1995; 771: 234–239.

108. Nibuya M, Morinobu S, Duman RS. Regulation of BDNF and trkB mRNA in rat brain by chronic electroconvulsive seizure and antidepressant drug treatments. J Neurosci 1995; 15: 7539–7547.

109. Dowlatshahi D, MacQueen GM, Wang JF, Young LT. Increased temporal cortex CREB concentrations and antidepressant treatment in major depression. Lancet 1998; 352: 1754–1755.

110. Conti AC, Cryan JF, Dalvi A, et al. cAMP response element-binding protein is essential for the upregulation of brain-derived neurotrophic factor transcription, but not the behavioral or endocrine responses to antidepressant drugs. J Neurosci 2002; 22: 3262–3268.

111. Lahteinen S, Pitkanen A, Koponen E, et al. Exacerbated status epilepticus and acute cell loss, but no changes in epileptogenesis, in mice with increased brain-derived neurotrophic factor signaling. Neuroscience 2003; 122: 1081–1092.

112. Eisch AJ, Bolanos CA, de Wit J, et al. Brain-derived neurotrophic factor in the ventral midbrain–nucleus accumbens pathway: a role in depression. Biol Psychiatry 2003; 54: 994–1005.

113. Chen B, Dowlatshahi D, MacQueen GM, et al. Increased hippocampal BDNF immunoreactivity in subjects treated with antidepressant medication. Biol Psychiatry 2001; 50: 260–265.

114. Riccio A, Ahn S, Davenport CM, et al. Mediation by a CREB family transcription factor of NGF-dependent survival of sympathetic neurons. Science 1999; 286: 2358–2361.

115. Bonni A, Brunet A, West AE, et al. Cell survival promoted by the Ras-MAPK signaling pathway by transcription-dependent and -independent mechanisms. Science 1999; 286: 1358–1362.

116. Yuan PX, Huang LD, Jiang YM, et al. The mood stabilizer valproic acid activates mitogen-activated protein kinases and promotes neurite growth. J Biol Chem 2001; 276: 31674–31683.

117. Duman RS, Heninger GR, Nestler EJ. A molecular and cellular theory of depression. Arch Gen Psychiatry 1997; 54: 597–606.

118. Takahashi M, Terwilliger R, Lane C, et al. Chronic antidepressant administration increases the expression of cAMP-specific phosphodiesterase 4A and 4B isoforms. J Neurosci 1999; 19: 610–618.

119. Siuciak JA, Lewis DR, Wiegand SJ, Lindsay RM. Antidepressant-like effect of brain-derived neurotrophic factor (BDNF). Pharmacol Biochem Behav 1997; 56: 131–137.

120. Steer RA, Brown GK, Beck AT, Sanderson WC. Mean Beck Depression Inventory-II scores by severity of major depressive episode. Psychol Rep 2001; 88: 1075–1076.

121. Willner P, Mitchell PJ. The validity of animal models of predisposition to depression. Behav Pharmacol 2002; 13: 169–188.

122. Porsolt RD. Animal models of depression: utility for transgenic research. Rev Neurosci 2000; 11: 53–582.

123. Haracz JL, Minor TR, Wilkins JN, Zimmermann EG. Learned helplessness: an experimental model of the DST in rats. Biol Psychiatry 1988; 23: 388–396.

124. King JA, Abend S, Edwards E. Genetic predisposition and the development of posttraumatic stress disorder in an animal model. Biol Psychiatry 2001; 50: 231–237.

125. Morris RG, Garrud P, Rawlins JN, O'Keefe J. Place navigation impaired in rats with hippocampal lesions. Nature 1982; 297: 681–683.

126. Brunson KL, Vishai-Eliner S, Hatalski CG, Baram TZ. Neurobiology of the stress response early in life: evolution of a concept and the role of corticotropin releasing hormone. Mol Psychiatry 2001; 6: 647–656.

127. Hatalski CG, Brunson KL, Tantayanubutr B, et al. Neuronal activity and stress differentially regulate hippocampal and hypothalamic corticotropin-releasing hormone expression in the immature rat. Neuroscience 2000; 101: 571–580.

128. Kalin NH, Shelton SE, Davidson RJ. The role of the central nucleus of the amygdala in mediating fear and anxiety in the primate. J Neurosci 2004; 24: 5506–5515.

129. Kalin NH, Shelton SE. Nonhuman primate models to study anxiety, emotion regulation, and psychopathology. Ann NY Acad Sci 2003; 1008: 189–200.

130. Kalin NH, Shelton SE, Rickman M, Davidson RJ. Individual differences in freezing and cortisol in infant and mother rhesus monkeys. Behav Neurosci 1998; 112: 251–254.

131. Mayberg HS. Limbic–cortical dysregulation: a proposed model of depression. J Neuropsychiatry Clin Neurosci 1997; 9: 471–481.

132. Liotti M, Mayberg HS. The role of functional neuroimaging in the neuropsychology of depression. J Clin Exp Neuropsychol 2001; 23: 121–136.

133. Drevets WC. Neuroimaging abnormalities in the amygdala in mood disorders. Ann NY Acad Sci 2003; 985: 420–444.

134. Blair HT, Tinkelman A, Moita MA, LeDoux JE. Associative plasticity in neurons of the lateral amygdala during auditory fear conditioning. Ann NY Acad Sci 2003; 985: 485–487.

135. Sheline YI, Barch DM, Donnelly JM, et al. Increased amygdala response to masked emotional faces in depressed subjects resolves with antidepressant treatment: an fMRI study. Biol Psychiatry 2001; 50: 651–658.

136. Drevets WC. Neuroimaging and neuropathological studies of depression: implications for the cognitive–emotional features of mood disorders. Curr Opin Neurobiol 2001; 11: 240–249.

137. Hall J, Thomas KL, Everitt BJ. Fear memory retrieval induces CREB phosphorylation and Fos expression within the amygdala. Eur J Neurosci 2001; 13: 1453–1458.

138. Manji HK, Drevets WC, Charney DS. The cellular neurobiology of depression. Nat Med 2001; 7: 541–547.

3

Brain structural changes associated with depression

Yvette I Sheline

INTRODUCTION

Both early-onset and late-onset depression have been associated with structural brain changes. Studies of late-onset depression associated with neurological disorders, often referred to as secondary depression, have revealed that, similar to early-onset depression, there are structural brain changes within a specific neuroanatomical circuit. This circuit, which was originally described by Nauta[1], has been called the limbic–cortical–striatal–pallidal–thalamic (LCSPT) tract[2]. The structures that comprise it are extensively interconnected[3] and include the hippocampus, amygdala, caudate, putamen and frontal cortex. There have also been reports of brain changes in this circuit associated with early-onset recurrent major depression or primary depression. This chapter first briefly reviews the evidence for brain changes occurring in this circuit in early-onset recurrent depression. Following this is a summary of depression-associated brain changes in stroke, Parkinson's disease, epilepsy and dementia of the Alzheimer's type, illnesses characterized by tissue loss in the hippocampus, amygdala, basal ganglia and frontal cortex, the same areas as reported in major depression without comorbid illness. Studies examining brain changes in aging are also described. Finally, potential mechanisms for structural brain loss in depression are explored, and the question of whether depression is the cause or effect of abnormalities in brain structure is addressed.

THE LIMBIC–CORTICAL–STRIATAL–PALLIDAL–THALMIC CIRCUIT

Both primary and secondary mood disorders often involve abnormalities in specific neuroanatomic circuits[2,4–11]. The prefrontal cortex has an important role in modulating activity in basal ganglia and limbic regions, and there are extensive cortical–subcortical interconnections. In addition, the prefrontal cortex subserves important cognitive functions, including the so-called 'executive functions'. The caudate and putamen (striatum) and globus pallidus are organized in parallel, to connect with limbic and cortical regions. It has been proposed that one arm of the LCSPT circuit consists of a limbic–thalamic–cortical branch composed of the amygdala and hippocampus, mediodorsal nucleus of the thalamus and medial and ventrolateral prefrontal cortex. A limbic–striatal–pallidal–thalamic branch has been proposed as the other arm of the circuit.

With multiple neurotransmitter systems and interconnections, there are many ways in which lesions within this system could result in depression. Underactive forebrain dopamine activity resulting in disinhibition of the limbic striatum is one hypothesis[2] to account for depressive symptoms. Disinhibition of the limbic striatum could then produce overinhibition of the ventral pallidum with a decreased inhibitory connection to the mediodorsal thalamus. This in turn could result in disinhibition of the excitatory loop involving the mediodorsal thalamus, prefrontal cortex and amygdala. This is clearly a simplification that might explain some behaviors but could not explain underactivation of some cognitive functions, such as decreased attention and, in some studies, impairment in executive control[12]. However, overall this hypothesis is attractive, since it can explain a number of characteristic emotional, cognitive and motor activities in depression; for example, guilty ruminations, motor slowing and recurrent thoughts of death. While it is difficult to explain all of the manifestations of depression, lesions within the LCSPT structures themselves or in the interconnections among them could result in malfunction predisposing to depression. Brain structural changes reported in major depression are summarized in Table 3.1.

FROM DEPRESSION TO STRUCTURAL BRAIN CHANGES

The frontal cortex

Frontal cortex volume reductions have been reported to range from 7% overall in major depression[13] to 48% in the subgenual prefrontal cortex[14]. A number of recent studies have examined more localized frontal cortex volume losses in major depressive disorder (MDD). One study, for example[15], found a smaller volume of bilateral middle frontal gyrus gray matter, and a smaller volume of the left anterior cingulate and right middle frontal gyrus white matter. Brambilla *et al.*[16] found no significant difference in the subgenual prefrontal cortex in either familial of non-familial unipolar or bipolar disorder. Volume loss has been described in the orbitofrontal cortex in middle-aged (mean age 37[17]) and older (mean age 68.7[18]) and mean age 66.2[19] depressed subjects. Other prefrontal subregion volume reductions in older subjects were found in the anterior cingulate and gyrus rectus[19] (Figure 3.1).

Postmortem study of the prefrontal cortex in major depression[20] found significant differences from controls in several prefrontal cortical areas. Abnormalities included rostral orbitofrontal cortex decreases in cortical thickness, neuronal size decrease and loss of glial cells in layers II–IV. Caudal orbitofrontal cortex abnormalities were reductions in glial cells in layers V–VI and decreases in neuronal sizes. Reductions in glial and neuronal cells throughout all layers as well as reduction in cell size were reported in the dorsolateral prefrontal cortex. Subgenual prefrontal cortex glial cell loss has also been reported in major depression[21]. Some of the magnetic resonance imaging (MRI) volumetric findings in the frontal cortex could be accounted for by neuropathological changes such as these. The prefrontal cortex is particularly important as a target of monoamine projections and abnormalities in monoamine receptors, transporters and second-messenger systems[22–26] are reported to occur in major depression. Overactivation in one part of the interconnected LCSPT neuroanatomical circuit is another possibility that may lead to overexcitation in the other components, resulting in excitotoxic damage. In particular, the orbitomedial prefrontal cortex has high concentrations of glucocorticoid receptors, potentially rendering it vulnerable to stress-mediated damage (see below).

Table 3.1 Brain structural changes reported in major depression

Author	*Brain region*	*Sample (number and diagnosis)*	*Age (mean ± SD)*	*Methods and resolution*	*Findings*
Frontal cortex					
Krishnan *et al.*[47]	Frontal lobe	50 NC 50 MDD	49.3 ± 18 48.3 ± 17	1.5 T 5 mm	Bifrontal distances smaller; bifrontal brain widths smaller
Coffey *et al.*[85]	Frontal lobe	76 NC 48 MDD 44 unipolar; 4 bipolar) ECT referred	62.4 ± 16.4 61.6 ± 15.9	1.5 T 5 mm 5.2 interval	Smaller frontal lobe volumes in major depression
Bell-McGinty *et al.*[15]	Frontal lobe	47 NC 30 MDD	66.9 ± 7.3 69.3 ± 5.7	1.5 T 8 mm	Smaller bilateral middle frontal gyrus gray matter volume in MDD Smaller left anterior cingulate and right middle frontal gyrus white matter volume
Almeida *et al.*[99]	Frontal lobe	37 NC 24 EOD 27 LOD	72.9 72.8 75.5	1.0 T 1.0 mm	Smaller right frontal lobe in LOD group than EOD and controls
Ballmaier *et al.*[19]	Anterior cingulate	19 NC 24 MDD	66.24 ± 7.25 65.85 ± 8.18	1.5 T 1.4 mm	Significant reduction in bilateral anterior cingulate volume in major depression
Drevets *et al.*[14]	Subgenual prefrontal cortex	33 NC 13 MDD 10 MDD, remitted	36.2 ± 8.9 33.6 ± 10.0 30.1 ± 7.8	1.5 T 1 mm	Smaller subgenual prefrontal cortex volumes in major depression
Brambilla et al.[16]	Subgenual prefrontal cortex	38 NC 18 MDD-unipolar 27 MDD-bipolar	37.0 ± 10.0 42.0 ± 10.0 35.0 ± 11.0	1.5 T 1.5 mm	No significant difference in the SGPFC in either familial or non-familial unipolar or bipolar disorder

continued

Table 3.1 *continued*

Author	*Brain region*	*Sample (number and diagnosis)*	*Age (mean ± SD)*	*Methods and resolution*	*Findings*
Ballmaier *et al.*[19]	Gyrus rectus	19 NC 24 MDD	66.24 ± 7.25 65.85 ± 8.18	1.5 T 1.4 mm	Significant reduction in bilateral gyrus rectus volume in major depression
Lee *et al.*[18]	Orbitofrontal cortex	41 NC 41 MDD	71.15 ± 6.25 68.73 ± 6.98	1.5 T 3 mm Acquired as 2 scans with a 3 mm gap to include all possible slices	Significant reduction in OFC volume in major depression
Ballmaier *et al.*[19]	Orbitofrontal cortex	19 NC 24 MDD	66.24 ± 7.25 65.85 ± 8.18	1.5 T 1.4 mm	Significant reduction in bilateral OFC volume in major depression
Lacerda *et al.*[17]	Orbitofrontal cortex	34 NC 31 MDD	37.03 ± 11.88 39.26 ± 11.9	1.5 T 1.5 mm	Significant reduction in right medial and left lateral OFC volume in major depression
Hippocampus					
Sheline *et al.*[31]	Hippocampus	10 NC 10 MDD, remitted	68.0 ± 9.5 68.5 ± 10.4	1.5 T 0.5 mm	Decreased hippocampal gray matter volume in major depression
Shah *et al.*[30]	Hippocampus	20 NC 20 MDD 20 TRD	49.3 ± 11.8 47.7 ± 9.9 48.9 ± 9.8	1.0 T 2 mm	Decreased hippocampal volume in treatment-resistant depression
Sheline *et al.*[32]	Hippocampus	24 NC 24 MDD, remitted	52.8 ± 17.8 52.8 ± 18.4	1.5 T 0.5 mm	Decreased hippocampal volume in major depression
Bremner *et al.*[27]	Hippocampus	16 NC 16 MDD (1 panic disorder)	45.0 ± 10.0 43.0 ± 8.0	1.5 T 3 mm	Decreased hippocampal volume in major depression

continued

Table 3.1 *continued*

Author	*Brain region*	*Sample (number and diagnosis)*	*Age (mean ± SD)*	*Methods and resolution*	*Findings*
Mervaala *et al.*[35]	Hippocampus	17 NC 34 MDD (6 bipolar, 28 monopolar)	42.1 ± 14.6 42.2 ± 12.2	1.5 T 3 mm	No significant difference
Vakili *et al.*[37]	Hippocampus	20 NC 38 MDD	40.3 ± 10.4 38.5 ± 10.0	1.5 T 3 mm	No significant difference
Frodl *et al.*[39]	Hippocampus	30 NC 30 MDD (FE)	40.6 ± 12.5 40.3 ± 12.6	1.5 T 1.5 mm	Smaller left hippocampal gray matter volume in males Female patients showed a tendency toward larger right hippocampal gray matter volumes Smaller bilateral hippocampal white matter volume in all subjects
MacQueen *et al.*[29]	Hippocampus	20 NC 20 MDD (FE) 17 NC 17 MDD (ME)	28.4 ± 11.5 28.4 ± 11.8 36.2 ± 11.9 35.9 ± 11.1	1.5 T 1.2 mm	No significant difference in the FE group Smaller bilateral hippocampal volume in the ME group
Posener *et al.*[36]	Hippocampus	42 NC 27 MDD	33.2 ± 10.8 33.0 ± 10.7	1.5 T 1.0 mm	No significant difference
MacMaster and Kusumakar[28]	Hippocampus	17 NC 17 MDD	16.23 ± 1.61 16.67 ± 1.83	1.5 T 1.4 mm	Significant reduction in left hippocampus volume in major depression

continued

Table 3.1 *continued*

Author	*Brain region*	*Sample (number and diagnosis)*	*Age (mean ± SD)*	*Methods and resolution*	*Findings*
Frodl *et al.*[40]	Hippocampus	30 NC 30 MDD (1-year follow-up)	45.7 ± 12.9 48.4 ± 13.4	1.5 T 1.5 mm	No significant difference between baseline and follow-up Significant reduction in bilateral hippocampus volume in non-remitted compared with those that remitted Significantly smaller right hippocampus volume in the remitted group compared with controls
Bell-McGinty *et al.*[15]	Hippocampus/ entorhinal cortex	47 NC 30 MDD	66.9 ± 7.3 69.3 ± 5.7	1.5 T 8 mm	Smaller right hippocampal gray matter volume Hippocampus/entorhinal cortex is inversely associated with the number of years since the first lifetime depressive episode
Amygdala					
Sheline *et al.*[44]	Amygdala	20 NC 20 MDD, remitted	53.8 ± 17.7 54.1 ± 18.1	1.5 T 0.5 mm	Decreased amygdala core nuclei volume in major depression
Tebartz van Elst *et al.*[100]	Amygdala	20 NC 38 TLE 12 TLE with dysthymia	36.5 32 30	1.5 T 1.5 mm	Increased amygdala volume in both the TLE and TLE with dysthymia groups
Bremner *et al.*[27]	Amygdala	16 NC 16 MDD (1 panic disorder)	45.0 ± 10.0 43.0 ± 8.0	1.5 T 3 mm	Increased right amygdala volume in major depression

continued

Table 3.1 *continued*

Author	*Brain region*	*Sample (number and diagnosis)*	*Age (mean ± SD)*	*Methods and resolution*	*Findings*
Mervaala *et al.*[35]	Amygdala	17 NC 34 MDD (6 bipolar, 28 monopolar)	42.1 ± 14.6 42.2 ± 12.2	1.5 T 3 mm	Significant asymmetry in amygdala volume (right smaller than left)
Tebartz van Elst *et al.*[101]	Amygdala	20 NC 47 TLE 17 TLE with dysthymia	36 33.3 32.8	1.5 T 1.5 mm	Increased amygdala volume in dysthymia females but not males and in the TLE with dysthymia group as a whole
Frodl *et al.*[43]	Amygdala	30 NC 30 MDD (hospitalized with FE)	40.6 ± 12.5 40.3 ± 12.6	1.5 T 1.5 mm	Increased bilateral amygdala volume in the depressed group
Frodl *et al.*[40]	Amygdala	30 NC 30 MDD (1-year follow-up)	45.7 ± 12.9 48.4 ± 13.4	1.5 T 1.5 mm	No significant difference between baseline and follow-up
Hippocampus–amygdala complex					
Axelson *et al.*[34]	Amygdala/ hippocampus complex	30 NC 19 MDD	46.7 ± 20.4 56.6 ± 19.1	1.5 T 5 mm	No significant difference
Pantel *et al.*[73]	Amygdala/ hippocampus complex	13 NC 19 MDD 27 AD	68.2 ± 5.3 72.4 ± 8.8 71.9 ± 8.0	1.5 T 1.25 mm	No significant difference
Ashtari *et al.*[33]	Amygdala/ hippocampus complex	46 NC 40 MDD	71.4 ± 0.3 74.3 ± 6.0	1.0 T 3.1 mm	No significant difference

continued

Table 3.1 *continued*

Author	*Brain region*	*Sample (number and diagnosis)*	*Age (mean ± SD)*	*Methods and resolution*	*Findings*
Basal ganglia					
Krishnan *et al.*[47]	Caudate	50 NC 50 MDD	49.3 ± 18 48.3 ± 17	1.5 T 5 mm	Decreased caudate volumes in major depression
Greenwald *et al.*[45]	Caudate	30 NC 36 MDD	72.8 ± 6.6 75.9 ± 6.7	1.0 T 3 mm	Decreased left caudate volume in major depression
Lacerda *et al.*[50]	Caudate	48 NC 25 MDD	35 ± 10 41 ± 11	1.5 T 1.5 mm	No significant difference
Husain *et al.*[46]	Putamen	44 NC 41 MDD	56.4 ± 19.2 55.3 ± 18.8	1.5 T 5 mm (2 patients with 7 mm)	Decreased putamen volume in major depression
Lacerda *et al.*[50]	Putamen	48 NC 25 MDD	35 ± 10 41 ± 11	1.5 T 1.5 mm	No significant difference
Dupont *et al.*[87]	Caudate and lenticular nucleus	26 NC 36 MDD-bipolar 30 MDD-unipolar	39.1 ± 9.4 36.6 ± 10.8 38.6 ± 10.6	1.5 T 5–2.5 mm gap	No significant difference
Lenze and Sheline[49]	Caudate and putamen	24 NC 24 MDD-remitted	52.8 ± 17.8 52.8 ± 18.4	1.5 T 0.5 mm	No significant difference in healthy 'supernormal population'
Lacerda *et al.*[50]	Globus pallidus	48 NC 25 MDD	35 ± 10 41 ± 11	1.5 T 1.5 mm	Significant decrease in asymmetrical globus pallidus volume in major depression

continued

Table 3.1 *continued*

Author	*Brain region*	*Sample (number and diagnosis)*	*Age (mean ± SD)*	*Methods and resolution*	*Findings*
Pituitary					
Axelson *et al.*[102]	Pituitary	21 MDD (1 bipolar; 1 adjustment disorder; 1 multi-infarct dementia; 1 schizo affective)	47.9 ± 18.4 39.6 ± 13.2	1.0 T 3 mm	Increase in pituitary volume in major depression

NC, normal control; MDD, major depressive disorder; ECT, electroshock therapy; EOD, early-onset depression; LOD, late-onset depression; T, Tesla; TRD, treatment resistant depression; AD, Alzheimer's disease; FE, first episode; ME, multiple episodes; PFC, prefrontal cortex; SGPFC, subgenual prefrontal cortex; OFC, orbitofrontal cortex; TLE, temporal lobe epilepsy

The hippocampus

Some studies that examined hippocampal volume in depression have found significant reductions[15,27–32]. Others[33–37] have not found significant reductions in hippocampal volumes in depression. The volume loss appears to have functional significance, with an association between acute depression and abnormalities of declarative memory[38] and recollection memory[29] as well as an association between depression in remission and lower scores on tests of verbal memory[32]. In one study[30], hippocampal atrophy was found in patients with chronic depression but not in patients with remitted depression. Vakili *et al.*[37] also observed correlations between depression severity and hippocampal volumes, although no group differences were found between depressed and control subjects. White matter changes were noted in one study[39], but no overall differences in hippocampal volume, although in a follow-up study non-remitted patients were found to have hippocampal volume reductions[40]. Most studies that assessed depression severity in unipolar subjects and used high-resolution MRI techniques found depression to be associated with hippocampal volume loss, ranging from 8 to 19%. Studies that measured only the hippocampus–amygdala complex found no differences. A postmortem study[41] has found glial cell loss in the dentate gyrus of the hippocampus as well as in the amygdala in major depression. In addition, a recent study has found increased neuronal and glial cell packing density[42], suggesting a decrease in the hippocampal neuropil in MDD.

The amygdala

Amygdala volume findings in major depression have been inconsistent. Studies have identified

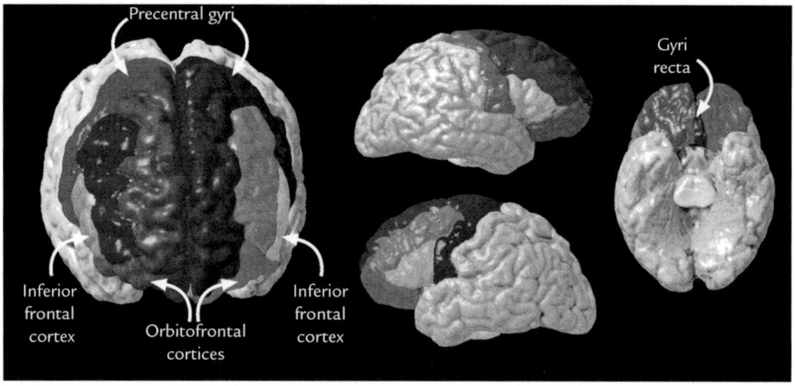

Figure 3.1 Volumetric studies of prefrontal cortex subregions have identified volume decreases in major depression. Reproduced from reference 19, with permission from the *American Journal of Psychiatry*, copyright 2004, American Psychiatric Association. (See also color image on p. xvii)

an increase in volume in the right amygdala[27], in the bilateral amygdala in first-episode subjects[43], loss of normal asymmetry[35] or reduction in the bilateral core nuclei[44]. In a 1-year follow-up study[40], no significant difference from baseline was identified in depressed subjects. The amygdala is a difficult structure to measure, since in many areas the 'cortical amygdala' merges with the surrounding cortex, and specific boundaries selected have varied greatly in different studies.

Basal ganglia

Decreased volumes have been found in many studies of basal ganglia structures in major depression, especially in late-onset depression[45–48]. Negative findings were reported in the caudate and putamen in depressed subjects who were otherwise physically healthy[49], a criterion not clearly present in other studies. Another study in middle-aged subjects[50] found no significant difference in the putamen but found significant asymmetry differences in the globus pallidus.

FROM CO-MORBID DISEASE TO DEPRESSION

High rates of depression are found in neurological illnesses associated with both cortical and subcortical atrophy. These include Huntington's disease[51], post-stroke syndromes[52], dementia of the Alzheimer's type[53], epilepsy[54] and Parkinson's disease[55]. These illnesses involve damage to brain structures critical in emotional functioning, namely the frontal cortex, hippocampus, thalamus, amygdala and basal ganglia. Further, these same brain structures are involved in more classical or early-onset major depression.

Epilepsy

In temporal lobe epilepsy (see subsequent chapters), not only is there hippocampal and amygdala sclerosis from the epilepsy itself, but postsurgical volume loss associated with depression has been reported in connected structures, the medial–dorsal thalamus and putamen[56].

Stroke

In groundbreaking work Robinson *et al.*[57] and Lipsey *et al.*[58] reported that ischemic lesions located in the anterior frontal cortex were associated with more severe depression. Subsequently, inconsistent results have been reported on the relationship between infarct site and depression after stroke, with systematic review of the numerous studies not supporting the hypothesis that stroke lesion location predicts depression[59]. A recent study in a large, well-defined series of patients with ischemic stroke, however, found a strong correlation between lesions affecting the prefrontosubcortical circuits (particularly on the left) and subsequent depression[60]. These issues are discussed in more detail in a subsequent chapter.

Parkinson's disease

Different subtypes of Parkinson's disease have neuronal loss in a variety of subcortical brain regions, with some evidence for an association with clinical syndromes in postmortem studies[61]. Decreased volumes of caudate and putamen have been demonstrated *in vivo* in Parkinson's disease[62]. Further, in depressed compared with non-depressed Parkinson's patients fluorodeoxyglucose metabolism studies using positron emission tomography (PET) have demonstrated selective hypometabolism in the caudate and orbital–inferior frontal lobe[63].

Dementia of the Alzheimer's type

Postmortem studies of patients with dementia of the Alzheimer's type have been shown to have profound atrophy of the hippocampus, and frontal and parietal cortex; atrophy in each of these areas is associated with a specific neuropsychological profile[64]. Depression is an important complication of dementia of the Alzheimer's type (DAT), affecting as many as 50% of subjects[65]. Studies of individuals with cognitive impairment have shown that baseline depression was associated with a three-fold increased risk of incident dementia. *In vivo* MRI studies[66] have shown that small left hippocampal size on neuroimaging predicts later dementia. In summary, neurological diseases have been associated with structural changes in the LCSPT tract.

Co-morbid medical illness

In addition to an association of specific neurological illnesses with increased rates of depression, late-onset depression is itself associated with an increased prevalence of structural brain changes in its own right. This is due at least in part to the increased prevalence of co-morbid illness with age. Depression onset in late age is characterized by lower familial frequency of affective disorders[67], greater medical morbidity and mortality[68], and higher rates of neuroradiological abnormalities, particularly white matter hyperintensities[69,70], compared with early-onset depression. Late-onset depression has also been found to be associated with higher rates of neuropsychological impairment and greater treatment refractoriness[71,72].

MRI and computed tomography (CT) have demonstrated diffuse cortical and subcortical atrophy and ventricular enlargement in many studies of late-life depression[9,73–75]. In addition to neurological illnesses, conditions that have been associated with brain atrophy include hypertension[76], Cushing's disease[77], diabetes[78] and alcohol abuse[79]. Brain atrophy can potentially be produced by any condition which produces neuronal ischemia or neurotoxicity. Some studies have not found evidence for generalized atrophy in addition to volume loss in structures of the LCSPT circuit. Kumar *et al.*[80] and Ballmaier *et al.*[19], for example, have found prefrontal lobe volume loss in late-onset depression in the absence of generalized atrophy, suggesting that, similarly to patients with early-onset depression, some subjects with late-onset depression may also have focal volume loss. Whether this focal volume loss occurs by the same etiological mechanisms as early-onset recurrent depression (EORD) remains to be determined.

White matter hyperintensities

An important phenomenon in late-life depression is the presence of hyperintensities seen on T2-weighted scans (T2H). Increased numbers of T2H[74,81–84] are well-replicated findings in elderly depressed subjects. Some studies have shown younger depressed subjects to have increased T2H[85,86], although negative findings have also been reported with younger groups[87,88]. It is important to note that etiological mechanisms for T2H are unknown, and that T2H also occur at rates of up to 60% in healthy elderly individuals[89], in whom their significance is also unknown[90]. One explanation is a higher rate of 'silent' cerebral infarctions (T2H) in late- compared to early-onset MDD[91,92]. Vascular risk factors, older age, neuropsychological impairment and late age of onset[72,85,93] are the clinical correlates of MRI-defined T2H in late-life depression. The concept of 'vascular depression' with increased cardiovascular disease (CVD) risk factors and increased T2H has been proposed[94]. Interestingly, in late-life depression, T2H have been found to cluster in the frontal cortex[95] as well as in subcortical areas.

This section has summarized the evidence for structural brain changes leading to depression; however, there are important qualifications. LCSPT structure impairment in patients with diseases causing structural impairment, and even in healthy persons with small structures, does not always lead to depression. Thus, a direct correspondence between structural impairment in LCPST structures and depression does not exist. Rather, it appears that a subset of people with such structural impairment have increased vulnerability to depression and that when depression occurs it may further contribute to additional damage.

BRAIN STRUCTURAL CHANGES ASSOCIATED WITH AGING

The central question in brain volumetric studies in aging is to what extent brain volume declines are expected and to what extent they are caused by other factors. However, this is a difficult question, since aging is accompanied by a definite increase in the amount of medical co-morbidity. A few studies have handled this confound by selecting an extremely healthy patient population in conducting the comparisons. In some cross-sectional studies of healthy aging[96], very old subjects (mean age 90) had significantly smaller brain volumes than younger healthy subjects (mean age 70). In addition, there was a significant difference in the gray matter/white matter volume ratio with old healthy elderly subjects having a higher ratio than younger healthy elderly subjects. This suggested that in the very old, the decline of white matter volume is disproportionately greater than the decline of gray matter volume. One study[97] estimated that there was approximately a 1526 mm^3 gain in ventricular volume over 1 year but no detectable changes in total or regional brain volumes. In a longitudinal study[98] that followed healthy subjects through the 11th decade and had examined subjects longitudinally over a 5-year follow-up, there was no detectable increase in rate of volume loss in the very old. The authors pointed out that most studies that have examined brain volume loss have posed the question cross-sectionally and may have identified subjects in a prodromal period prior to the onset of dementia.

DEPRESSION – CAUSE OR EFFECT OF STRUCTURAL BRAIN ABNORMALITIES?

Whether depression is the cause or result of brain structural changes is not known. As discussed above and represented in Figure 3.2, there is evidence that medical illnesses such as Parkinson's disease, Alzheimer's disease and stroke produce brain changes that are associated with the onset of depression. Stress also can produce neuronal damage and structural changes in animal models and potentially in humans. Depression *per se* may produce structural brain changes, but this is difficult to establish in the face of higher rates of co-morbid illness in depression. Potential mediators of depression include genetic predisposition, stress and illness. Thus, it could be that depression acts through these other mechanisms rather than having an independent effect on brain structure. Alternatively, depression may have an independent effect in producing changes in brain structure (represented in Figure 3.2 by

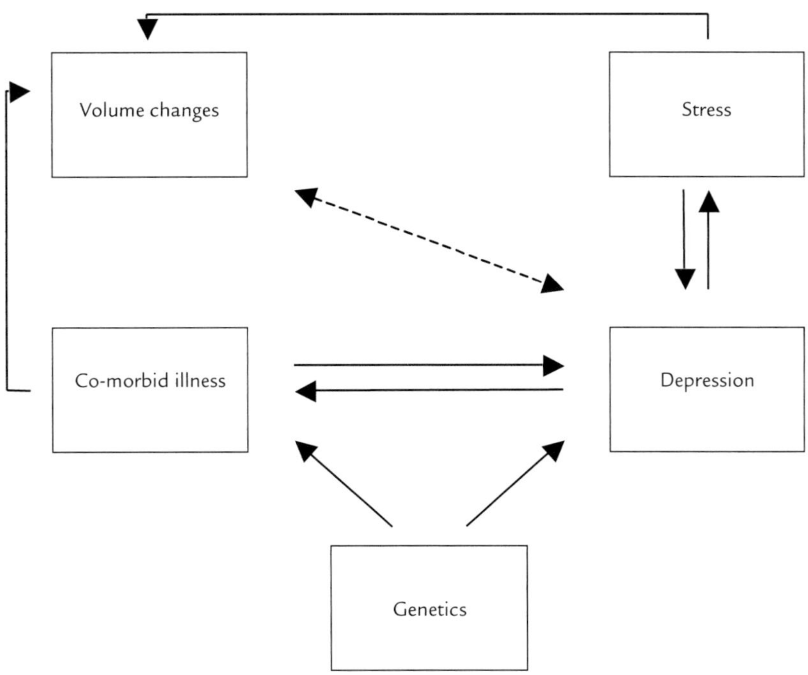

Figure 3.2 Brain changes associated with depression

the dashed arrow). The converse question, how brain structural changes might produce depression, is also highly speculative. Since the structures in the LCSPT circuit are all involved in emotional regulation, damage to any portion of this circuit could potentially produce depression, either directly by neuronal damage or indirectly by changes in neurotransmitter balance (again, represented by the dashed arrow). How it is that the symptoms produced would be similar, with disturbance in mood, sleep, appetite, energy, etc. from structural changes in disparate brain regions is not clear. Further, it is not apparent how episodic disturbances would be produced by structural changes. Finally, not all patients who have such structural changes have depression.

REFERENCES

1. Nauta WJ. Neural associations of the frontal cortex. Acta Neurobiol Exp 1972; 32: 125–140.
2. Swerdlow NR, Koob GF. Dopamine, schizophrenia, mania and depression: toward a unified hypothesis of cortico-striato-pallido-thalamic function. Behav Brain Sci 1987; 10: 197–245.
3. Price J, Russchen F, Amaral D. The limbic region. II: The amygdaloid complex. In: Bjorkland A, Hokfelt T, Swanson L, eds. Handbook of Chemical Neuroanatomy, vol 5. New York: Elsevier, 1987; 279–388.
4. Nauta WJ, Domesick V. Afferent and efferent relationships of the basal ganglia. In: Evered D, O'Conner M, eds. CIBA Foundation Symposium 107. Function of the Basal Ganglia. London: Pitman Press, 1984; 3–29.
5. Drevets WC, Videen TO, Price JL, et al. A functional anatomical study of unipolar depression. J Neurosci 1992; 12: 3628–3641.
6. McDonald WM, Krishnan KR. Magnetic resonance in patients with affective illness. Eur Arch Psychiatry Clin Neurosci 1992; 241: 283–290.
7. Mayberg HS. Frontal lobe dysfunction in secondary depression. J Neuropsychiatry Clin Neurosci 1994; 6: 428–442.
8. Mega MS, Cummings JL. Frontal–subcortical circuits and neuropsychiatric disorders. J Neuropsychiatry Clin Neurosci 1994; 6: 358–370.
9. Soares JC, Mann JJ. The anatomy of mood disorders – review of structural neuroimaging studies. Biol Psychiatry 1997; 41: 86–106.
10. Sheline Y. 3D MRI studies of neuroanatomic changes in unipolar major depression: the role of stress and medical comorbidity. Biol Psychiatry 2000; 48: 791–800.
11. Sheline Y. Neuroimaging studies of mood disorder effects on the brain. Biol Psychiatry 2003; 54: 338–352.
12. Degl'Innocenti A, Agren H, Backman L. Executive deficits in major depression. Acta Psychiatr Scand 1998; 97: 182–188.
13. Coffey CE, Wilkinson WE, Parashos IA, et al. Quantitative cerebral anatomy of the aging human brain: a cross-sectional study using magnetic resonance imaging. Neurology 1992; 42: 527–536.
14. Drevets WC, Price JL, Simpson JR Jr, et al. Subgenual prefrontal cortex abnormalities in mood disorders. Nature 1997; 386: 824–827.
15. Bell-McGinty S, Butters MA, Meltzer CC, et al. Brain morphometric abnormalities in geriatric depression: long-term neuro-

biological effects of illness duration. Am J Psychiatry 2002; 159: 1424–1427.

16. Brambilla P, Nicoletti MA, Harenski K, et al. Anatomical MRI study of subgenual prefrontal cortex in bipolar and unipolar subjects. Neuropsychopharmacology 2002; 27: 792–799.

17. Lacerda A, Keshavan MS, Hardan AY, et al. Anatomic evaluation of the orbitofrontal cortex in major depressive disorder. Biol Psychiatry 2004; 55: 353–358.

18. Lee SH, Payne ME, Steffens DC, et al. Subcortical lesion severity and orbitofrontal cortex volume in geriatric depression. Biol Psychiatry 2003; 54: 529–533.

19. Ballmaier M, Toga AW, Blanton RE, et al. Anterior cingulate, gyrus rectus, and orbitofrontal abnormalities in elderly depressed patients: an MRI-based parcellation of the prefrontal cortex. Am J Psychiatry 2004; 161: 99–108.

20. Rajkowska G, Miguel-Hidalgo JJ, Wei J, et al. Morphometric evidence for neuronal and glial prefrontal cell pathology in major depression. Biol Psychiatry 1999; 45: 1085–1098.

21. Ongur D, Drevets WC, Price JL. Glial reduction in the subgenual prefrontal cortex in mood disorders. Proc Natl Acad Sci USA 1998; 95: 13290–13295.

22. Arango V, Underwood MD, Gubbi AV, Man JJ. Localized alterations in pre- and postsynaptic serotonin binding sites in the ventrolateral prefrontal cortex of suicide victims. Brain Res 1995; 688: 121–133.

23. Biver F, Wikler D, Lotstra F, et al. Serotonin 5-HT_2 receptor imaging in major depression: focal changes in orbito-insular cortex. Br J Psychiatry 1997; 171: 444–448.

24. Duman RS. Novel therapeutic approaches beyond the serotonin receptor. Biol Psychiatry 1998; 44: 324–335.

25. Mintun MA, Sheline YI, Moerlein SM, et al. Decreased hippocampal 5-HT2A receptor binding in major depressive disorder: in vivo measurement with [18F] Altanserin positron emission tomography. Biol Psychiatry 2004; 55: 217–224.

26. Price JL. Prefrontal cortical networks related to visceral function and mood. Ann NY Acad Sci 1999; 877: 383–396.

27. Bremner JD, Narayan M, Anderson ER, et al. Hippocampal volume reduction in major depression. Am J Psychiatry 2000; 157: 115–118.

28. MacMaster FP, Kusumakar V. Hippocampal volume in early onset depression. BMC Med 2004; 2: 2–7.

29. MacQueen GM, Campbell S, McEwen B, et al. Course of illness, hippocampal function and hippocampal volume in major depression. Proc Natl Acad Sci USA 2003; 100: 1387–1392.

30. Shah PJ, Ebmeier KP, Glabus MF, Goodwin GM. Cortical grey matter reductions associated with treatment resistant chronic unipolar depression. Controlled magnetic resonance imaging study. Br J Psychiatry 1998; 172: 527–532.

31. Sheline Y, Wang P, Gado M, et al Hippocampal atrophy in recurrent major depression. Proc Natl Acad Sci USA 1996; 93: 3908–3913.

32. Sheline Y, Sanghavi M, Mintun M, Gado M. Depression duration but not age predicts hippocampal volume loss in women with recurrent major depression. J Neurosci 1999; 19: 5034–5043.

33. Ashtari M, Greenwald BS, Kramer-Ginsberg E, et al. Hippocampal/amygdala volumes in geriatric depression. Psychol Med 1999; 29: 629–638.

34. Axelson DS, Doraiswamy PM, McDonald WM, et al. Hypercortisolemia and hippocampal changes in depression. Psychiatry Res 1993; 47: 163–173.

35. Mervaala E, Fohr J, Kononen M, et al. Quantitative MRI of the hippocampus and amygdala in severe depression. Psychol Med 2000; 30: 117–25.

36. Posener JA, Wang L, Price JL, et al. High-dimensional mapping of the hippocampus in depression. Am J Psychiatry 2003; 160: 83–89.

37. Vakili K, Pillay SS, Lafer B, et al. Hippocampal volume in primary unipolar major depression: a magnetic resonance imaging study. Biol Psychiatry 2000; 47: 1087–90.

38. Burt DB, Zembar MJ, Niederehe G. Depression and memory impairment: a meta-analysis of the association, its pattern, and specificity. Psychol Bull 1995; 117: 285–305.

39. Frodl T, Meisenzahl E, Zetzsche T, et al. Hippocampal changes in patients with a first episode of major depression. Am J Psychiatry 2002; 159: 1112–1118.

40. Frodl T, Meisenzahl E, Zetzsche T, et al. Hippocampal and amygdala changes in patients with major depressive disorder and healthy controls during a 1-year follow-up. J Clin Psychiatry 2004; 65: 492–499.

41. Bowley MP, Drevets WC, Ongur D, Price JL. Glial changes in the amygdala and entorhinal cortex in mood disorders. Soc Neurosci Abstr 2000; 26: 876.10.

42. Stockmeier C, Mahajan G, Konick L, et al. Neuronal and glial density is increased and neuronal soma size is decreased in the hippocampus in major depressive disorder (MDD). Society for Biological Psychiatry Annual Meeting. San Francisco, CA, 2003: abstr 650713.

43. Frodl T, Meisenzahl E, Zetzsche T, et al. Enlargement of the amygdala in patients with a first episode of major depression. Biol Psychiatry 2002; 51: 708–714.

44. Sheline YI, Gado MH, Price JL. Amygdala core nuclei volumes are decreased in recurrent major depression. Neuroreport 1998; 9: 2023–2028.

45. Greenwald BS, Kramer-Ginsberg E, Bogerts B, et al. Qualitative magnetic resonance imaging findings in geriatric depression. Possible link between later-onset depression and Alzheimer's disease? Psychol Med 1997; 27: 421–31.

46. Husain MM, McDonald WM, Doraiswamy PM, et al. A magnetic resonance imaging study of putamen nuclei in major depression. Psychiatry Res 1991; 40: 95–99.

47. Krishnan KR, McDonald W, Escalona P, et al. MRI of the caudate nuclei in depression. Arch Gen Psychiatry 1992; 49: 553–557.

48. Steffens DC, Krishnan KR. Structural neuroimaging and mood disorders: recent findings, implications for classification, and future directions. Biol Psychiatry 1998; 43: 705–12.

49. Lenze E, Sheline Y. Absence of striatal volume differences between healthy depressed subjects and matched comparisons. Am J Psychiatry 1999; 156: 1989–1991.

50. Lacerda A, Nicoletti MA, Brambilla P, et al. Anatomical MRI study of basal ganglia in major depressive disorder. Psychiatry Res Neuroimag 2003; 124: 129–140.

51. Folstein SE, Abbott MH, Chase GA, et al The association of affective disorder with Huntington's disease in a case series and in families. Psychol Med 1983; 13: 537–542.

52. Starkstein SE, Robinson RG. Affective disorders and cerebral vascular disease. Br J Psychiatry 1989; 154: 170–182.

53. Burns A, Jacoby R, Levy R. Psychiatric phenomena in Alzheimer's disease III: disorders of mood. Br J Psychiatry 1990; 157: 81–86.

54. Sawrie SM, Martin RC, Knowlton R, et al. Relationships among hippocampal volumetry, proton magnetic resonance spectroscopy, and verbal memory in temporal lobe epilepsy. Epilepsia 2001; 42: 1403–1407.

55. Cummings JL. Depression and Parkinson's disease: a review. Am J Psychiatry 1992; 149: 443–454.

56. Parashos IA, Oxley SL, Boyko OB, Krishnan KR. In vivo quantitation of basal ganglia and thalamic degenerative changes in two temporal lobectomy patients with affective disorder. J Neuropsychiatry Clin Neurosci 1993; 5: 337–341.

57. Robinson RG, Kubos KL, Starr LB, et al. Mood changes in stroke patients: relationship to lesion location. Compr Psychiatry 1983; 24: 555–566.

58. Lipsey JR, Robinson RG, Pearlson GD, et al. Mood change following bilateral hemisphere brain injury. Br J Psychiatry 1983; 143: 266–267.

59. Carson JA, MacHale S, Allen K, et al. Depression after stroke and lesion location: a systematic review. Lancet 2000; 356: 122–127.

60. Vataja R, Pohjasvaara T, Leppavuori A, et al. Magnetic resonance imaging correlates of depression after ischemic stroke. Arch Gen Psychiatry 2001; 58: 925–931.

61. Jellinger KA. Post mortem studies in Parkinson's disease – is it possible to detect brain areas for specific symptoms? J Neural Transm Suppl 1999; 56: 1–29.

62. Lisanby SH, McDonald WM, Massey EW, et al. Diminished subcortical nuclei volumes in Parkinson's disease by MR imaging. J Neural Transm Suppl 1993; 40: 13–21.

63. Mayberg HS, Starkstein SE, Sadzot B, et al. Selective hypometabolism in the inferior frontal lobe in depressed patients with Parkinson's disease. Ann Neurol 1990; 28: 57–64.

64. Kanne SM, Balota DA, Storandt M, et al. Relating anatomy to function in Alzheimer's disease: neuropsychological profiles predict regional neuropathology 5 years later. Neurology 1998; 50: 979–985.

65. Lyketsos CG, Olin J. Depression in Alzheimer's disease: overview and treatment. Biol Psychiatry 2002; 52: 243–252.

66. Steffens DC, Payne ME, Greenberg DL, et al. Hippocampal volume and incident dementia in geriatric depression. Am J Geriatr Psychiatry 2002; 10: 62–71.

67. Baron M, Mendlewicz J, Klotz J. Age-of-onset and genetic transmission in affective disorders. Acta Psychiatr Scand 1981; 64: 373–380.

68. Jacoby RJ, Levy R, Bird JM. Computed tomography and the outcome of affective disorder: a follow-up study of elderly patients. Br J Psychiatry 1981; 139: 288–292.

69. Coffey C, Figiel G, Djang W. Leukoencephalopathy in elderly depressed patients referred for ECT. Biol Psychiatry 1988; 24: 143–161.

70. Figiel GS, Krishnan KRR, Doraiswamy PM, et al. Subcortical hyperintensities on brain magnetic resonance imaging: a comparison between late age onset and early onset elderly depressed subjects. Neurobiol Aging 1991; 12: 245–247.

71. Alexopoulos GS, Meyers B, Young R, et al. Recovery in geriatric depression. Arch Gen Psychiatry 1996; 53: 305–312.

72. Simpson S, Baldwin RC, Jackson A, Burns AS. Is subcortical disease associated with a poor response to antidepressants? Neurological, neuropsychological and neuroradiological findings in late-life depression. Psychol Med 1998; 28: 1015–26.

73. Pantel J, Schroder J, Essig M, et al. Quantitative magnetic resonance imaging in geriatric depression and primary degenerative dementia. J Affect Disord 1997; 42: 69–83.

74. Rabins PV, Pearlson GD, Aylward E, et al Cortical magnetic resonance imaging changes in elderly inpatients with major depression. Am J Psychiatry 1991; 148: 617–620.

75. Rothschild AJ, Benes F, Hebben N, et al. Relationships between brain CT scan findings and cortisol in psychotic and nonpsychotic depressed patients. Biol Psychiatry 1989; 26: 565–575.

76. Kobayashi S, Okada K, Yamashita K. Incidence of silent lacunar lesion in normal adults and its relation to cerebral blood flow and risk factors. Stroke 1991; 22: 1379–1383.

77. Starkman MN, Gebarski SS, Berent S, Schteingart DE. Hippocampal formation volume, memory dysfunction, and cortisol levels in patients with Cushing's syndrome. Biol Psychiatry 1992; 32: 756–765.

78. Aronson S. Intracranial vascular lesions in patients with diabetes mellitus. J Neuropathol Exp Neurol 1973; 32: 183–196.

79. Charness ME. Brain lesions in alcoholics. Alcohol Clin Exp Res 1993; 17: 2–11.

80. Kumar A, Jin Z, Bilker W, et al. Late-onset minor and major depression: early evidence for common neuroanatomical substrates detected by using MRI. Proc Natl Acad Sci 1998; 95: 7654–7658.

81. Coffey CE, Figiel GS, Djang WT, Weiner RD. Subcortical hyperintensity on magnetic imaging: a comparison of normal and depressed elderly subjects. Am J Psychiatry 1990; 147: 187–189.

82. Howard RJ, Beats B, Forstl H, et al. White matter changes in late onset depression: a magnetic resonance imaging study. Int J Geriatr Psychiatry 1993; 8: 183–185.

83. Krishnan KR, McDonald WM, Doraiswamy PM. Neuroanatomical substrates of depression in the elderly. Eur Arch Psychatry Clin Neurosci 1993; 243: 41–46.

84. Lesser IM, Miller BL, Boone KB, et al. Brain injury and cognitive function in late-onset psychotic depression. J Neuropsychiatry Clin Neurosci 1991; 3: 33–40.

85. Coffey CE, Wilkinson WE, Weiner RD, et al. Quantitative cerebral anatomy in depression: a controlled magnetic resonance imaging study. Arch Gen Psychiatry 1993; 50: 7–16.

86. Hickie I, Scott E, Mitchell P, et al. Subcortical hyperintensities on magnetic resonance imaging: clinical correlates and prognostic significance in patients with severe depression. Biol Psychiatry 1995; 37: 151–160.

87. Dupont RM, Jernigan TL, Heindel W, et al. Magnetic resonance imaging and mood disorders – localization of white matter and other subcortical abnormalities. Arch Gen Psychiatry 1995; 52: 747–755.

88. Guze BH, Szuba MP. Leukoencephalopathy and major depression: a preliminary report. Psychiatry Res 1992; 45: 169–175.

89. Fazekas F, Kleinert R, Offenbacher H, et al. The morphologic correlate of incidental punctate white matter hyperintensities on MR images. Am J Nucl Radiol 1991; 12: 915–921.

90. Mirsen TR, Lee DH, Wong CJ, et al. Clinical correlates of white-matter changes on magnetic resonance imaging scans of the brain. Arch Neurol 1991; 48: 1015–21.

91. Fujikawa T, Yamawaki S, Touhouda Y. Incidence of silent cerebral infarction in patients with major depression. Stroke 1993; 24: 1631–1634.

92. Fujikawa T, Yamawaki S, Touhouda Y. Background factors and clinical symptoms of major depression with silent cerebral infarction. Stroke 1994; 25: 798–801.

93. Krishnan KR, Goli V, Ellinwood EH, et al. Leukoencephalopathy in patients diagnosed as major depressive. Biol Psychiatry 1988; 23: 519–522.

94. Krishnan K, Hays J, Blazer D. MRI-defined vascular depression. Am J Psychiatry 1997; 154: 497–501.

95. MacFall JR, Payne ME, Provenzale JE, Krishnan KR. Medial orbital frontal lesions in late-onset depression. Biol Psychiatry 2001; 49: 803–806.

96. Salat DH, Kaye JA, Janowsky JS. Prefrontal gray and white matter volumes in healthy aging and Alzheimer disease. Arch Neurol 1999; 56: 338–344.

97. Resnick S, Goldszal AF, Davatzikos C, et al. One-year age changes in MRI brain volumes in older adults. Cereb Cortex 2000; 10: 464–472.

98. Mueller EA, Moore MM, Kerr DCR, et al. Brain volume preserved in healthy elderly through the eleventh decade. Neurology 1998; 51: 1555–1562.

99. Almeida OP, Burton EJ, Ferrier N, et al. Depression with late onset is associated with right frontal lobe atrophy. Psychol Med 2003; 33: 675–681.

100. Tebartz van Elst L, Woermann FG, Lemieux L, Trimble MR. Amygdala enlargement in dysthymia – a volumetric study of patients with temporal lobe epilepsy. Biol Psychiatry 1999; 46: 1614–1623.

101. Tebartz van Elst L, Woermann FG, Lemieux L, Trimble MR. Increased amygdala volumes in female and depressed humans. A quantitative magnetic resonance study. Neurosci Lett 2000; 281: 103–106.

102. Axelson DS, Doraiswamy PM, Boyko OB, et al. In vivo assessment of pituitary volume with magnetic resonance imaging and systematic stereology: relationship to dexamethasone suppression test results in patients. Psychiatry Res 1992; 44: 63–70.

4

Functional imaging in secondary depressions

Andrei G Vlassenko, Yvette I Sheline and Mark A Mintun

Depression is a frequently associated syndrome in many neurological and psychiatric disorders including neurodegenerative, neuroinflammatory and vascular diseases. Specific neural pathways, neurophysiological changes and neurotransmitter abnormalities have been proposed for different types of secondary depression, although there are commonalities as well. Clinical manifestations of secondary depressions may vary from moderate mood disturbances and apathy to full-blown depression. In recent decades, an increasing number of investigators have begun to study the pathophysiology of secondary depression using modern methods of functional brain imaging including positron emission tomography (PET) and single-photon emission tomography (SPECT). In this chapter we mention briefly the findings in major depressive disorder and discuss in more detail the imaging characteristics of secondary depressions that are often comparable to those in major depressive disorder, but may have distinct patterns depending on the primary disease.

Imaging data in primary depression, or major depressive disorder, most commonly demonstrate frontal and cingulate abnormalities, and often other limbic–paralimbic and subcortical changes have been reported (Figure 4.1)[1–5]. Frontal and cingulate abnormalities involve multiple but distinct areas including the dorsolateral and ventrolateral prefrontal cortex, dorsomedial and ventromedial frontal cortex, and dorsal, rostral and subgenual cingulate cortex. Not all of these areas are demonstrated in all studies, and the direction and severity of cerebral blood flow (CBF) and glucose metabolism (^{18}F-fluorodeoxyglucose; FDG) changes vary between studies[6].

Repeated observations from independent studies of patients with different primary diseases and depression suggest that these functional abnormalities most probably have underlying structural changes within specific neuroanatomical circuits. Several pathways have been proposed, including orbitofrontal–basal, ganglia–thalamic, basotemporal limbic and limbic–cortical–striatal–pallidal–thalamic circuits[7–10]. One of the popular and well-accepted hypotheses is that depressed mood correlates with changes in stereotypic brain regions and is state-specific rather than disease-specific[8]. As depression is believed to be the result of disruption of multisystem functional interactions between a highly integrated network of limbic and cortical brain areas, a limbic–cortical depression model has been proposed, which may help to integrate the existing knowledge of the pathogenesis of depressive disorders and improve treatment strategies (Figure 4.2)[4,6,11]. Serotonin, dopamine and norepinephrine systems have been implicated in behavioral changes in depression; altered radioligand binding to serotonin and dopamine receptors have been repeatedly demonstrated using PET and SPECT[5,12,13].

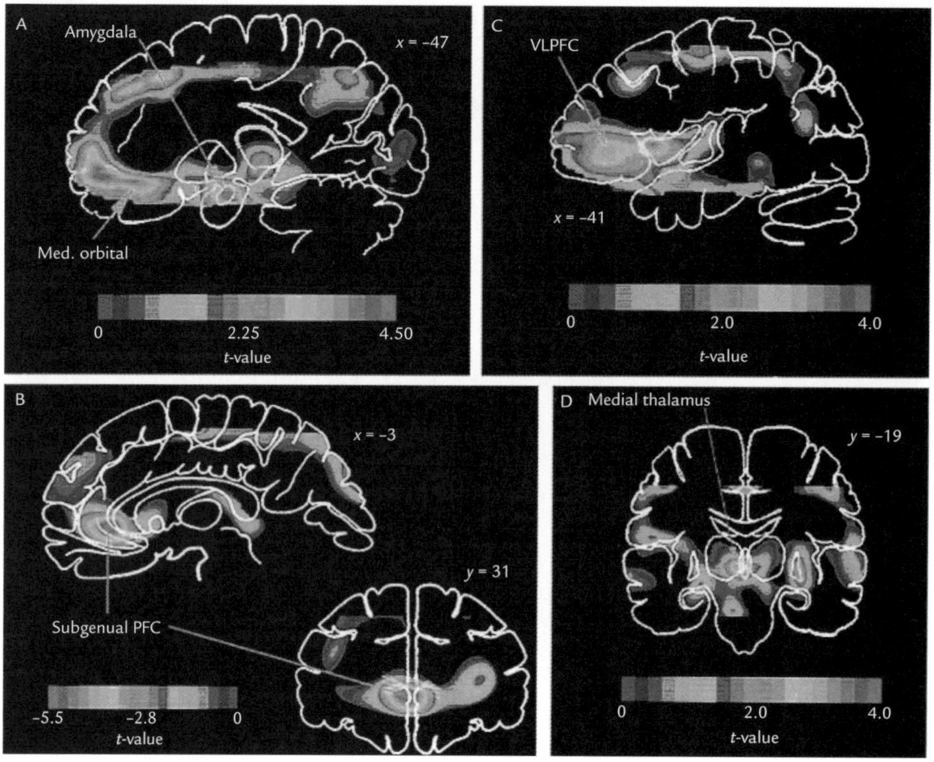

Figure 4.1 (A) Areas of abnormally increased cerebral blood flow (CBF) in familial major depressive disorder (MDD) shown in an image of *t* values, produced by a voxel-by-voxel computation of the unpaired *t* statistic to compare CBF between depressed and control samples. The positive *t* values in this sagittal section at 17 mm left of midline ($x = -17$) show areas of increased CBF in depressives relative to control subjects in the amygdala and medial orbital cortex. Anterior is to the left. (B) Coronal (31 mm anterior to the anterior commissure, or $y = 31$) and sagittal (3 mm left of midline, or $x = -3$) sections showing negative voxel *t* values where glucose metabolism is decreased in depressives relative to control subjects. Anterior is to the left. (C) Sagittal *t* image section 41 mm left of midline ($x = -41$) shows an area of increased CBF in depressives relative to healthy control subjects in the left ventrolateral prefrontal cortex (VLPFC), lateral orbital cortex and anterior insula. (D) Coronal *t* image section 19 mm caudal to the anterior commissure (i.e. $y = -19$) showing the area corresponding to increased CBF in the left medial thalamus of depressives relative to control subjects. Reproduced from reference 3: Drevets WC. Neuroimaging studies of mood disorders. *Biol Psychiatry* 2000; 48: 813–829, with permission from the Society of Biological Psychiatry, copyright 2000. (See also color image on p. xviii)

STROKE

The prevalence of post-stroke depression has been considered highly variable, ranging from 20 to 60%[14–17]. The relationship between the location of the lesion and the occurrence of post-stroke depression has been repeatedly studied; different locations (especially left frontal and basal ganglia lesions) were associated with mood changes after stroke[14,18–21]. It was also suggested that mood changes may result not solely from the stroke lesion but also

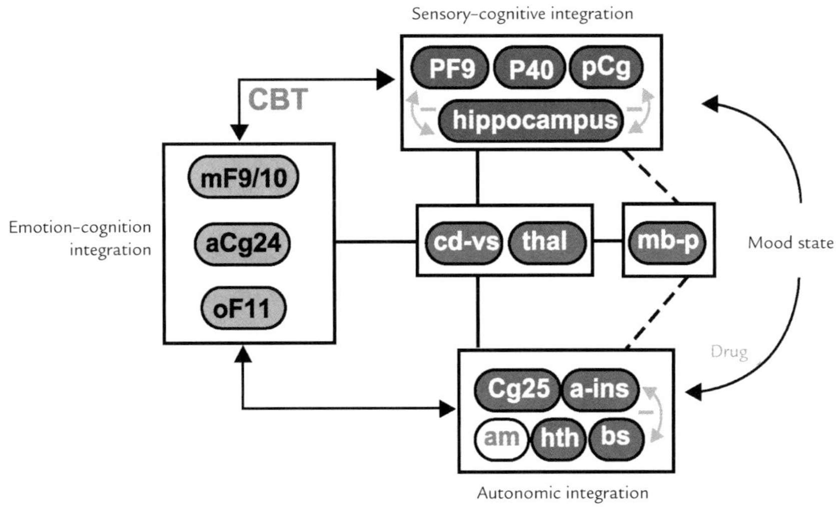

Figure 4.2 Limbic–cortical dysregulation model. Regions with known anatomical interconnections are grouped into four main behavioral compartments. The cortical limbic (dorsal–ventral) segregation within each 'compartment' additionally identifies those brain regions where an inverse relationship is seen across the different PET paradigms. Sadness and depressive illness are both associated with decreases in dorsal neocortical regions (sensory–cognitive compartment) and relative increases in ventral limbic and paralimbic areas (autonomic compartment). The model, in turn, proposes that illness remission occurs when there is appropriate modulation of dysfunctional limbic–cortical interactions (small gray arrows) – an effect facilitated by various forms of treatment (purple). mF, medial prefrontal; aCg, rostral anterior cingulate; oF, orbital frontal; cd-vs, caudate–ventral striatum; thal, thalamus; mb-p, midbrain-pons; Cg25, subgenual cingulate; a-ins, anterior insula; am, amygdala, hth, hypothalamus; bs, brain stem; PF, dorsolateral prefrontal; P, parietal; pCg, posterior cingulate. Numbers: Brodmann designations. Reproduced from reference 6: Seminowicz DA, Mayberg HS, McIntosh AR, *et al.* Limbic–frontal circuitry in major depression: a path modeling metanalysis. *Neuroimage* 2004; 22: 409–418, with permission from Elsevier, copyright 2004. (See also color image on p. xix)

from interruption of subcortical–cortical or cortical–cortical connections (so-called 'remote diaschisis'[22–25]. This hypothesis explains the strong association of both basal ganglia and frontal lesions with depression.

FDG changes in stroke patients with depression are often localized in the same areas as in patients with primary depression (Figures 4.3 and 4.4). CBF has been studied in post-stroke depression using the ^{133}Xenon inhalation method[26] and ^{99m}Tc hexamethylpropyleneamine oxime (^{99m}Tc-HMPAO) SPECT imaging[27,28]. In these studies there was a significant correlation between the size of the CBF defect and the severity of depression[26,27]. A longitudinal study demonstrated that improvement of test scores in post-stroke patients with depression was associated with an increase in regional CBF[26]. In patients with subcortical stroke, neither location nor laterality of lesions correlated with the occurrence of depression; however, the mesial temporal cortex ipsilateral to the subcortical ischemia was significantly less perfused in depressed patients compared to non-depressed patients[28].

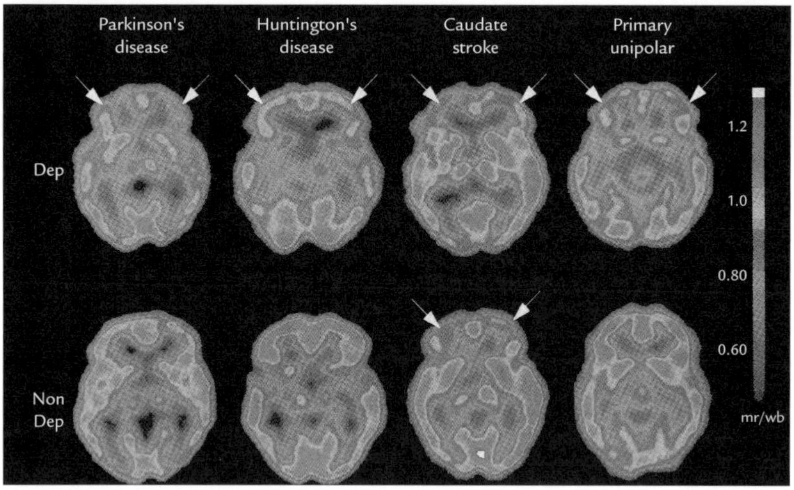

Figure 4.3 ^{18}F-fluorodeoxyglucose (FDG) positron emission tomography (PET) images in primary and secondary depressions – basal ganglia level. There is bilateral frontal lobe hypometabolism in all depressed patients (Dep), independent of disease etiology (arrows, top row). Frontal cortex metabolism is normal in non-depressed patients (Non Dep) except those with strokes (arrows, bottom row). Scale: relative metabolic rate (regional absolute metabolic rate/whole brain metabolic rate). Reproduced from reference 8, with permission from the *American Journal of Geriatric Psychiatry*, copyright 1994, American Psychiatric Association. (See also color image on p. xx)

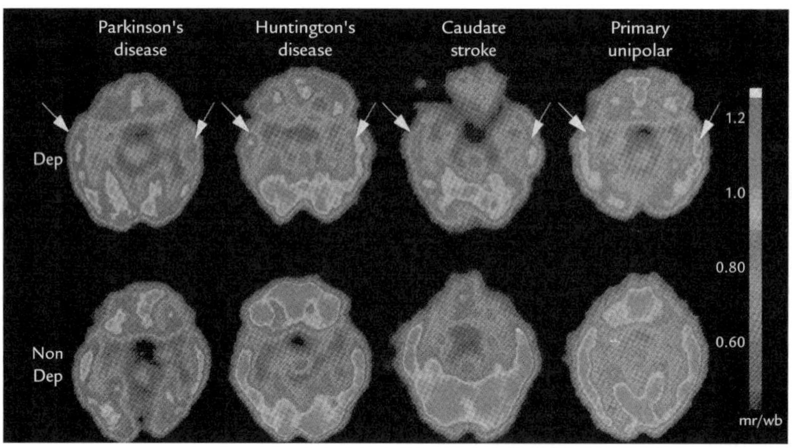

Figure 4.4 ^{18}F-fluorodeoxyglucose (FDG) positron emission tomography (PET) images in primary and secondary depressions – temporal lobe level. There is symmetric bitemporal hypometabolism in all depressed patients (Dep), independent of disease etiology (arrows, top row). Temporal cortex metabolism is normal in all non-depressed patients (Non Dep). Scale: relative metabolic rate (regional absolute metabolic rate/whole brain metabolic rate). Reproduced from reference 8, with permission from the *American Journal of Geriatric Psychiatry*, copyright 1994, American Psychiatric Association. (See also color image on p. xx)

Regional CBF has also been studied in patients with post-stroke subcortical lesions and apathy compared to patients without mood disturbances[29]. The apathetic group showed a significantly reduced regional CBF in the right dorsolateral frontal and left frontotemporal regions, and the apathy score for all patients was significantly negatively correlated with the CBF in the same regions. The authors suggested that vascular damage to frontal–subcortical circuits (including those in the anterior cingulate, dorsolateral prefrontal and cholinergic projection to the temporal cortex) is responsible for producing apathy in subcortical infarctions[29].

EPILEPSY

Depression is a relatively frequent concomitant of epilepsy, which significantly increases the rate of suicide in epileptic patients[30–35]. Laterality of ictal onset and degree of interictal temporal lobe hypometabolism were suggested as interdependent factors contributing to the risk of depression in patients with complex partial seizures[32].

Bromfield *et al.*[36] studied depressed patients with complex partial seizures using FDG PET and reported that patients with left temporal lobe seizure foci were more likely to exhibit depressive symptoms than those with right temporal lobe lesions, and that the neurophysiological correlate for these symptoms may be bilateral in the inferior frontal rather than the temporal lobe. The similarity of metabolic abnormalities in depressions associated with different neurological and psychiatric disorders may suggest that the depression itself may be a correlate, if not a consequence, of frontal hypometabolism independent of the underlying illness[36]. This hypothesis was later supported by Schmitz *et al.*[37] who reported that bilateral frontal hypoperfusion as assessed using ^{99m}Tc-HMPAO SPECT was associated with a significantly higher score on the Beck Depression Inventory.

Ring *et al.*[38] compared patients with temporal epilepsy and depression to non-depressive epileptic patients using ^{99m}Tc-HMPAO SPECT. In contrast to previous studies[36,37], they found widespread regions of relative hyperactivity across left frontal, temporal, parietal and subcortical structures in depressed compared with non-depressed patients. The authors suggested that similar functional circuits may be involved in the manifestation of depression in epilepsy, but they are dysfunctional by virtue of being more active, rather than less active, and the mechanism underlying the development of epilepsy-related depression may be distinct from that occurring in primary or movement disorder-related depression[38].

MULTIPLE SCLEROSIS

Among psychiatric disorders in multiple sclerosis, depression is the most frequent[39]; as many as 60% of patients with multiple sclerosis experience depression on a life-time basis[40]. Mechanisms underlying depression in multiple sclerosis are unknown, with several hypothetical explanations proposed, including relation to the location of the disease lesions and reactive character of mood changes due to high disability and the unpredictable course of the disease[41,42].

In spite of the lack of differences in regional area of cerebral lesions between depressed and non-depressed patients with multiple sclerosis, a significant association between CBF asymmetry in the limbic system and depression was demonstrated using SPECT with ^{99m}Tc-HMPAO[43]. Moreover, the degree of CBF asymmetry in the limbic system significantly correlated with the severity of the depression. It was suggested that changes in the activity of the limbic system might reflect cortical suppression or deactivation after disconnection from subcortical structures, owing to white matter lesions of multiple sclerosis.

PARKINSON'S DISEASE

In Parkinson's disease (PD), depression occurs in 40–50% of patients and often precedes motor symptoms[44–46]. Depressed patients with PD have faster motor impairment and greater decline in cognitive and self-care functions compared to non-depressed patients. However, depression in PD may be qualitatively different from major depressive disorder, with fewer suicides and less guilt and discouragement[45,47,48]. The cause of depression in PD remains unknown; however, the loss of mesocortical and mesolimbic dopamine connections to the frontal lobe and disruption of monoaminergic afferents from the mesencephalon to the limbic frontal cortex have been proposed as a hypothetical mechanism of depression in patients with PD, and this hypothesis has been supported by postmortem data[49–51].

The relationship between depression and PD abnormalities has been characterized by functional imaging methods including CBF SPECT and glucose metabolism PET studies[52,53]. Mayberg *et al.*[52] reported bilateral caudate and orbital–inferior frontal hypometabolism (Brodmann areas 10, 11, 12 and 47) in depressed PD patients compared to both non-depressed patients and control subjects. The magnitude of hypometabolism in the inferior prefrontal–orbital frontal region inversely correlated with depression severity as measured by the Hamilton Rating Scale for Depression, but not with changes in cognitive performance. Later, Jagust *et al.*[53] assessed regional cerebral perfusion using SPECT with [^{123}I]iodoamphetamine in patients with PD including a subgroup of depressed patients, and also suggested that the effects of depression on frontal lobe performance are mediated by the orbitofrontal cortex.

Ring *et al.*[54] studied resting CBF changes using PET and a statistical parametric mapping technique in patients with PD and depression, in patients with PD alone, in normal subjects and in patients with primary depression. Bilateral decreases in regional CBF were observed in anteromedial regions of the medial frontal cortex and the cingulate cortex (Brodmann areas 9 and 32) in the depressed PD group, compared with those with PD alone and compared with normal controls. There were no significant differences in these regions between the depressed PD group and the patients with primary depression. As similar medial prefrontal brain regions have been shown to be involved in primary depression[1], Ring *et al.*[54] suggested that the depression of PD shares a common biological substrate with that of primary depression.

The hypothesis that primary and secondary depressions, regardless of disease etiology, have comparable patterns of abnormal metabolism involving brain areas with limbic connections has been tested by Mayberg *et al.*[8,55] (Figures 4.3 and 4.4). They demonstrated that depressed mood correlated with changes in stereotypic brain regions and was state-specific rather than disease-specific.

In PD patients with varying degrees of dysphoria but not major depression, Mentis *et al.*[56] found bilaterally low FDG metabolism in the anterior cingulate cortex, orbitofrontal cortex, and lateral and medial frontal lobes (Figure 4.5). As this pattern included areas involved in major depression in PD[52–54], it was suggested that abnormal connectivity among prefrontal and limbic structures underlies not only major depression but also dysphoria in PD[56].

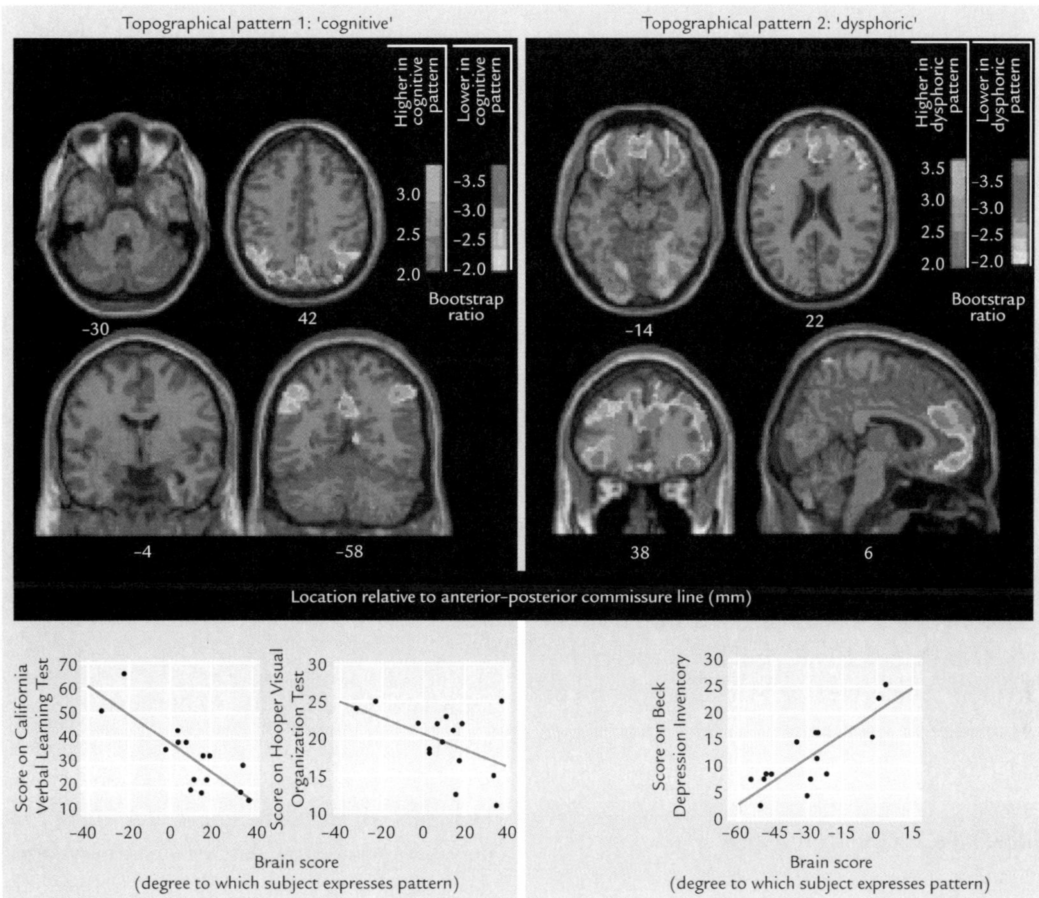

Figure 4.5 Two orthogonal topographic patterns of brain metabolism in 15 patients with Parkinson's disease and doubly dissociated correlations with cognitive and mood measures. The orthogonal topographic patterns were derived from glucose metabolic scans at rest. The bootstrap ratio is the ratio of voxel salience value to its standard error (estimated from 100 bootstrap samples). It is a measure of the stability of high (red scale) and low (blue scale) metabolic values that constitute the topographic pattern. Scatterplots of relationship between behavior and brain score are presented at the bottom row. Brain–behavior correlations for dysphoria and cognition were doubly dissociated, whereas those for the two cognitive measures were similar. Reproduced from reference 56, with permission from the *American Journal of Psychiatry*, copyright 1971, American Psychiatric Association. (See also color image on p. xxi)

In contrast to motor disability[57,58], mood and cognitive performance in PD patients was found to be unrelated to the degree of striatal dopaminergic depletion[59]. In this study, however, depression was found only in advanced PD and it did not segregate with any variables of principal component analysis including the striatal [^{18}F]-6-fluoro-L-dopa rate constant measured with PET[59].

Serotonergic and noradrenergic mechanisms have also been implicated, because depressed patients with PD respond favorably to antidepressant therapy[60,61] and pilot serotonin receptor studies using carbonyl-[^{11}C]-*N*-[2-[4-(2-methoxyphenyl)-1-piperazinyl]ethyl]-*N*-(2-pyridinyl)cyclohexane carboxamide (^{11}C-WAY) PET showed significant 17.5% reduction in cortical postsynaptic 5-hydroxytryptamine (5-HT)$_{1A}$ binding in depressed PD patients in contrast to unaltered binding in non-depressed PD patients[62]. These preliminary data suggested that the presence of depression in PD is associated with a reduced cortical responsiveness to serotonin along with a loss of serotonin output. More studies with serotonin receptors would be helpful in improving our knowledge in this field.

HUNTINGTON'S DISEASE

In Huntington's disease (HD), depression affects approximately 40% of patients and may precede choreiform movements and cognitive symptoms[63,64]. Many PET studies in patients with HD have reported decreases in striatal, frontal and cingulate cortex glucose metabolism. However, only a few of them have examined mood disorders and specific metabolic patterns of depression in HD[55,65–69].

Mayberg *et al.*[55] studied brain glucose metabolism in depressed and non-depressed patients with HD using PET. They found that caudate, putamen and cingulate metabolism was significantly lower in the patients with HD compared to the control subjects, and orbital–inferior prefrontal cortex and thalamic hypometabolism differentiated depressed patients from non-depressed patients and normal controls. They hypothesized that disruption of pathways linking the paralimbic frontal cortex and the basal ganglia is integral to the development of depression in HD[8,55]. Several potential explanations were suggested for the frontal cortical hypometabolism in depressed patients with HD, including primary dysfunction or degeneration of neurons in the frontal cortex, anterograde or retrograde degeneration of pathways linking dorsomedial caudate, dorsomedial thalamus and orbitofrontal–inferior prefrontal cortex, or changes in basotemporal limbic pathways linking the orbitofrontal cortex with the amygdala, temporal pole and dorsomedial thalamus (Figures 4.3 and 4.4)[8].

DEMENTIA OF THE ALZHEIMER'S TYPE

Dementia of the Alzheimer's type (DAT) may be accompanied by different neuropsychiatric disturbances including hallucinations, personality alterations and mood disorders. Depression is a common symptom, affecting 30–50% of patients with DAT[70–73].

Hirono *et al.*[74] reported that, in DAT patients, depression was associated with glucose (FDG) hypometabolism in the bilateral superior gyri and left anterior cingulate gyrus and that, without reaching statistical significance, other frontal regions showed a similar trend.

Glucose hypometabolism measured using FDG PET was also observed in the parietotemporal area in patients with DAT and depression[75]. Severity of depressive symptoms was associated with decreased perfusion in frontotemporal and basal ganglia regions of interest as reported by Ebmeier *et al.*[76] based on SPECT data in depressed DAT patients.

Analysis of statistical parametric mapping showed revealing differences in ^{99m}Tc-HMPAO SPECT images in AD patients with and without depression[77]. The depressed group demonstrated selective hypoperfusion in the bilateral anterior and posterior cingulate gyri and precuneus. An inverse correlation was found between cerebral perfusion and the severity of depression assessed by the Hamilton Rating Scale for Depression[77].

Sultzer *et al.*[78] studied glucose metabolism using FDG PET in patients with DAT and reported that the relationship between cortical hypometabolism and psychiatric symptoms in DAT was regionally specific, the anxiety/depression score in the Neurobehavioral Rating Scale correlating with the parietal metabolic rate, but not with frontal glucose metabolism.

Starkstein *et al.*[79] compared regional changes in CBF between DAT patients with depression, patients with dysthymia and DAT patients without depression, using ^{99m}Tc-HMPAO. They found that DAT patients with depression had significant CBF deficits in specific brain regions compared to dysthymic and non-depressed DAT patients. Compared to non-depressed DAT patients, DAT patients with depression showed significant CBF decrease in the left temporal superior and left parietal areas. When compared to DAT patients with dysthymia, these CBF changes also involved the left frontal lobe, right temporal superior area and the right basal ganglia[79]. It was suggested that dysthymia and depression in DAT patients have different mechanisms. Dysthymia may be an emotional reaction of the subgroup of DAT patients when confronted by the progressive cognitive decline. A partial support of this hypothesis is provided by the findings that patients with dysthymia had no significant CBF differences as compared to non-depressed patients. In depressed DAT patients, a similar prevalence of depression across the progressive stages of the illness and poor awareness about their cognitive decline were noted, suggesting, together with CBF findings, that depression in DAT may be related primarily to biological factors[79].

The prevalence of apathy in DAT may be high (up to 61%[80]). Regional differences in CBF were studied using ^{99m}Tc-HMPAO SPECT in patients with DAT with apathy and compared with those in DAT patients without apathy[81]. The presence of apathy was associated with more severe prefrontal and anterior temporal hypoperfusion. Craig *et al.*[81] suggested that intact neuronal networks rather than single cortical regions are important for preservation of the complex personality traits of drive, initiative and interest. Appearance of apathy early in DAT may be primarily related to loss of nucleus basalis of Meynert cholinergic input to prefrontal (primary anterior cingulate) and subcortical (thalamic) structures, and later impairment may be related both to more profound cholinergic denervation and to pathological involvement of network structures (prefrontal and anterior and medial temporal)[81]. In another SPECT study, apathy was correlated with diminished blood flow in the right temporoparietal region[82]. Diminished regional CBF in the right dorsolateral prefrontal cortex and basal ganglia was demonstrated in a patient with DAT and apathy with a ^{15}O water PET scan[83].

Benoit *et al.*[84] studied the relationship between regional cerebral perfusion measured with ^{99m}Tc-bicisate SPECT and apathy assessed using the Neuropsychiatric Inventory (NPI) and Mini-Mental State Examination (MMSE) in patients with DAT. The apathy NPI score was correlated with right cingulate gyrus hypoperfusion, whereas the highest correlation for the MMSE was with the left temporoparietal area. Later, the same investigators[85] presented the results of comparison of changes in brain perfusion between two groups of elderly subjects differentiated by the presence or absence of apathy. The apathy group included DAT patients and also patients with other cognitive disorders such as mild cognitive impairment or organic personality disorders. Using statistical parametric mapping (SPM96) they demonstrated that the common feature of the apathetic subgroups (whole population, demented subjects only or non-demented subjects only) was always the anterior cingulate hypoperfusion. They suggested that apathy is related to anterior cingulate region hypoperfusion and this dysfunction is not disease-specific[85].

In summary, secondary depressions, common concomitants of various brain diseases, are associated with a wide range of metabolic abnormalities, as shown by functional imaging methods. These abnormalities may be caused not only by local damage of regions directly involved in the pathophysiology of depression, but also by disruption of subcortical–cortical or cortical–cortical circuits. This may be one reason why secondary depressions of different origins have common patterns of functional abnormality. For example, the areas of metabolic abnormalities most often seen in depressions associated with stroke, PD and HD (the orbital inferior frontal cortex or the basotemporal cortex) are involved in the orbitofrontal–basal ganglia–thalamic pathway and the basotemporal limbic pathway that links the orbitofrontal cortex and anterior temporal cortex via the uncinate fasciculus[8,86,87]. The anterior cingulate area appears to be most often involved in depressed patients with diseases of different pathophysiological and pathological origin, such as stroke, PD, HD and DAT[29,54–56,74,77,85]. Abnormal metabolism and diffusion in this area in depression, irrespective of the etiology of the primary disease, suggests a state-specific rather than a disease-specific character of the dysfunctions. Improving the quality and accuracy of both structural and functional imaging techniques together with the development of new radioligands and tracers may significantly contribute to the identification of the causes and mechanisms of secondary depression as well as specific markers of the disease and targets for treatment.

REFERENCES

1. Bench CJ, Friston KJ, Brown RG, et al. The anatomy of melancholia – focal abnormalities of cerebral blood flow in major depression. Psychol Med 1983; 22: 607–615.
2. Drevets WC, Videen TO, Price JL, et al. A functional anatomical study of unipolar depression. J Neurosci 1992; 12: 3628–3641.
3. Drevets WC. Neuroimaging studies of mood disorders. Biol Psychiatry 2000; 48: 813–829.
4. Mayberg HS. Positron emission tomography imaging in depression: a neural systems perspective. Neuroimag Clin North Am 2003; 13: 805–815.

5. Mintun MA, Sheline YI, Moerlein SM, et al. Decreased hippocampal 5-HT2A receptor binding in major depressive disorder: in vivo measurement with [18F]altanserin positron emission tomography. Biol Psychiatry 2004; 55: 217–224.

6. Seminowicz DA, Mayberg HS, McIntosh AR, et al. Limbic–frontal circuitry in major depression: a path modeling metanalysis. Neuroimage 2004; 22: 409–418.

7. Nauta WJ. Neural associations of the frontal cortex. Acta Neurobiol Exp (Wars) 1972; 32: 125–140.

8. Mayberg HS. Frontal lobe dysfunction in secondary depression. J Neuropsychiatry Clin Neurosci 1994; 6: 428–442.

9. Sheline YI. 3D MRI studies of neuroanatomic changes in unipolar major depression: the role of stress and medical comorbidity. Biol Psychiatry 2000; 48: 791–800.

10. Sheline YI. Neuroimaging studies of mood disorder effects on the brain. Biol Psychiatry 2003; 54: 338–352.

11. Mayberg HS. Limbic-cortical dysregulation: a proposed model of depression. J Neuropsychiatry Clin Neurosci 1997; 9: 471–481.

12. D'haenen HA, Bossuyt A. Dopamine D2 receptors in depression measured with single photon emission computed tomography. Biol Psychiatry 1994; 35: 128–132.

13. Biver F, Wikler D, Lotstra F, et al. Serotonin 5-HT2 receptor imaging in major depression: focal changes in orbito-insular cortex. Br J Psychiatry 1997; 171: 444–448.

14. Robinson RG, Szetela B. Mood change following left hemispheric brain injury. Ann Neurol 1981; 9: 447–453.

15. Robinson RG, Price TR. Post-stroke depressive disorders: a follow-up study of 103 patients. Stroke 1982; 13: 635–641.

16. Astrom M, Adolfsson R, Asplund K. Major depression in stroke patients. A 3-year longitudinal study. Stroke 1993; 24: 976–982.

17. Burvill PW, Johnson GA, Jamrozik KD, et al. Prevalence of depression after stroke: the Perth Community Stroke Study. Br J Psychiatry 1995; 166: 320–327.

18. Robinson RG, Kubos KL, Starr LB, et al. Mood disorders in stroke patients. Importance of location of lesion. Brain 1984; 107: 81–93.

19. Sinyor D, Jacques P, Kaloupek DG, et al. Poststroke depression and lesion location. An attempted replication. Brain 1986; 109: 537–546.

20. Starkstein SE, Robinson RG, Price TR. Comparison of cortical and subcortical lesions in the production of poststroke mood disorders. Brain 1987; 110: 1045–1059.

21. Morris PL, Robinson RG, de Carvalho ML, et al. Lesion characteristics and depressed mood in the stroke data bank study. J Neuropsychiatry Clin Neurosci 1996; 8: 153–159.

22. Baron JC, D'Antona R, Pantano P, et al. Effects of thalamic stroke on energy metabolism of the cerebral cortex. A positron tomography study in man. Brain 1986; 109: 1243–1259.

23. Baron JC. Depression of energy metabolism in distant brain structures: studies with positron emission tomography in stroke patients. Semin Neurol 1989; 9: 281–285.

24. Perani D, Di Piero, V, Lucignani G, et al. Remote effects of subcortical cerebrovascular lesions: a SPECT cerebral perfusion

study. J Cereb Blood Flow Metab 1988; 8: 560–567.

25. Pappata S, Mazoyer B, Tran DS, et al. Effects of capsular or thalamic stroke on metabolism in the cortex and cerebellum: a positron tomography study. Stroke 1990; 21: 519–524.

26. Yamaguchi S, Kobayashi S, Koide H, Tsunematsu T. Longitudinal study of regional cerebral blood flow changes in depression after stroke. Stroke 1992; 23: 1716–1722.

27. Schwartz JA, Speed NM, Mountz JM, et al. 99mTc-hexamethylpropyleneamine oxime single photon emission CT in poststroke depression. Am J Psychiatry 1990; 147: 242–244.

28. Grasso MG, Pantano P, Ricci M, et al. Mesial temporal cortex hypoperfusion is associated with depression in subcortical stroke. Stroke 1994; 25: 980–985.

29. Okada K, Kobayashi S, Yamagata S, et al. Poststroke apathy and regional cerebral blood flow. Stroke 1997; 28: 2437–2441.

30. Barraclough B. Suicide and epilepsy. In: Reynolds EH, Trimble MR, eds. Epilepsy and Psychiatry. Edinburgh: Churchill Livingstone, 1981: 72–76.

31. Edeh J, Toone BK. Antiepileptic therapy, folate deficiency, and psychiatric morbidity: a general practice survey. Epilepsia 1985; 26: 434–440.

32. Victoroff JI, Benson F, Grafton ST, et al. Depression in complex partial seizures. Electroencephalography and cerebral metabolic correlates. Arch Neurol 1994; 51: 155–163.

33. Gilliam F, Hecimovic H, Sheline Y. Psychiatric comorbidity, health, and function in epilepsy. Epilepsy Behav 2003; 4 (Suppl 4): S26–S30.

34. Hecimovic H, Goldstein JD, Sheline YI, Gilliam FG. Mechanisms of depression in epilepsy from a clinical perspective. Epilepsy Behav 2003; 4 (Suppl 3): S25–S30.

35. Jones JE, Hermann BP, Barry JJ, et al. Rates and risk factors for suicide, suicidal ideation, and suicide attempts in chronic epilepsy. Epilepsy Behav 2003; 4 (Suppl 3): S31–S38.

36. Bromfield EB, Altshuler L, Leiderman DB, et al. Cerebral metabolism and depression in patients with complex partial seizures. Arch Neurol 1992; 49: 617–623.

37. Schmitz EB, Moriarty J, Costa DC, et al. Psychiatric profiles and patterns of cerebral blood flow in focal epilepsy: interactions between depression, obsessionality, and perfusion related to the laterality of the epilepsy. J Neurol Neurosurg Psychiatry 1997; 62: 458–463.

38. Ring HA, Acton PD, Scull D, et al. Patterns of brain activity in patients with epilepsy and depression. Seizure 1999; 8: 390–397.

39. Whitlock FA, Siskind MM. Depression as a major symptom of multiple sclerosis. J Neurol Neurosurg Psychiatry 1980; 43: 861–865.

40. Lamberg L. Psychiatric symptoms common in neurological disorders. JAMA 2001; 286: 154–156.

41. Baretz RM, Stephenson GR. Emotional responses to multiple sclerosis. Psychosomatics 1981; 22: 117–126.

42. Honer WG, Hurwitz T, Li DK, et al. Temporal lobe involvement in multiple sclerosis patients with psychiatric disorders. Arch Neurol 1987; 44: 187–190.

43. Sabatini U, Pozzilli C, Pantano P, et al. Involvement of the limbic system in multi-

ple sclerosis patients with depressive disorders. Biol Psychiatry 1996; 39: 970–975.

44. Mayeux R, Stern Y, Rosen J, Leventhal J. Depression, intellectual impairment, and Parkinson disease. Neurology 1981; 31: 645–650.

45. Taylor AE, Saint-Cyr JA, Lang AE, Kenny FT. Parkinson's disease and depression. A critical re-evaluation. Brain 1986; 109: 279–292.

46. Cummings JL. Depression and Parkinson's disease: a review. Am J Psychiatry 1992; 149: 443–454.

47. Gotham AM, Brown RG, Marsden CD. Depression in Parkinson's disease: a quantitative and qualitative analysis. J Neurol Neurosurg Psychiatry 1986; 49: 381–389.

48. Brown RG, MacCarthy B, Gotham AM, et al. Depression and disability in Parkinson's disease: a follow-up of 132 cases. Psychol Med 1988; 18: 49–55.

49. Scatton B, Javoy-Agid F, Rouquier L, et al. Reduction of cortical dopamine, noradrenaline, serotonin and their metabolites in Parkinson's disease. Brain Res 1983; 275: 321–328.

50. Jellinger K. Overview of morphological changes in Parkinson's disease. Adv Neurol 1987; 45: 1–18.

51. Torack RM, Morris JC. The association of ventral tegmental area histopathology with adult dementia. Arch Neurol 1988; 45: 497–501.

52. Mayberg HS, Starkstein SE, Sadzot B, et al. Selective hypometabolism in the inferior frontal lobe in depressed patients with Parkinson's disease. Ann Neurol 1990; 28: 57–64.

53. Jagust WJ, Reed BR, Martin EM, et al. Cognitive function and regional cerebral blood flow in Parkinson's disease. Brain 1992; 115: 521–537.

54. Ring HA, Bench CJ, Trimble MR, et al. Depression in Parkinson's disease. A positron emission study. Br J Psychiatry 1994; 165: 333–339.

55. Mayberg HS, Starkstein SE, Peyser CE, et al. Paralimbic frontal lobe hypometabolism in depression associated with Huntington's disease. Neurology 1992; 42: 1791–1797.

56. Mentis MJ, McIntosh AR, Perrine K, et al. Relationships among the metabolic patterns that correlate with mnemonic, visuospatial, and mood symptoms in Parkinson's disease. Am J Psychiatry 2002; 159: 746–754.

57. Brooks DJ, Salmon EP, Mathias CJ, et al. The relationship between locomotor disability, autonomic dysfunction, and the integrity of the striatal dopaminergic system in patients with multiple system atrophy, pure autonomic failure, and Parkinson's disease, studied with PET. Brain 1990; 113: 1539–1552.

58. Vingerhoets FJ, Schulzer M, Calne DB, Snow BJ. Which clinical sign of Parkinson's disease best reflects the nigrostriatal lesion? Ann Neurol 1997; 41: 58–64.

59. Broussolle E, Dentresangle C, Landais P, et al. The relation of putamen and caudate nucleus 18F-Dopa uptake to motor and cognitive performances in Parkinson's disease. J Neurol Sci 1999; 166: 141–151.

60. Andersen J, Aabro E, Gulmann N, et al. Anti-depressive treatment in Parkinson's disease. A controlled trial of the effect of nortriptyline in patients with Parkinson's disease treated with L-DOPA. Acta Neurol Scand 1980; 62: 210–219.

61. Stark P, Hardison CD. A review of multicenter controlled studies of fluoxetine vs. imipramine and placebo in outpatients with

major depressive disorder. J Clin Psychiatry 1985; 46: 53–58.

62. Doder M, Rabiner EA, Turjanski N, et al. Brain serotonin 1A receptors in Parkinson's disease with and without depression measured by positron emission tomography with 11C-WAY 100635. Mov Disord 2000; 15 (Suppl 3): 213

63. Caine ED, Shoulson I. Psychiatric syndromes in Huntington's disease. Am J Psychiatry 1983; 140: 728–733.

64. Folstein S, Abbott MH, Chase GA, et al. The association of affective disorder with Huntington's disease in a case series and in families. Psychol Med 1983; 13: 537–542.

65. Kuhl DE, Phelps ME, Markham CH, et al. Cerebral metabolism and atrophy in Huntington's disease determined by 18FDG and computed tomographic scan. Ann Neurol 1982; 12: 425–434.

66. Hayden MR, Martin WR, Stoessl AJ, et al. Positron emission tomography in the early diagnosis of Huntington's disease. Neurology 1986; 36: 888–894.

67. Young AB, Penney JB, Starosta-Rubinstein S, et al. PET scan investigations of Huntington's disease: cerebral metabolic correlates of neurological features and functional decline. Ann Neurol 1986; 20: 296–303.

68. Berent S, Giordani B, Lehtinen S, et al. Positron emission tomographic scan investigations of Huntington's disease: cerebral metabolic correlates of cognitive function. Ann Neurol 1988; 23: 541–546.

69. Kuwert T, Lange HW, Langen KJ, et al. Cerebral glucose consumption measured by PET in patients with and without psychiatric symptoms of Huntington's disease. Psychiatry Res 1989; 29: 361–362.

70. Wragg RE, Jeste DV. Overview of depression and psychosis in Alzheimer's disease. Am J Psychiatry 1989; 146: 577–587.

71. Cummings JL, Ross W, Absher J, et al. Depressive symptoms in Alzheimer disease: assessment and determinants. Alzheimer Dis Assoc Disord 1995; 9: 87–93.

72. Lee HB, Lyketsos CG. Depression in Alzheimer's disease: heterogeneity and related issues. Biol Psychiatry 2003; 54: 353–362.

73. Lyketsos CG, Lee HB. Diagnosis and treatment of depression in Alzheimer's disease. A practical update for the clinician. Dement Geriatr Cogn Disord 2004; 17: 55–64.

74. Hirono N, Mori E, Ishii K, et al. Frontal lobe hypometabolism and depression in Alzheimer's disease. Neurology 1998; 50: 380–383.

75. Grady CL, Haxby JV, Schapiro MB, et al. Subgroups in dementia of the Alzheimer type identified using positron emission tomography. J Neuropsychiatry Clin Neurosci 1990; 2: 373–384.

76. Ebmeier KP, Prentice N, Ryman A, et al. Temporal lobe abnormalities in dementia and depression: a study using high resolution single photon emission tomography and magnetic resonance imaging. J Neurol Neurosurg Psychiatry 1997; 63: 597–604.

77. Liao YC, Liu RS, Lee YC, et al. Selective hypoperfusion of anterior cingulate gyrus in depressed AD patients: a brain SPECT finding by statistical parametric mapping. Dement Geriatr Cogn Disord 2003; 16: 238–244.

78. Sultzer DL, Mahler ME, Mandelkern MA, et al. The relationship between psychiatric symptoms and regional cortical metabolism in Alzheimer's disease. J Neuropsychiatry Clin Neurosci 1995; 7: 476–484.

79. Starkstein SE, Vazquez S, Migliorelli R, et al. A SPECT study of depression in Alzheimer's disease. Neuropsychiatry Neuropsychol Behav Neurol 1995; 8: 38–43.

80. Bozzola FG, Gorelick PB, Freels S. Personality changes in Alzheimer's disease. Arch Neurol 1992; 49: 297–300.

81. Craig AH, Cummings JL, Fairbanks L, et al. Cerebral blood flow correlates of apathy in Alzheimer disease. Arch Neurol 1996; 53: 1116–1120.

82. Ott BR, Noto RB, Fogel BS. Apathy and loss of insight in Alzheimer's disease: a SPECT imaging study. J Neuropsychiatry Clin Neurosci 1996; 8: 41–46.

83. Lopez OL, Zivkovic S, Smith G, et al. Psychiatric symptoms associated with cortical–subcortical dysfunction in Alzheimer's disease. J Neuropsychiatry Clin Neurosci 2001; 13: 56–60.

84. Benoit M, Dygai I, Migneco O, et al. Behavioral and psychological symptoms in Alzheimer's disease. Relation between apathy and regional cerebral perfusion. Dement Geriatr Cogn Disord 1999; 10: 511–517.

85. Migneco O, Benoit M, Koulibaly PM, et al. Perfusion brain SPECT and statistical parametric mapping analysis indicate that apathy is a cingulate syndrome: a study in Alzheimer's disease and nondemented patients. Neuroimage 2001; 13: 896–902.

86. Nauta WJ. The problem of the frontal lobe: a reinterpretation. J Psychiatr Res 1971; 8: 167–187.

87. Robinson RG, Chemerinski E, Jorge R. Pathophysiology of secondary depressions in the elderly. J Geriatr Psychiatry Neurol 1999; 12: 128–136.

5

Co-morbidity of neurological and affective disorders: the model of the genetically epilepsy-prone rat

Phillip C Jobe and Richard J Weber

INTRODUCTION

Co-morbidity between the affective disorders and the epilepsies is a topic of current interest among neurologists, psychiatrists and experimental neuroscientists[1–15]. People with epilepsy are more likely to suffer from affective disorder than are individuals in the general population. Moreover, as summarized recently, the co-morbidity between epilepsy and affective disorders in humans appears to be 'bidirectional', with epilepsy patients having an elevated risk of affective disorder, and with affective disorder patients having an elevated risk of epilepsy[3].

The linkage between epilepsy and affective disorders may arise via one or more general processes. First, seizures may initiate biological mechanisms that at some distant point in time cause or increase the propensity for the occurrence of dysfunctional affective episodes. Second, dysfunctional affective episodes may initiate processes that at some extended point in time contribute to the appearance of seizures. Third, neither may cause the other, although both may be caused by common underlying pathogenic mechanisms. Fourth, these alternatives may act together to cause co-morbidity.

CO-MORBIDITY AND THE GENETICALLY EPILEPSY-PRONE RAT

Behavioral linkages between the epilepsies and affective disorders, first noted in humans, have also emerged in studies of the genetically epilepsy-prone rat (GEPR). Such cross-species linkages would be anticipated if the biological linkage mechanisms between human epilepsy and affective disorder were evolutionarily preserved without the superimposition of prohibitive compensatory processes. A vast body of evidence shows that the neuronal systems that participate as determinants of affective disorders and the epilepsies[11–13,16,17] are present in the brains of different mammalian species, including mice, rats, cats, dogs, monkeys, apes and humans[18–31].

Corresponding abnormalities in these neuronal systems across species would be expected to underlie the expression of affective disorder and epilepsy equivalents in each species. Evidence obtained in studies of the GEPR demonstrate that these mammals are predisposed both to the expression of seizures and to dysfunctional affective episodes, and that they are characterized by neuronal defects and other abnormalities that putatively underlie epilepsy and affective disorder in humans[12,32]. Two independently derived GEPR strains have been developed. They are known by the acronyms GEPR-3 and GEPR-9. At this time, epilepsy and affective disorder co-morbidity has been more extensively studied in GEPR-3 than in GEPR-9. Evidence suggests that the GEPR-3 models co-morbidity. Behavioral experiments with GEPR-9 are planned. Nevertheless, many of the pathophysiological determinants of co-morbidity have been identified in both strains.

SEIZURE PREDISPOSITION IN GENETICALLY EPILEPSY-PRONE RATS

The biological bases of seizure predisposition, expression and termination have been frequent investigative themes in the moderately epileptic GEPR-3 and the more severely epileptic GEPR-9[33–50]. GEPRs are characterized by genetically determined predisposition to generalized tonic–clonic seizures (GTCSs) and limbic seizures. The GTCSs of the GEPR mimic behaviorally and electrographically this type of seizure in humans[48,51,52]. A typical generalized electroencephalogram (EEG) seizure with its corresponding tonic–clonic behaviors is shown in Figure 5.1.

GTCSs in GEPRs may occur spontaneously or they may be ignited by stimuli that do not cause such episodes in normal mammals. GEPRs also exhibit GTCSs secondarily in response to limbic seizures, as do some humans with epilepsy[51]. Moreover, the forebrain seizure circuitry of GEPRs shows a sharp acceleration of kindling from the amygdala.

Spontaneous seizures appear to occur more frequently in developing GEPRs than in their adult counterparts[44] (Jobe, unpublished observations). Some of these spontaneous seizures develop into convulsive status epilepticus with death as the ultimate short-term consequence. Because of the absence of exogenous precipitating events, the absence of known endogenous seizure foci, and because of their rarity, these cases of spontaneous seizures are believed to occur in response to infrequently occurring seizure-evoking alterations in endogenous neuronal events, coupled with an underlying and persistent seizure predisposition. The predisposition may enhance intracerebral conditions necessary for the generation of the extreme neuronal impulse rates sufficient to drive GTCSs. Evidence supports the concept that the tonic flexor and extensor components of GTCSs occur in response to an extremely high frequency of impulses per unit time being delivered by the brain to the spinal cord[53–56]. Interestingly, in field potential recordings, the tonic extensor components of GTCSs occur during an interval of marked EEG amplitude attenuation, a phenomenon denoted in the clinical literature as the electrodecremental interval or response[57]. This electrographic event occurs throughout the rostral–caudal extent of the GEPR-9 brain. Intuitively it appears to be a condition associated with temporal desynchronization of the neuronal action potentials[50,51,58,59]. Recent observations support the concept that 'asynchrony is necessary to maintain a high level of activity in neuronal networks for sustained periods of time and that synchrony may disrupt such activity.'[60] Experiments in GEPR-9s demonstrate that the electrodecremental interval (or neuronal asynchrony) occurs during the tonic flexor or extensor components of GTCSs and that such events are followed by increasing synchrony, as represented by spike-wave complexes that temporally increase duration and amplitude (Figure

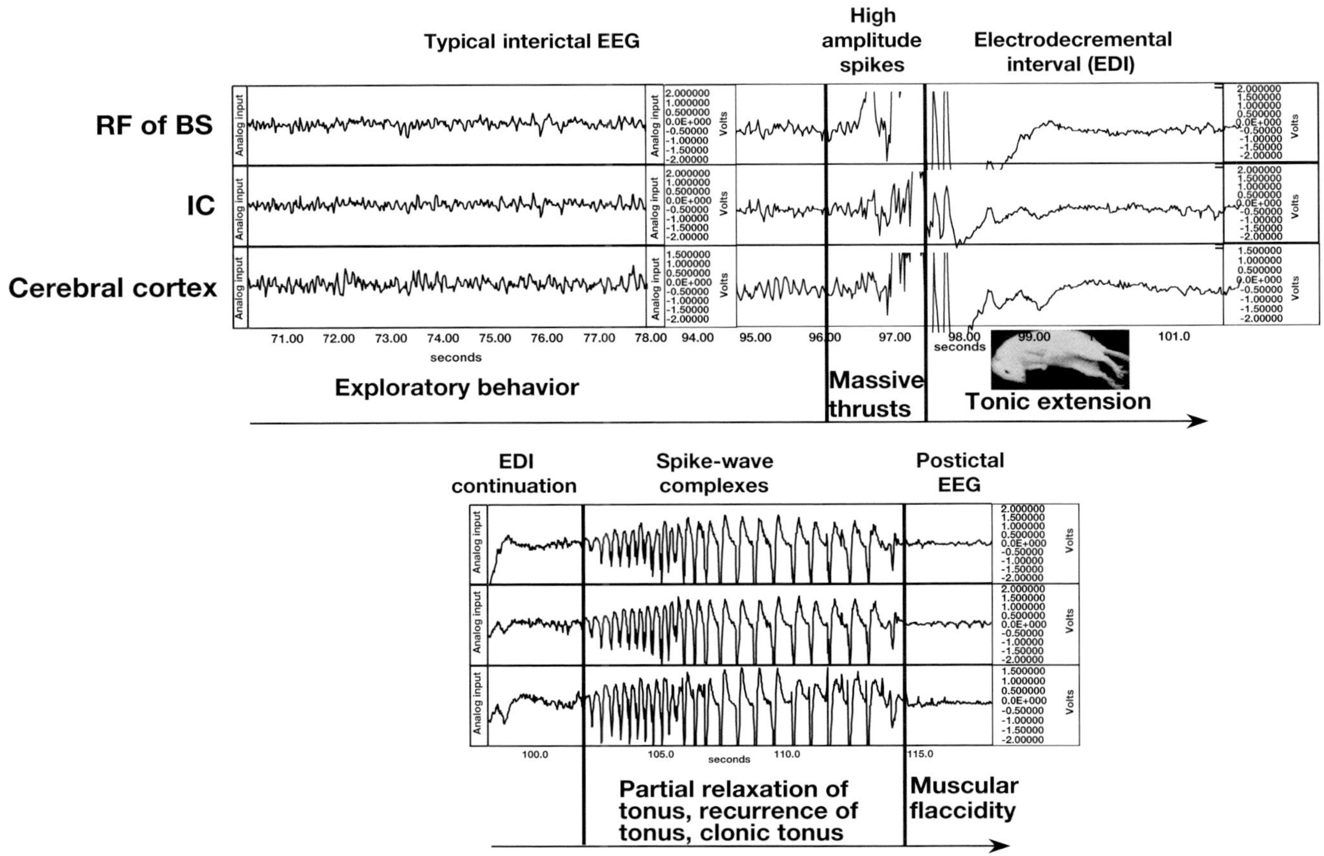

Figure 5.1 Typical electroencephalogram (EEG) of the GEPR-9 generalized tonic–clonic seizure ignited by sound. The close similarities of this EEG seizure and human generalized tonic–clonic seizures have been previously documented. Data were previously published (reference 52). RF of BS, reticular formation of the brain stem; IC, inferior colliculus

5.1). Seizure termination occurs with the end of the spike-wave EEG morphology.

Exogenous events that precipitate seizures in GEPRs, but not in adult non-epileptic rats, include sound and hyperthermia[46,61,62]. Much less dependably, handling and postural alterations may also ignite seizures in GEPRs[49]. Sound is effective both in developing and in adult GEPRs[63]. Hyperthermia has been investigated only in adult GEPRs. Stimuli that ignite seizures both in GEPRs and in non-epileptic subjects include electrical current[42], pentylenetetrazol[42], flurothyl[64], aminophylline[65] and high atmospheric pressure[66], but at lower thresholds or with greater severity than in non-epileptic subjects.

SEIZURE PREDISPOSITION IN HUMANS WITH EPILEPSY

All adult mammals, including humans, have the capacity to experience seizures. Accordingly, people without epilepsy will exhibit seizures in response to an electrical stimulus or convulsant drug of sufficient magnitude[67,68]. Therapeutically induced seizures are used in psychiatric patients with and without a diagnosis of epilepsy[1,69]. Convulsions in people have been reported to occur in response to accidentally ingested strychnine, camphor, cocaine, amphetamine and methamphetamine[70]. An enormous body of experimental data in many different mammalian species has provided solid evidence for the convulsant properties of these drugs.

In contrast, people with epilepsy experience seizures in response to exogenous stimuli or endogenous events that do not produce seizures in normal individuals. In this regard, seizure predisposition in humans appears to correspond to predisposition in GEPRs. Like GEPRs, humans with epilepsy are susceptible to endogenous and exogenous stimuli that fail to produce seizures in normal people, regardless of dose or magnitude of exposure. Also, humans with epilepsy may exhibit an exaggerated seizure responsiveness to stimuli that also provoke less severe seizures in non-epileptic humans. The concept of spontaneous seizures in humans has achieved such prominence in clinical medicine that definitions of epilepsy have tended to exclude or minimize the importance of disorders wherein seizures occur in response to environmental alterations. Engel states, 'In general, epileptic seizures can be defined as being brief, intermittent, spontaneous, and paroxysmal...'[71] Underlying this approach is the idea that 'Epileptic seizures can be described as spontaneous to emphasize the presence of an inherent seizure generator.'

Many different environmental stimuli are known to elicit seizures in some humans. Flickering visual stimulation is relatively common among triggers for sensory-precipitated seizures in humans[72]. Thinking, reading, listening to music, typing, speaking and eating have also been reported to ignite seizures in humans. 'Thinking-triggers' include mental arithmetic and games of cards or checkers. In reading-induced seizures, comprehension of the concepts conveyed by the written material may, in some cases, be a crucial component of the trigger. In others, visual patterns created by the written material may play a role in seizure initiation.

As is well known, febrile seizures occur in humans[73]. In medical practice, the diagnosis of febrile seizures is often limited to children who express febrile seizures between the ages of 3 months and 6 years. These seizures are approximately three times more common among family members of affected children than in the general population. Interestingly, an elevated incidence of afebrile epilepsy has been found in first-degree relatives of patients undergoing febrile seizures. Moreover, investigators are also reporting cases of children exhibiting seizures with fever in early childhood, and who continue to exhibit febrile seizures beyond the age of 6 years[74,75]. Some of these older individuals experience afebrile tonic–clonic seizures.

Some authors argue that 'Despite the implicit relationship between fever and seizure activation, temperature itself probably does not lower the seizure threshold.'[73] Nevertheless, adult GEPRs with broadly represented seizure predisposition exhibit seizures in response to elevated body temperature, whereas neurologically normal adult rats fail to exhibit this trait[77]. These data support the concept that elevated body temperature does indeed ignite seizures in mammals with innately deficient seizure thresholds. The question of whether elevated body temperatures cause decreases in seizure threshold is not answered by these data.

As in GEPRs, spontaneous seizures reportedly occur in people with environmentally evoked seizures[72]. For example, approximately one-third of patients with photosensitive epilepsy have been noted to exhibit spontaneous seizures. Similarly, patients with seizures induced by thinking also have been reported to experience spontaneous seizures. In addition to susceptibility to spontaneous and environmentally evoked seizures, humans with epilepsy are characterized by an exaggerated seizure responsiveness when exposed to stimuli that also provoke seizures in non-epileptic humans (Figure 5.2). Both behavioral and EEG seizures

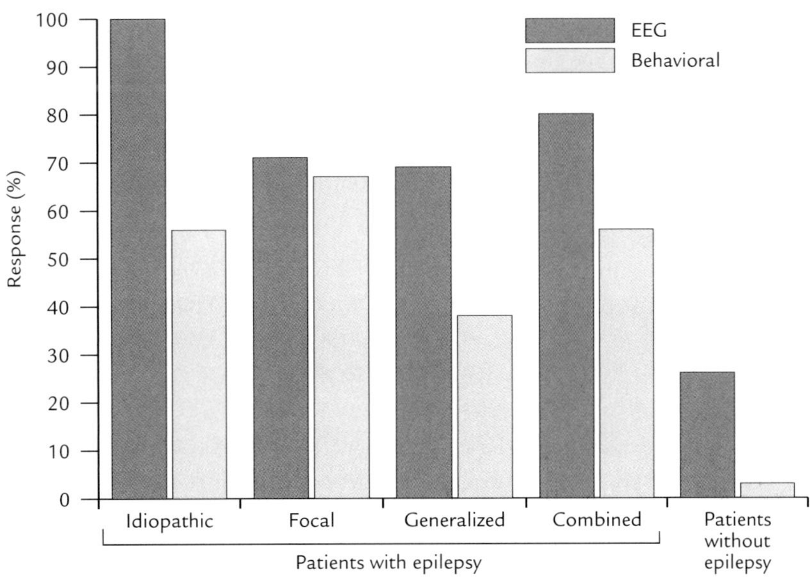

Figure 5.2 Exaggerated seizure responsiveness to intravenous pentylenetetrazol (maximum dose 400 mg) in humans with epilepsy versus people without epilepsy. (From reference 76)

are more likely to occur in pentylenetetrazol-treated patients with epilepsy than in patients without epilepsy. For example, pentylenetetrazol-induced seizures occurred in 100% of patients with idiopathic epilepsy versus approximately 25% of patients without epilepsy[76].

PREDISPOSITION TO DYSFUNCTIONAL AFFECTIVE EPISODES IN HUMANS WITH A DIAGNOSIS OF AN AFFECTIVE DISORDER

Normal humans experience elements of depressive episodes if negative life events are of sufficient magnitude[16]. Normal people who undergo conjugal bereavement or the loss of an intimate friend may experience symptoms that mimic many of those exhibited in a depressive episode. Thus, many or perhaps all of the mechanisms of dysfunctional affective episode initiation, continuation and termination are seemingly inherent properties of the mammalian brain (including the human brain). In contrast, because of an underlying biologically determined predisposition, many, if not most, people with a diagnosis of affective disorder diverge from normality physiologically even before their first dysfunctional affective episode. However, with regard to bipolar disorder specifically, Rush points out 'Until complete genetic information of this disease is available, it remains unclear whether an environmental insult alone could induce expression of bipolar disorder in a patient with no apparent predisposing genetic susceptibility.'[77] Although few if any causality hypotheses are completely unassailable, evidence supports the hypothesis that multiple genes may underlie a gradient of liability to experimentally induced affective dysfunction. Concepts that clinical manifestations of affective disorders are triggered by interactions of genetic determinants with environment influences hold a position of prominence in biological psychiatry[12,16,78–84]. As demonstrated in rats, genetic predisposition sharply accelerates disease progression caused by limbic kindling[51,77,85–88].

The biology of sensitization and adaptation provides an intellectually attractive basis for the idea that diagnosable affective disorders and/or epilepsy may never occur in individuals if genetically determined biological defenses are sufficient and if traumatic events during the course of a lifetime, perhaps especially during development, are not excessively severe and too frequently encountered[13]. In accordance with this concept, deficiencies in endogenous defense mechanisms underlie predisposition to neurological and psychiatric disease, whereas experientially induced emergence of neurobiological alterations in brain function represent the pathophysiological components of disease progression, some of which may include further deterioration of the initial genetically determined biological defenses against illness (Figure 5.3).

PREDISPOSITION TO MANIFESTATIONS OF DYSFUNCTIONAL AFFECTIVE EPISODES IN GENETICALLY EPILEPSY-PRONE RATS

Several indices of affective disorder can be measured in mammals other than humans. Because of its relevance to depressive episodes, inescapable stress induced by forced swimming is utilized by investigators as an index of experimentally induced helplessness or behavioral despair[89,90]. During the course of such testing, rats discontinue swimming and become

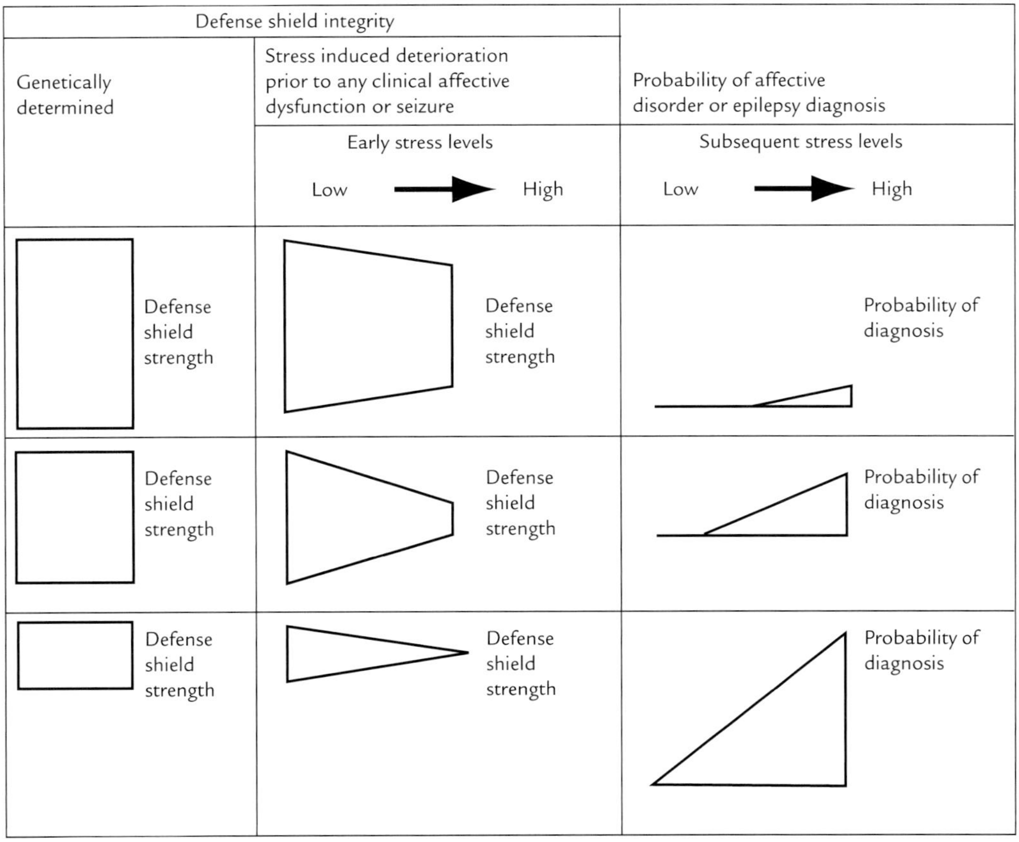

Figure 5.3 Magnitude of defense shield integrity versus probability of affective disorder or epilepsy diagnosis. Magnitude of genetically determined defense shield strength is indicated by geometric figures in the left column, with decreasing strength from the top to the bottom row. Environmentally induced defense shield strengths are indicated in the middle column. Within each cell of this middle column, progressive stress-induced deficits are indicated by the decreasing vertical height of each geometric shape from left to right. Also, at the left of each cell in the middle column, the initial strength of the defense shield is determined by the strength of the genetically derived shield. Furthermore, the rate of deterioration of the shield is determined genetically. Probability of affective dysfunction or seizure being sufficient to lead to a diagnosis as a psychiatric or neurological disorder is indicated in column 3 by progressive increases in the vertical height of each geometric shape

immobile. In our laboratory, the time of immobility was measured[89] twice, both in GEPR-3s and in neurologically normal rats, once on day 0 of testing and again on day 1[91]. As shown in Figure 5.4 for day 0, the duration of immobility per minute of testing for GEPR-3s was substantially greater than for non-epileptic control rats. During the second test period on the next day of the experiment, GEPR-3s appeared to experience a longer duration (30%) of

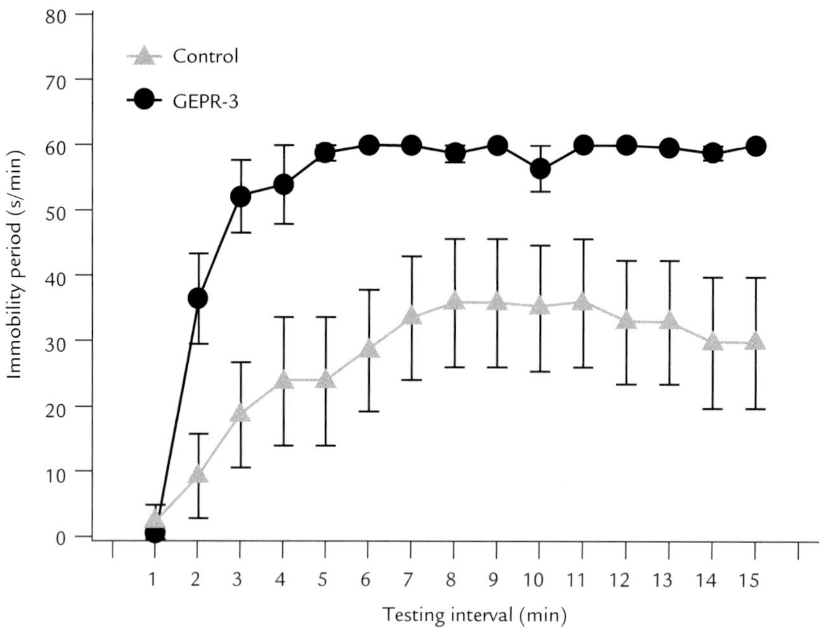

Figure 5.4 Time in immobility (seconds ± SEM) during each 1-minute interval of the initial swim test conducted on day 0 of study: GEPR-3s (10 females) verus non-epileptic control rats (10 females)

immobility than they did during the initial test. Moreover, these GEPR-3s continued to exhibit exaggerated immobility compared to neurologically normal controls. These observations further support the concept that GEPR-3s model predisposition to behavioral despair, and that disease progression occurs as a consequence of exposure stimuli.

ANHEDONIA IN GENETICALLY EPILEPSY-PRONE RATS AND IN HUMANS WITH AFFECTIVE DISORDER

Anhedonia, a feature of depressive disorder, has been modeled in rodents by measurement of reductions in and preferences for the consumption of saccharin solution[92]. In a preliminary study, saccharin solution consumption was determined in male GEPR-3s versus control rats following a protocol previously described[93]. When given a choice between saccharin and water, GEPR-3s consumed 13% less saccharin than did the control subjects (unpublished observations, Weber, Jobe and colleagues). Animals each had two sources of liquid: 10 mmol/l saccharin solution and tap water. In contrast, GEPR-3s ate more food than control rats.

DIURNAL ABNORMALITIES IN MOTOR ACTIVITIES

In a recent study, unmedicated inpatients who met the DSM-IV criteria for a major depressive disorder (subtype unipolar) exhibited an abnormal diurnal pattern of motor activity[94]. During daylight hours (activity cycle for most humans), the depressed patients exhibited less motor activity than did control subjects. In contrast, during night hours (resting cycle for most humans), depressed individuals exhibited more activity than controls. Similar findings of abnormal diurnal activities in depressed inpatients and outpatients have been previously reported[95–97].

Major diurnal motor abnormalities of the GEPR appear largely analogous to those of people with depressive disorder. Accordingly, during the daily inactivity period for rats, the per cent elevation of GEPRs (compared to non-epileptic control rats was 191%, 279% and 226%, for horizontal, rearing and grooming activities, respectively (Figure 5.5). During the rat activity cycle, rearing activity was significantly ($p = 0.014$) less than for controls. However, in grooming activity, GEPR-3s exhibited a 31% elevation in grooming time ($p < 0.001$).

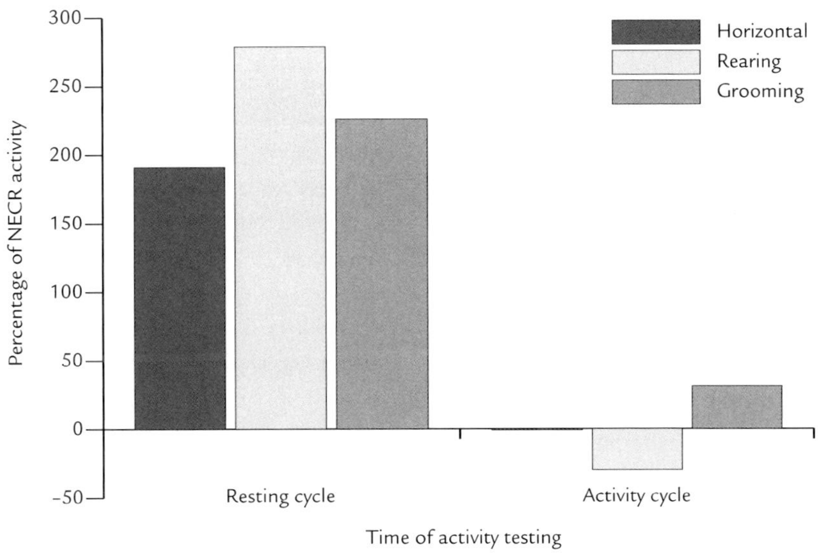

Figure 5.5 GEPR-3 activity as percentage difference from the activity of non-epileptic control rats (NECR). GEPR-3s evaluated during a 3-hour interval (09.00 to 12.00) of their daily 12-hour resting period (lights on) exhibited significantly ($p < 0.001$) greater activity in horizontal, rearing and grooming behaviors in an open field than did NECRs. During the normal rat activity cycle, the GEPR exhibited less rearing behavior and more grooming behavior than did NECRs. Motor activities were determined with a Digiscan Animal Activity Monitor System using a previously described protocol[100]. Animals were monitored singly in open-field boxes. Data were collected from female and male rats by a Digiscan Analyser, connected to a Hewlett-Packard Vectra 386s/20 computer running Oasis v2.4 (Omnitech Electronics, Columbus, OH). The light period activity was assessed 3 hours beginning at 09.00, whereas the dark period activity was assessed for 3 hours beginning at 21.00. Rats were maintained on a circadian 'on/off' cycle with light activation and inactivation at 06:00 and 18.00, respectively. Observations not previously published by Weber and Jobe

ENDOCRINOLOGICAL ABNORMALITIES

GEPRs share several endocrinological abnormalities with people suffering from affective disorder. A complete assessment of these shared defects cannot be made until more comprehensive investigations are undertaken in GEPRs. Nevertheless, the known endocrinological similarities between the GEPR and human depression provide a basis for future studies.

Hypothyroidism

Humans with depressive disorder and GEPRs exhibit abnormalities in the hypothalamic–pituitary–thyroid axis. GEPR-9s are characterized by evidence of hypothyroidism[99,100]. Concentrations of serum thyroxine and tri-iodothyronine are deficient, and levels of thyroid-stimulating hormone are excessive when comparisons are made with non-epileptic control rats. GEPR-3s exhibit low concentrations of thyroxine plus high levels of tri-iodothyronine. Similarly, many authorities have associated subtle hypothyroidism with depression, and hyperthyroidism, in some instances, with mania[101–106]. Some investigators point out that hypothyroidism may give rise to frank depression. Investigators report that thyroid hormone administration shortens the onset of action of antidepressant therapy, or enhances the efficacy of antidepressant drug therapy in refractory depression[107–109].

Corticosterone

A recent unpublished investigation shows that the plasma of GEPR-3s is characterized by elevated concentrations of corticosterone[91], the major glucocorticoid in rats. This finding is noteworthy in the present context, because some patients with depression exhibit high resting plasma cortisol concentrations as well as refractoriness to dexamethasone-induced suppression of cortisol[110]. The capacity of dexamethasone to suppress corticosterone levels in GEPRs has not been examined.

Growth hormone

Abnormalities in growth hormone concentrations have been reported both for people with depression and for GEPRs. The concentrations of this hormone are deficient in developing GEPRs[111]. Also, during the interval of time (early hours of the night) in which pulses of growth hormone secretion are the highest in normal people, secretion is deficient in depressed patients[112]. Also, people with depression show a lower response than control patients to increments in growth hormone produced by growth hormone releasing hormone[113].

IMMUNE DYSFUNCTION

Immune dysfunctions have been detected in GEPRs[91,114–118] in people with epilepsy[119–125], in other mammalian models of depression[126], and in people with depression[127–137]. Deficiencies in the present understanding of psycho- and neuroimmunology and the complexity of the interactions between underlying genetic factors, stress and other environmental influences that give rise to variations in the emergent properties of epilepsy and depression prohibit a unified hypothesis regarding commonalties in immune function in these two disorders. Nevertheless, the existence of such abnormalities in epilepsy and depression supports the concept of a linkage between these disorders and of the hypothesis that the GEPR is a co-morbidity model. Additional studies are required to reach a decision regarding this possibility.

NORADRENERGIC AND SEROTONERGIC DEFICIENCIES IN AFFECTIVE DISORDERS AND EPILEPSY

Innate deficits

Serotonergic and noradrenergic deficits have been proposed as biological linkages between the epilepsies and affective disorders[11,13,16,32]. Such deficits – whether genetically determined and/or experientially derived – serve to facilitate the occurrence and the severity of the dysfunctional episodes of both the epilepsies and the affective disorders.

The results of some but not all pertinent studies support the concept of innate noradrenergic and serotonergic deficiencies existing in humans with affective disorder and/or epilepsy[11,13,16,32]. Similar insufficiencies characterize the GEPR brain.

Numerous experimental approaches have been used to investigate the indices of noradrenergic and serotonergic transmission in affective disorders. Deficits in serotonergic transmission in humans with affective disorder appear to arise at least partially from a paucity of serotonergic innervation of its terminal areas, and from over- and under-representation of presynaptic and postsynaptic serotonin receptors, respectively. Deficient innervation is supported by the observation that affective disorder and/or suicidality is associated with a scarcity of serotonin levels in brain tissue, plasma and platelets[138–142], and with a deficit in serotonin transporter binding sites in the postmortem human brain[143–152]. Serotonin stores and transporter protein are important components of serotonin terminals, so that a combined deficit is a plausible indicator of reduced axonal branching and synapse formation. Serotonergic transmission may also be deficient in affective disorder as a result of excessive presynaptic serotonin$_{1D}$ autoreceptors[153].

Upon release of presynaptic serotonin molecules into the extracellular fluid, serotonin (the neurotransmitter) is confronted with a deficit in the density or affinity of postsynaptic serotonin$_{1A}$ receptors in the hippocampus and amygdala (drug-free suicides)[154]. Also, in people with affective disorder, serotonergic transmission is impaired by defects in the dorsal raphe nuclei wherein serotonergic cell bodies are located. An excessive density of serotonergic somatodendritic impulse suppressing serotonin$_{1A}$ autoreceptors has been detected in this nucleus in suicide victims with major depression[155].

Discoveries of noradrenergic deficiencies in affective disorder/suicidality have proceeded with scientific approaches that are similar to those used to identify serotonergic deficits[16,156–165]. Though more limited than for affective disorders in numbers of experiments and participating laboratories, the published literature on noradrenergic and serotonergic deficits in the human epilepsies provides valuable insights[166–179]. As has been established for epilepsy in GEPRs, the defects in humans with epilepsy appear to include deficiencies in serotonergic$_{1A}$ and α_1 noradrenergic receptors, and 5-HIAA (serotonin metabolite) and norepinephrine levels.

Noradrenergic deficiencies in GEPRs appear to result from deficient arborization of neurons arising from the locus coeruleus[180,181] coupled with excessive presynaptic suppression of stimulated norepinephrine release in the terminal fields[182] and lack of postsynaptic compensatory up-regulation[16]. These determinants are partially offset by a deficient inhibition of firing by somatodendritic noradrenergic receptors within the locus coeruleus[16]. Evidence also supports the hypothesis that the GEPR brain is characterized by deficits in serotonergic arborization coupled with deficient postsynaptic serotonin$_{1A}$-receptor density and excessive serotonin$_{1B}$ autoreceptor-mediated feedback inhibition of the serotoninergic neuron[183].

A recent observation relates immunological abnormalities to serotonergic abnormalities in epilepsy. This concept is pertinent to the monoaminergic hypotheses of depression and epilepsy. Anti-serotonin$_{1A}$ receptor antibodies have been found in sera of children with developmental disorders[171]. These findings are believed to support the existence of an autoimmune component in epilepsy. Other observations suggest that serotonergic abnormalities in people with depressive disorder are determined genetically[184].

Experimentally induced deficits

The observations of innate noradrenergic and serotonergic deficits described above do not provide sufficient evidence that such deficits mechanistically underlie affective disorder/suicidality and epilepsy. Nevertheless, the question of causality has been addressed, for example, by determining the effects on epilepsy and affective disorder of experimental interventions with known noradrenergic and serotonergic mechanisms[16,49,185–197]. These investigations of affective manifestations in humans, and seizures in GEPRs, epileptic baboons (*Papio papio*) and non-epileptic rats and other seizure models have produced remarkably similar results. The results will be exemplified by focusing on studies undertaken in GEPRs. With regard to noradrenergic and serotonergic reduction paradigms, storage vesicle inactivators (reserpine, tetrabenazine or RO 4-1284) exacerbate seizures in GEPRs, epileptic baboons and other mammals, as the norepinephrine false transmitter α-methyl-*m*-tyrosine is also known to do in GEPRs. When administered to people in the high psychiatric doses of former times, the storage vesicle inactivators also appear to have the potential to increase the likelihood of depression, although this insight is subject to equivocal interpretation, because it relies on observations that lack rigorous experimental design[16]. The norepinephrine synthesis inhibitor (α-methyl-ρ-tyrosine) exacerbates seizures in GEPRs exposed to protracted stress and produces depressive symptoms in medication-free, euthymic people with a history of major depression. Moreover, this drug contributes to the reversal of effectiveness of the anticonvulsant properties of antidepressant drugs in GEPRs and reverses summer remission of the beneficial effects of light therapy in seasonal affective disorder. α-Methyl-ρ-tyrosine also exacerbates seizures in GEPRs under treatment with lithium[198] and causes transient relapse of hypomanic symptoms in bipolar patients in remission with lithium therapy[199]. Inhibition of serotonin synthesis via restriction of the ingestion of the dietary serotonin precursor (tryptophan) exacerbates depressive symptoms in humans[200–202]. Likewise, the serotonin synthesis inhibitor ρ-chlorophenylalanine intensifies seizures in GEPRs[203].

Although they are not ethically subject to investigation in humans, noradrenergic and serotonergic neurotoxins have been evaluated in GEPRs. Selective neurotoxic deletion or destruction of noradrenergic or serotonergic neurons exacerbates seizures in GEPRs[204,205]. Also, in experiments in GEPRs with experimental destruction of serotonergic neurons by treatment with a serotonergic neurotoxin, fetal raphe (but not neocortical) implants into the third ventricle are anticonvulsant[206], as are fetal locus coeruleus implants following the noradrenergic destructive effects of a noradrenergic neurotoxin[207].

Experimentally induced augmentation

Noradrenergic and serotonergic intensification produces effects that are opposite to those of monoaminergic reduction paradigms. The serotonin precursor 5-HTP is anticonvulsant in GEPRs when combined with a monoamine oxidase inhibitor[198]. A meta-analysis suggests but provides no conclusive evidence for antidepressive effects of 5-HTP in humans[208]. If given in doses sufficient to increase brain norepinephrine concentration, L-dopa, a precursor to dopamine and norepinephrine, is anticonvulsant in the GEPR. Effectiveness is lost when its biological conversion to norepinephrine is disallowed by inhibition of dopamine-β-hydroxylase, the enzyme that converts dopamine to norepinephrine[209].

ANTIDEPRESSANTS AND ANTICONVULSANTS

The anticonvulsant effects of the tricyclic antidepressant drugs were demonstrated in GEPRs many years ago[210]. Moreover, through studies in GEPRs, the anticonvulsant properties of other antidepressants including the specific serotonin reuptake inhibitors have been documented in multiple investigations[49,211–218].

Speculations that the anticonvulsant effects of antidepressants were species/strain specific to the GEPR have been soundly disproved[219]. Anticonvulsant effects of the antidepressants have been demonstrated in non-epileptic rats and a multiplicity of other species, including epileptic and non-epileptic mice, cats, dogs, baboons and rhesus monkeys[220–223].

In humans with epilepsy or epilepsy and depression, anticonvulsant effects of antidepressant medications have been reported in case studies, retrospective analyses, unblinded, open-label add-on trials, and a double-blind crossover investigation[224–233]. The evaluated antidepressants include amoxapine, citalopram, clomipramine, desipramine, doxepin, fluoxetine, imipramine, protriptyline and trazadone in myoclonic–astatic seizures, absence seizures, temporal lobe seizures and secondary or primary GTCSs.

In a double-blind crossover study, imipramine therapy caused a significant reduction in seizure frequency in patients with absence and myoclonic–astatic epilepsy, despite the discontinuation of other antiepileptic medications[227]. The most recent of the investigations in humans was an open, multicenter, uncontrolled study of depressed epileptic patients on antiepileptic medications[233]. Baseline values for seizure frequency and depressive symptoms (Montgomery Asberg Depression Rating Scale (MADRS) and Zung Self-Rating Depression Scale (Zung-SDS) were determined during the initial 4-month interval when patients were being treated with antiepileptic medications but not the specific serotonin reuptake inhibitor citalopram. Upon the addition of citalopram to the treatment regimen, reductions occurred both in seizure frequency and in depressive symptoms. With regard to seizure type, significant reductions in seizure frequency were observed in people with complex partial seizures and partial seizures with secondary generalization, but not in those with simple partial seizures.

Mean plasma levels of concomitant carbamazepine, phenobarbital and valproate did not change during citalopram treatment when compared with baseline concentrations.

Although these studies provide appealing evidence for an anticonvulsant effect of the antidepressant drugs in human epilepsy, one study with fluoxetine[234] and another with citalopram[235] failed to show anticonvulsant effects. However, as pointed out in the preceding paragraph, the most recently published investigation provides a measure of confidence that citalopram does have anticonvulsant properties in humans[233]. Also, the negative fluoxetine study[234] is challenged by other observations that support an anticonvulsant effect for fluoxetine[225]. Also, the few negative studies of the antidepressants sharply contrast with the large body of evidence of anticonvulsant effects obtained from studies of both epileptic and non-epileptic mammals, ranging from mice to primates.

What interpretations best fit the data? Are antidepressant medications also anticonvulsants? Only one double-blind crossover study has been published on this subject, and it provides an affirmative answer to the question. The two studies that failed to show anticonvulsant effects in humans are offset by analogous studies that do show anticonvulsant effects for the very same antidepressants. Furthermore, the retrospective analyses that have been published as well as several case studies and various open-label trials provide an affirmative answer. Finally, anticonvulsant effects of the antidepressants have been detected in a large number of mammalian species, in both epileptic and non-epileptic subjects. No other drugs known, in the medical sciences, to be anticonvulsant in epileptic and non-epileptic mice, rats, dogs, cats, rhesus monkeys and baboons have failed to exert anticonvulsant effects in humans. In other words, *in toto* this large battery of mammalian models has not to our knowledge produced false positives. The adverse effects of some drugs that have been identified as anticonvulsant compounds via screening with such a large battery of tests have been too pronounced to be of human use. However, none have failed to exert anticonvulsant effects in humans unless toxicities have prohibited adequate dose exploration.

Thus, if the question is whether the antidepressant drugs that exist today have anticonvulsant properties in mammals, the answer is yes. If the question is whether they have anticonvulsant properties in humans, the answer is yes, with little doubt. If the question is whether they will have a clinically acceptable anticonvulsant therapeutic index in people with epilepsy, the answer is not known. Finally, whether they will prove superior to existing antiepileptic medications is again unknown.

In summary, the antidepressant drugs have anticonvulsant properties in mammals. No persuasive evidence exists to reject the idea that the antidepressants may be useful in the treatment of epilepsy in patients with affective disorders. Also, the existing body of evidence supports the proposition that novel antiepileptic drugs can be developed using molecules that are inhibitors of norepinephrine and serotonin reuptake. The progress made in the past 20 years in developing increasingly selective reuptake inhibitors heralds opportunities for greater progress. Dosage escalation needed to produce clinically and commercially attractive anticonvulsant effects of these drugs in humans may become feasible with treatments that are selective pharmacodynamically and that elevate norepinephrine and/or serotonin levels selectively in brain regions that are relevant to epilepsy and affective disorders.

CONVULSIONS IN RESPONSE TO SOME ANTIDEPRESSANT TREATMENTS

Some antidepressant treatments that produce therapeutic effects at least partially via noradrenergic and serotonergic augmentation also have convulsant properties in humans with epilepsy and in GEPRs[219]. On different occasions, investigators have incorrectly imputed convulsant properties to noradrenergic and/or serotonergic elevation. However, as has been demonstrated in GEPRs, the anticonvulsant properties – not the convulsant properties – of these drugs are caused by noradrenergic and/or serotonergic elevations[212–216,218,236].

Indeed, the convulsant and anticonvulsant actions of the norepinephrine/serotonin reuptake inhibitor antidepressants occur at distinctly different doses[211]. The anticonvulsant doses are evident at relatively small doses, whereas the convulsant actions are apparent at large doses. Results from a study with the tricyclic antidepressant drug imipramine are summarized in Figure 5.6.

Moreover, the anticonvulsant doses are mechanistically linked to the reuptake properties of the antidepressant drug[182,236–238]. One experiment demonstrated that anticonvulsant doses of desipramine lost effectiveness if administered to GEPR-3s pretreated to deplete norepinephrine stores (Figure 5.7).

Further, intracerebral studies with microdialysis have demonstrated the crucial dependency of antidepressant drug-induced anticonvulsant effects on the capacity of these drugs to cause increases in extracellular norepinephrine and/or serotonin concentrations[49,182,212,214,215,237,238]. An example of such observations is set forth in Figure 5.8[215,218].

Additional studies have demonstrated that the inverse correlation between extracellular fluid concentrations and seizure severity is causal rather than coincidental. If serotonin increments are responsible for the anticonvulsant effects of specific serotonin reuptake inhibitors, pretreatment to deplete storage vesicles of serotonin content should sharply reduce antidepressant drug-induced serotonin

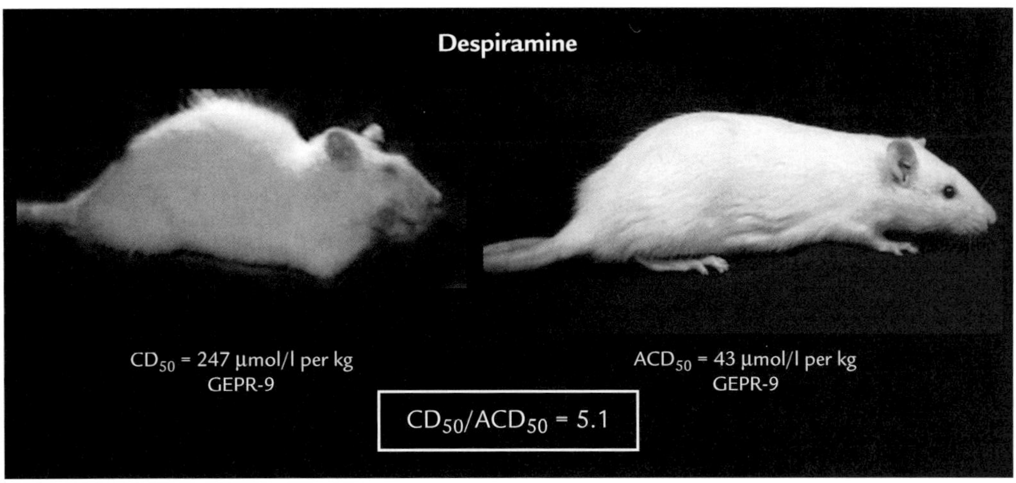

Figure 5.6 Convulsant dose fifty (CD_{50}) versus the anticonvulsant dose fifty (ACD_{50}) of desipramine in GEPR-9s. The CD_{50} is approximately five times greater than the ACD_{50}

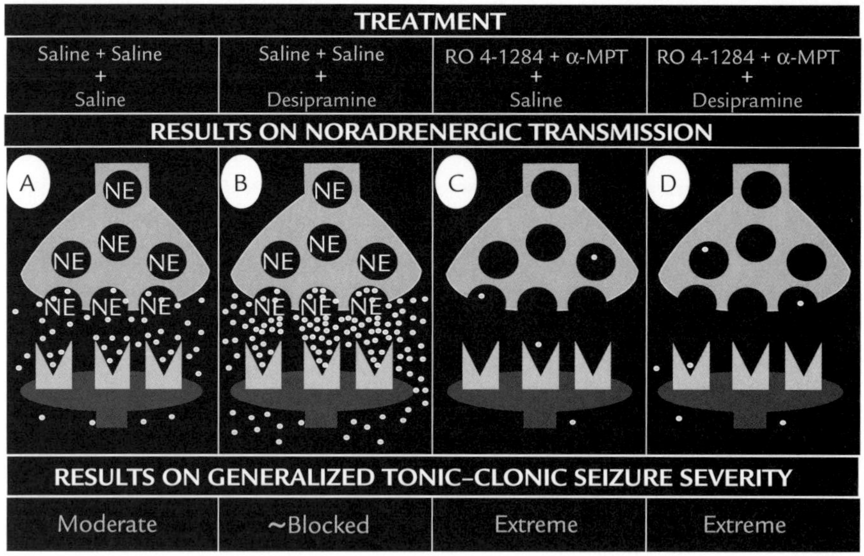

Figure 5.7 Effect of norepinephrine (NE) depletion on the severity of GTCSs induced by sound in GEPR-3s. , noradrenergenic terminals; , noradrenergic storage vesicles; norepinephrine and white dots, norepinephrine molecules; , noradrenergic postsynaptic receptors; , postsynaptic neuron. Three injections were administered to all animals as shown (groups A–D). Group A received saline each time; group B received two injections of saline followed by desipramine; group C received RO 4-1284 (a norepinephrine storage vesicle inactivator) followed by α-methyl-ρ-tyrosine (α-MPT, a norepinephrine synthesis inhibitor) plus saline; group D received RO 4-1284, then, α-MPT, and finally desipramine. These data are summarized from a previous publication (reference 236). (See also color image on p. xxii)

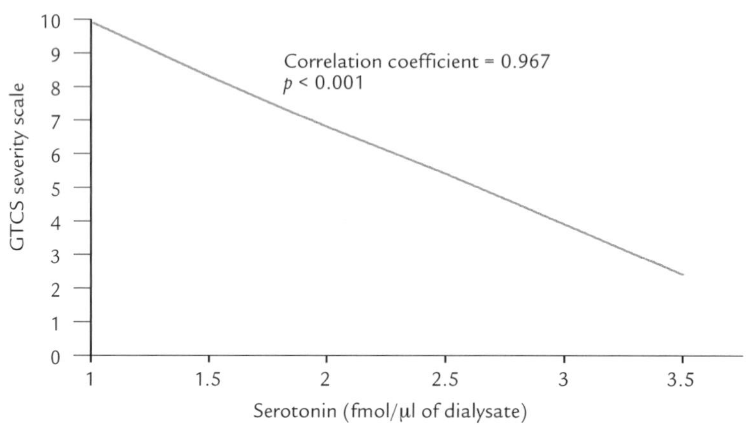

Figure 5.8 Correlation between means of extracellular serotonin concentrations in the thalamus and means of classification scores for generalized tonic–clonic seizures (GTCSs) in GEPR-9s after the intraperitoneal injection of specific serotonin reuptake inhibitor sertraline. Modified graphically from published material (reference 215)

increments in the extracellular fluid and should also block anticonvulsant effectiveness. Accordingly, when GEPR-9s are pretreated with a serotonin synthesis inhibitor (ρ-chlorophenylalanine, PCPA) serotonin stores are depleted; moreover, antidepressant drug-induced extracellular serotonin increments and anticonvulsant effects are markedly reduced (Figure 5.9).

Several antidepressant drugs are both anticonvulsant and convulsant[211,219,239]. Evidence supports the concept that antidepressant drugs suppress seizures when blood and brain levels are within a therapeutic range, whereas some may precipitate seizures when administered in overdoses or when doses thought to be in the therapeutic range are excessive for those rare individuals who may be slow metabolizers[239]. As indicated in the preceding section, the antidepressant monoamine oxidase inhibitors and serotonin and norepinephrine reuptake inhibitors are known to be anticonvulsant in genetically epileptic and in non-epileptic mammals.

As recognized recently and as depicted in Table 5.1, the relative risk for antidepressant drug-induced seizures appears to be equivalent in GEPRs and in humans[32]. In order of descending seizure liability in humans, antidepressants have been ranked as follows: (1) amoxapine; (2) imipramine; (3) desipramine; and (4) fluoxetine[1]. This same rank order was discovered for GEPRs upon determination of the ratio of the convulsant and anticonvulsant dose fifties[211] (Jobe, unpublished data). Interestingly, amoxapine is the only antidepressant to cause the most severe form of GTCSs as previously defined[51,240], whereas fluoxetine appears to have minimal to no convulsant properties.

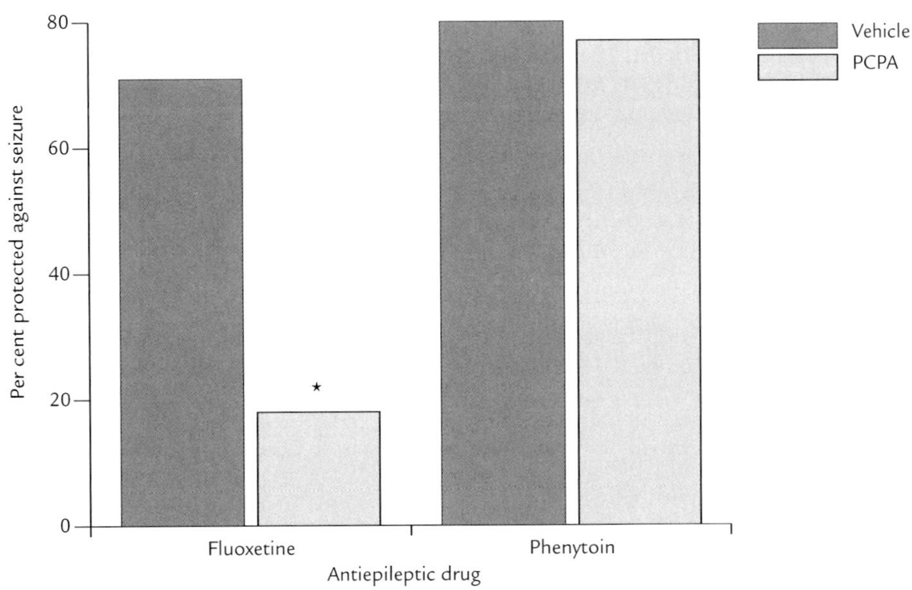

Figure 5.9 Effect of serotonin synthesis inhibition on the anticonvulsant effects of fluoxetine and phenytoin in GEPR-9s[213,216] ★, significant difference at the 0.01 level (χ^2 test). PCPA, ρ-chlorophenylalanine

Table 5.1 Rank order of convulsant liability of antidepressant drugs in humans[1,32] and in genetically epilepsy-prone rats (GEPRs)[211] (Jobe, unpublished data)

Human	*Seizure*	*GEPR*
Amoxepine	High	Amoxepine
Imipramine	Intermediate	Imipramine
Desipramine	Low	Desipramine
Fluoxetine	Lowest	Fluoxetine

In general terms, seizure liability has been inaccurately assigned to the antidepressant drugs. For example, the idea that therapeutic doses of fluoxetine have seizure liability is based on the observation that patients taking fluoxetine have a greater risk of seizures than do those in the general population. Unfortunately, this comparison provides no guidance on the issue of seizure liability. Without doubt, people at high risk for epilepsy, such as those with affective disorder, have an elevated risk of seizures when compared with those in the general population. Similarly, people taking antiepileptic drugs would be expected to have a greater risk for exhibiting seizures than those in the general population. After all, a large fraction of people taking antiepileptic drugs are those who have epilepsy. The risk ratio for epilepsy in patients taking tricyclic antidepressants is reportedly 1.5, compared with a risk ratio of 3.7 for people with depression[241]. As previously noted, it would be more consistent with experimental evidence to conclude that people taking antidepressants therapeutically for affective disorder may be at lower risk for seizures than people with depressive disorder in the absence of antidepressants. Indeed, people with affective disorder are at risk for epilepsy co-morbidity and, as a result, have an elevated risk for seizures because of their illness.

Discoveries of the usefulness of some, but not all, anticonvulsants in affective disorders have highlighted the possibility that antiepileptics may share one or more pharmacodynamic properties with antidepressants. Anticonvulsant properties of antidepressant treatments reinforce this concept. Stabilization of ion channels, adenosine receptor interactions, increases in GABAergic transmission and reductions in excitatory neurotransmission have been advanced as sources of clinical benefit in epileptology[242]. Serotonergic and/or noradrenergic activation now also appears to contribute appreciably to the anticonvulsant effects of antiepileptic drugs. Monoaminergic increments are known to occur in response to antiepilepsirine (a medication used clinically in China)[243], carbamazepine[243–246], lamotrigine[247], loreclezole[248], valproate[249] and zonisamide[250]. Several observations support the concept that these monoaminergic effects are therapeutically relevant. For example, valproate-induced increments in extracellular fluid serotonin in non-epileptic rats occur in response to doses of the drug that in other experiments are in the anticonvulsant range. Also, carbamazepine- and antiepilepsirine-induced increments in extracellular serotonin are tightly correlated with anticonvulsant effects[213,243,244,246], and depletion of serotonin stores sharply reduces the anticonvulsant effectiveness of these two antiepileptic drugs. The dose-dependent effect of carbamazepine on extracellular hippocampal serotonin concentrations is depicted in Figure 5.10. The capacity of serotonin depletion to reduce the anticonvulsant effectiveness of carbamazepine sharply is shown in Figure 5.11.

The serotonergic and noradrenergic effects of antiepileptic drugs are particularly interesting in light of the enormous body of evidence linking the therapeutic mechanisms of antidepressant drugs to these same neurotransmitter systems. The tricyclic antidepressants, such as amitriptyline (Elavil®) and desipramine (Norpramin®), the monoamine oxidase inhibitors, such as phenelzine (Nardil®) and tranylcypromine (Parnate®) and the selective serotonin reuptake inhibitors (SSRIs) all have the capacity to elevate noradrenergic and/or serotonergic transmission. Moreover, pretreatment-induced decrements in norepinephrine and/or serotonin stores have been shown to reduce the anticonvulsant effects of antidepressant drugs in GEPRs[213,216,236] as well as the antidepressant effects of light therapy in humans with seasonal affective disorder[197]. Moreover, both the anticonvulsant properties and the elevations of extracellular serotonin concentrations produced by fluoxetine are enhanced by combined treatment with the precursor 5-HTP[214].

GLUTAMATERGIC AND GABAERGIC ABNORMALITIES

Serotonergic and noradrenergic deficits are only two of the pathophysiological links between epilepsies and affective disorders in GEPRs and in humans. GABAergic interneurons are an internal component of corticolimbic circuitry. These and other GABAergic neurons appear to play important roles in seizure initiation, continuation and termination in the GEPRs and also in other epileptic and non-epileptic mammals[34,37,45,251–262]. Numerous, but not all investigations support the concept that GABAergic deficiencies contribute to seizure mechanisms and to epileptogenesis in humans[263–267].

Both GABAergic excess and deficiency hypotheses of mood disorders are being investigated[16,31,268–270]. Also, as previously suggested, GABAergic deficits may contribute to mood disorders and the epilepsies through interactions with serotonergic and/or noradrenergic systems[32]. GABA interacting directly or indirectly with noradrenergic terminals elicits norepinephrine release. Thus, GABAergic deficiencies in the vicinity of noradrenergic terminals could exacerbate innate noradrenergic deficits. Interactions relevant to mechanisms of affective disorder and epilepsy predisposition also occur between GABAergic and glutamatergic neurons both in the GEPR and in humans.

SUMMARY AND CONCLUSIONS

GEPRs model prominent aspects of epilepsy and affective disorder co-morbidity. Both GEPR-9s and GEPR-3s exhibit genetically determined seizure predisposition. This trait includes a marked predilection to seizures that closely mimic GTCSs in humans – behaviorally, electrographically and pharmacologically. In GEPR-3s these exhibited GTCSs are less severe than those of the GEPR-9.

Spontaneous seizures infrequently occur in both GEPR strains. Both are susceptible to GTCSs evoked by several types of environ-

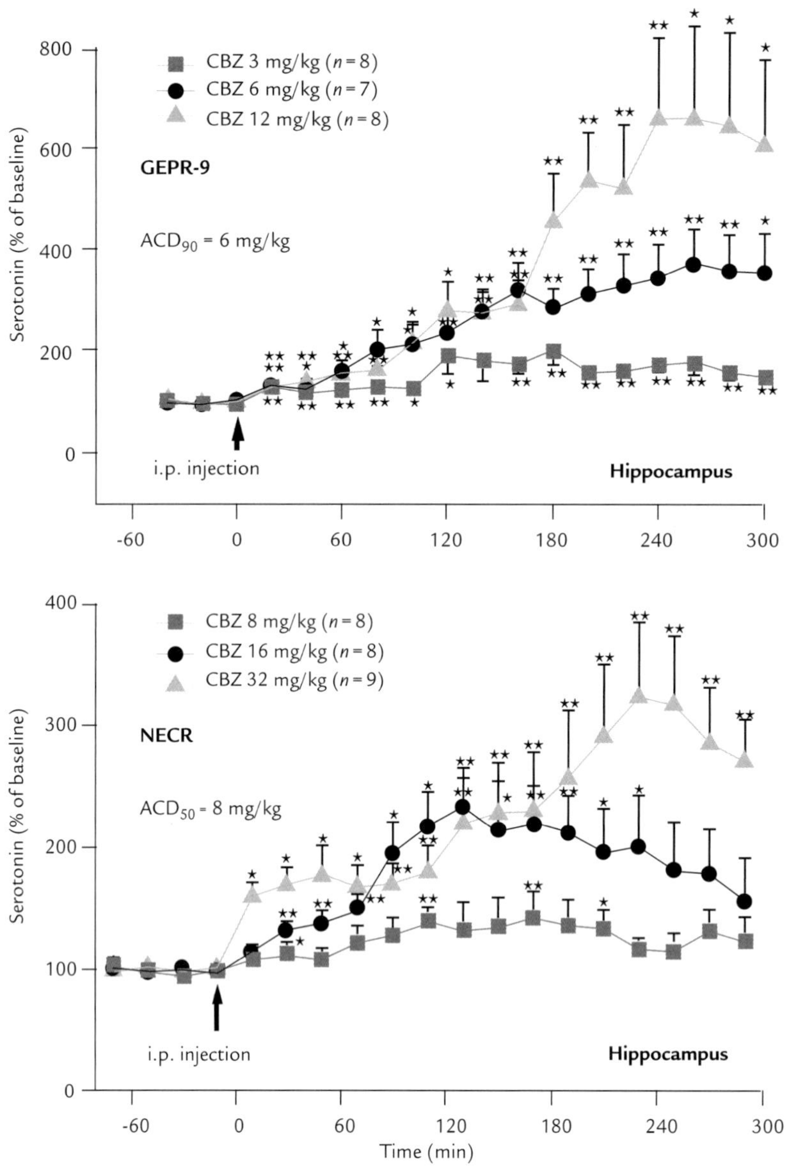

Figure 5.10 Effect of anticonvulsant doses of carbamazepine (intraperitoneal injection) on hippocampal serotonin concentrations in GEPRs and in non-epileptic control rats (NECR). The intraperitoneal (i.p.) doses of carbamazepine are shown in each graph. $^{\star}p < 0.01$; $^{\star\star}p < 0.001$. Reprinted from Dailey *et al.* Anticonvulsant doses of carbamazepine increase hippocampal extracellular serotonin in genetically epilepsy-prone rats: dose response relationships. *Neurosci Lett* 1997; 227: 13–16 and Dailey *et al.* Carbamazepine increases extracellular serotonin concentration: lack of antagonism by tetrodotoxin or zero Ca^{2+}. *Eur J Pharmacol* 1997; 328: 153–162, with permission from Elsevier (references 244 and 246)

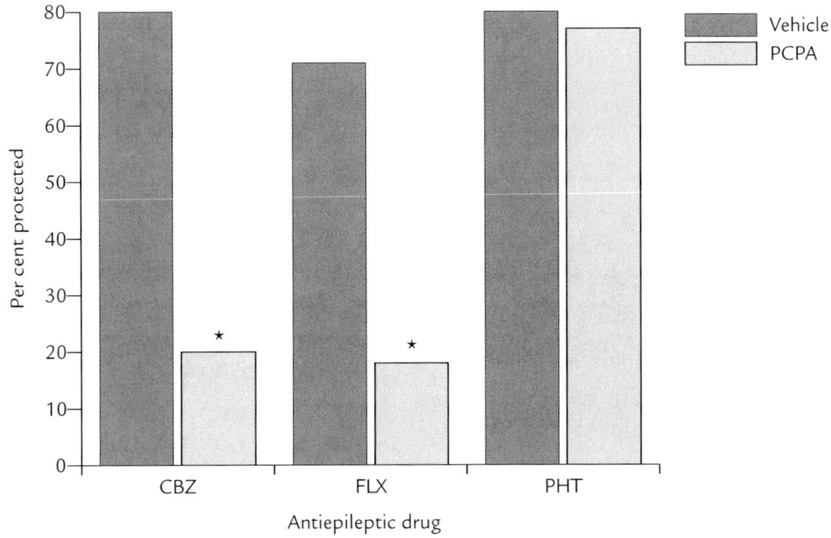

Figure 5.11 Effect of serotonin depletion on the anticonvulsant properties of carbamazepine (CBZ) in GEPR-9s. Serotonin stores were depleted by pretreatment with the serotonin synthesis inhibitor ρ-chlorophenylalanine (PCPA) and the effects on CBZ-induced protection against seizures is shown (CBZ with only saline pretreatment). FLX, fluoxetine; PHT, phenytoin; ★,$p < 0.01$. The data were previously published in reference 243

mental stimuli that do not ignite seizures in non-epileptic subjects. Moreover, both have abnormally low thresholds to seizures evoked by stimuli that also ignite seizures in non-epileptic subjects.

GEPR-3s and GEPR-9s model components of partial seizure epilepsy in humans. Both strains are characterized by predisposition to electrical limbic kindling, with this trait being more exaggerated in GEPR-9s. In addition, GEPR-9s model the process of secondary generalization of partial seizures to tonic–clonic seizures. Partial seizure with this type of secondary generalization is a feature exhibited by many humans with epilepsy.

Analog manifestations of affective disorder predisposition and expression have been examined in GEPR-3s. Corresponding information from GEPR-9s would be useful for additional hypothesis testing. The behavioral tests that have been undertaken with GEPR-9s may be useful in designing future experiments, but are not sufficiently advanced for testing affective disorder hypotheses. However, data from GEPR-3 experiments, with the inescapable stress induced by forced swimming, support the concept that these epileptic animals are predisposed to analog behavioral manifestations of dysfunctional affective episodes. Moreover, data from additional behavioral experiments support the concept that GEPR-3s exhibit anhedonia, and diurnal locomotor and endocrinological abnormalities that correspond to aberrations observed in people diagnosed with affective disorders.

Perhaps of equal importance, both GEPR-3s and GEPR-9s are characterized by predisposition to epileptogenesis or epilepsy disease progression. Evidence supports the concept that experientially and temporally dependent disease progression plays a prominent role both in the human epilepsies and in affective disorders. Ultimately, seizure-induced seizure predisposition in the GEPR combines with genetically determined seizure predisposition to produce an advanced state of neurological dysfunction. Analogous progressive events may characterize affective dysfunction in the GEPR-3.

The neuronal abnormalities that characterize GEPR-3s and GEPR-9s correspond closely to those that have been reported to underlie human epilepsy and affective disorders. Prominent among these are genetically derived deficits in noradrenergic and serotonergic transmission. These are widely distributed neuroanatomically, and play prominent roles as determinants of the severity of predisposition to seizures in both strains of GEPRs: the greater the defect, the greater the predisposition. GEPR-9s – having a more pronounced noradrenergic deficit than GEPR-3s – exhibit the most severe degree of seizure predisposition.

Deficits in noradrenergic and serotonergic transmission are only two of the prominent neurological abnormalities that may link the affective disorders and the epilepsies in humans, and that may underlie the co-morbidity expressed by the GEPR. In this context, aberrations in glutamatergic and GABAergic neurons may also play mechanistic roles in co-morbidity in both species.

REFERENCES

1. Barry JJ, Lembke A, Huynh N. Affective disorders in epilepsy. In: Ettinger AB, Kanner AM, eds. Psychiatric Issues in Epilepsy: A Practical Guide to Diagnosis and Treatment. Philadelphia, PA: Lippincott, Williams & Wilkins, 2001: 45.
2. Barry JJ. The recognition and management of mood disorders as a comorbidity of epilepsy. Epilepsia 2003; 44 (Suppl 4): 30–40.
3. Kanner AM, Barry JJ. The impact of mood disorders in neurological diseases: should neurologists be concerned? Epilepsy Behav 2003; 4: S3–S13.
4. Jones JE, Hermann BP, Barry JJ, et al. Rates and risk factors for suicide, suicidal ideation, and suicide attempts in chronic epilepsy. Epilepsy Behav 2003; 4: S31–S38.
5. Barry JJ, Lembke A, Bullock KD. Current status of the utilization of antiepileptic treatments in mood, anxiety and aggression: drugs and devices. Clin EEG Neurosci 2004; 35: 4–13.
6. Hauser WA, Hesdorffer DC. Psychosis, depression, and epilepsy epidemiologic considerations. In: Ettinger AB, Kanner AM, eds. Psychiatric Issues in Epilepsy: A Practical Guide to Diagnosis and Treatment. Philadelphia: Lippincott, Williams & Wilkins, 2001: 7.
7. Kanner AM, Balabanov A. Depression and epilepsy: how closely related are they? Neurology 2002; 58: S27–S39.
8. Kanner AM. Depression in epilepsy: prevalence, clinical semiology, pathogenic mechanisms, and treatment. Biol Psychiatry 2003; 54: 388–398.

9. Kanner AM. Depression in epilepsy: a frequently neglected multifaceted disorder. Epilepsy Behav 2003; 4: 11–19.

10. Kanner AM, Nieto JCR. Depressive disorders in epilepsy. Neurology 1999; 53: S26–S32.

11. Jobe PC. Affective disorder and epilepsy comorbidity: implications for development of treatments, preventions and diagnostic approaches. Clin Electroencephalogr 2004; 35: 53–68.

12. Jobe PC. Current and future therapeutic opportunities in the comorbidity between the epilepsies and affective disorders. Clin EEG Neurosci 2004; 35: 1–3.

13. Jobe PC. Shared mechanisms of antidepressant and antiepileptic treatments: drugs and devices. Clin Electroencephalogr 2004; 35: 25–37.

14. Hecimovic H, Goldstein JD, Sheline YI, Gilliam FG. Mechanisms of depression in epilepsy from a clinical perspective. Epilepsy Behav 2003; 4: S25–S30.

15. Agoub M, El Kadiri M, Chihabeddine K, et al. Depressive disorders among epileptic patients attending a specialised outpatient clinic. Encephale 2004; 30: 40–45.

16. Jobe PC, Dailey JW, Wernicke JF. A noradrenergic and serotonergic hypothesis of the linkage between epilepsy and affective disorders. Crit Rev Neurobiol 1999; 13: 317–356.

17. Jobe PC. Are there specific anatomical and/or transmitter systems (cortical or subcortical) that should be targeted? Int Rev Neurobiol 2002; 49: 221–252.

18. Wu MF, John J, Boehmer LN, et al. Activity of dorsal raphe cells across the sleep-waking cycle and during cataplexy in narcoleptic dogs. J Physiol 2004; 554: 202–215.

19. Levine ES, Litto WJ, Jacobs BL. Activity of cat locus coeruleus noradrenergic neurons during the defense reaction. Brain Res 1990; 531: 189–195.

20. Robertson HA, Leslie RA. Noradrenergic alpha 2 binding sites in vagal dorsal motor nucleus and nucleus tractus solitarius: autoradiographic localization. Can J Physiol Pharmacol 1985; 63: 1190–1194.

21. Quintero EM, Willis LM, Zaman V, et al. Glial cell line-derived neurotrophic factor is essential for neuronal survival in the locus coeruleus–hippocampal noradrenergic pathway. Neuroscience 2004; 124: 137–146.

22. Melchitzky DS, Lewis DA. Tyrosine hydroxylase- and dopamine transporter-immunoreactive axons in the primate cerebellum. Evidence for a lobular- and laminar-specific dopamine innervation. Neuropsychopharmacology 2000; 22: 466–472.

23. Smiley JF, Subramanian M, Mesulam MM. Monoaminergic–cholinergic interactions in the primate basal forebrain. Neuroscience 1999; 93: 817–829.

24. Moragues N, Ciofi P, Tramu G, Garret M. Localisation of GABA(A) receptor epsilon-subunit in cholinergic and aminergic neurones and evidence for co-distribution with the theta-subunit in rat brain. Neuroscience 2002; 111: 657–669.

25. Luque JM, Bleuel Z, Hendrickson A, Richards JG. Detection of MAO-A and MAO-B mRNAs in monkey brainstem by cross-hybridization with human oligonucleotide probes. Brain Res Mol Brain Res 1996; 36: 357–360.

26. Lavoie B, Parent A. Pedunculopontine nucleus in the squirrel monkey: distribution of cholinergic and monoaminergic neurons in the mesopontine tegmentum with evidence for the presence of glutamate in

cholinergic neurons. J Comp Neurol 1994; 344: 190–209.

27. Aston-Jones G, Rajkowski J, Kubiak P, Alexinsky T. Locus coeruleus neurons in monkey are selectively activated by attended cues in a vigilance task. J Neurosci 1994; 14: 4467–4480.

28. Rajkowski J, Kubiak P, Aston-Jones G. Locus coeruleus activity in monkey: phasic and tonic changes are associated with altered vigilance. Brain Res Bull 1994; 35: 607–616.

29. Shouse MN, Staba RJ, Saquib SF, Farber PR. Monoamines and sleep: microdialysis findings in pons and amygdala. Brain Res 2000; 860: 181–189.

30. Benes FM, Berretta S. GABAergic interneurons: implications for understanding schizophrenia and bipolar disorder. Neuropsychopharmacology 2001; 25: 1–27.

31. Heckers S, Stone D, Walsh J, et al. Differential hippocampal expression of glutamic acid decarboxylase 65 and 67 messenger RNA in bipolar disorder and schizophrenia. Arch Gen Psychiatry 2002; 59: 521–529.

32. Jobe PC. Common pathogenic mechanisms between depression and epilepsy: an experimental perspective. Epilepsy Behav 2003; 4: 14–24.

33. Chakravarty DN, Faingold CL. Differential roles in the neuronal network for audiogenic seizures are observed among the inferior colliculus subnuclei and the amygdala. Exp Neurol 1999; 157: 135–141.

34. Faingold CL. Neuronal networks in the genetically epilepsy-prone rat. Adv Neurol 1999; 79: 311–321.

35. Faingold CL, Randall ME. Neurons in the deep layers of superior colliculus play a critical role in the neuronal network for audiogenic seizures: mechanisms for production of wild running behavior. Brain Res 1999; 815: 250–258.

36. Faingold CL. Role of GABA abnormalities in the inferior colliculus pathophysiology – audiogenic seizures. Hear Res 2002; 168: 223–237.

37. N'Gouemo P, Faingold CL. The periaqueductal grey is a critical site in the neuronal network for audiogenic seizures: modulation by GABA(A), NMDA and opioid receptors. Epilepsy Res 1999; 35: 39–46.

38. Faingold CL. Neuronal networks, epilepsy, and the action of antiepileptic drugs. In: Faingold CL, Fromm GH, eds. Drugs for Control of Epilepsy: Actions on Neuronal Networks Involved in Seizure Disorders. Boca Raton, FL: CRC Press, 1992: 1.

39. Faingold CL, Naritoku DK. The genetically epilepsy-prone rat: neuronal networks and actions of amino acid neurotransmitters. In: Faingold CL, Fromm GH, eds. Drugs for the Control of Epilepsy: Actions on Neuronal Networks Involved in Seizure Disorders. Boca Raton, FL: CRC Press, 1992: 277.

40. Browning RA, Nelson DK, Mogharreban N, et al. Effect of midbrain and pontine tegmental lesions on audiogenic seizures in genetically epilepsy-prone rats. Epilepsia 1985; 26: 175–183.

41. Browning RA, Wang C, Nelson DK, Jobe PC. Effect of precollicular transection on audiogenic seizures in genetically epilepsy-prone rats. Exp Neurol 1999; 155: 295–301.

42. Browning RA, Wang C, Lanker ML, Jobe PC. Electroshock- and pentylenetetrazol-induced seizures in genetically epilepsy-prone rats (GEPRs): differences in threshold and pattern. Epilepsy Res 1990; 6: 1–11.

43. Buchhalter JR. Animal models of inherited epilepsy. Epilepsia 1993; 34 (Suppl 3): S31–S41.
44. Dailey JW, Reigel CE, Mishra PK, Jobe PC. Neurobiology of seizure predisposition in the genetically epilepsy-prone rat. Epilepsy Res 1989; 3: 3–17.
45. Evans MS, Viola-McCabe KE, Caspary DM, Faingold CL. Loss of synaptic inhibition during repetitive stimulation in genetically epilepsy-prone rats (GEPR). Epilepsy Res 1994; 18: 97–105.
46. Reigel CE, Dailey JW, Jobe PC. The genetically epilepsy-prone rat: an overview of seizure-prone characteristics and responsiveness to anticonvulsant drugs. Life Sci 1986; 39: 763–774.
47. Ribak CE, Morin CL. The role of the inferior colliculus in a genetic model of audiogenic seizures. Anat Embryol 1995; 191: 279–295.
48. Jobe PC, Dailey JW. Genetically epilepsy-prone rats (GEPRs) in drug research. CNS Drug Rev 2000; 6: 241–260.
49. Jobe PC, Mishra PK, Browning RA, et al. Noradrenergic abnormalities in the genetically epilepsy-prone rat. Brain Res Bull 1994; 35: 493–504.
50. Jobe PC, Mishra PK, Adams-Curtis LE, et al. The genetically epilepsy-prone rat (GEPR). Ital J Neurol Sci 1995; 16: 91–99.
51. Coffey LL, Reith MEA, Chen NH, et al. Amygdala kindling of forebrain seizures and the occurrence of brainstem seizures in genetically epilepsy-prone rats. Epilepsia 1996; 37: 188–197.
52. Jobe PC, Mishra PK, Dailey JW, et al. Genetic predisposition to partial (focal) seizures and to generalized tonic/clonic seizures: interactions between seizure circuitry of the forebrain and brainstem. In: Berkovic SF, Genton P, Hirsch E, Picard F, eds. Genetics of Focal Epilepsies. Avignon, France: John Libbey & Company, 1999: 251.
53. Jobe PC. Spinal seizures induced by electrical stimulation. In: Fromm GH, Faingold CL, Browning RA, Burnham WM, eds. Epilepsy and the Reticular Formation: The Role of the Reticular Core in Convulsive Seizures. New York: Alan R Liss, 1987: 81.
54. Esplin DW, Freston JW. Physiological and pharmacological analysis of spinal cord convulsions. J Pharmacol Exp Ther 1960; 130: 68–80.
55. Esplin DW. Spinal cord convulsions. Arch Neurol 1959; 1: 485–490.
56. Vernadakis A. Spinal cord convulsions in developing rats. Science 1962; 137: 532.
57. Engel J. Epileptic seizures. In: Engel J, ed. Seizures and Epilepsy. Philadelphia: FA Davis, 1989: 137.
58. Jobe PC, Moraes MFD, Chavali M, Garcia-Cairasco N. Artifact-free video-EEG recordings of the generalized tonic–clonic seizure of GEPR-9S and GEPR3s. Society for Neuroscience Abstracts 24 [Part 1]. 1998.
59. Mishra PK, Moraes MFD, Chavali M, et al. Seizure EEG of naive and acoustically stimulated GEPR-9S: factors important for evolution of seizure-experienced phenotype. Society for Neuroscience Abstracts 24 [Part 1]. 1998.
60. Netoff TI, Schiff SJ. Decreased neuronal synchronization during experimental seizures. J Neurosci 2002; 22: 7297–7307.
61. Jobe PC. Pharmacology of audiogenic seizures. In: Brown RD, Daigneault EA, eds. Pharmacology of Hearing: Experimental and Clinical Bases. New York: John Wiley & Sons, 1981: 271.

62. Jobe PC, Woods TW, McNatt LE, et al. Genetically epilepsy prone rats (GEPR), a model for febrile seizures? Fed Proc 1984; 41: 1560.

63. Reigel CE, Jobe PC, Dailey JW, Savage DD. Ontogeny of sound-induced seizures in the genetically epilepsy-prone rat. Epilepsy Res 1989; 4: 63–71.

64. Franck JE, Ginter KL, Schwartzkroin PA. Developing genetically epilepsy-prone rats have an abnormal seizure response to flurothyl. Epilepsia 1989; 30: 1–6.

65. DeSarro A, Grasso S, Zappala M, et al. Convulsant effects of some xanthine-derivatives in genetically epilepsy-prone rats. Naunyn Schmiedebergs Arch Pharmacol 1997; 356: 48–55.

66. Millan MH, Wardley-Smith B, Durmuller N, Meldrum BS. The high pressure neurological syndrome in genetically epilepsy prone rats: protective effect of 2-amino-7-phosphono heptanoate. Exp Neurol 1991; 112: 317–320.

67. Toman JEP, Loewe S, Goodman L. Physiology and therapy of convulsive disorders. Arch Neurol and Psychiatr 1947; 58: 312–324.

68. Goodman L, Gilman A. Metrazol, coramine and camphor. In: Goodman L, Gilman A, eds. The Pharmacological Basis of Therapeutics. New York: MacMillan, 1941: 267–273.

69. Weiner RD, American Psychiatric Association Task Force on Electroconvulsive Therapy. The Practice of Electroconvulsive Therapy: Recommendations for Treatment, Training, and Privileging: A Task Force Report of the American Psychiatric Association. Washington, DC: American Psychiatric Association, 1990.

70. Baselt RC. Disposition of Toxic Drugs and Chemical in Man, 5th edn. Foster City, California: Chemical Toxicology Institute, 2000.

71. Engel J. Terminology and classifications. In: Engel J, ed. Seizures and Epilepsy. Philadelphia: FA Davis, 1989: 3.

72. Zifkin B, Andermann F. Epilepsy with reflex seizures. In: Wyllie B, ed. The Treatment of Epilepsy: Principles and Practice. Philadelphia: Lippincott Williams & Wilkins, 2001: 537.

73. Duchowny M. Febrile seizures. In: Wyllie E, ed. The Treatment of Epilepsy: Principles and Practice. Philadelphia: Lippincott Williams & Wilkins, 2001: 601.

74. Singh R, Scheffer IE, Crossland K, Berkovic SF. Generalized epilepsy with febrile seizures plus: a common childhood-onset genetic epilepsy syndrome. Ann Neurol 1999; 45: 75–81.

75. Scheffer IE, Berkovic SF. Generalized epilepsy with febrile seizures plus. A genetic disorder with heterogeneous clinical phenotypes. Brain 1997; 120: 479–490.

76. Cure D, Rasmussen T, Jasper H. Activation of seizures and electroencephalographic disturbances in epileptic and in control subjects with 'metrazol'. Arch Neurol Psychiatr 1948; 59: 691–717.

77. Rush AJ. Toward an understanding of bipolar disorder and its origin. J Clin Psychiatry 2003; 64 (Suppl 6) : 4–8.

78. Heim C, Nemeroff CB. The role of childhood trauma in the neurobiology of mood and anxiety disorders: preclinical and clinical studies. Biol Psychiatry 2001; 49: 1023–1039.

79. Huber MT, Braun HA, Krieg JC. On the impact of episode sensitization on the

course of recurrent affective disorders. J Psychiatr Res 2001; 35: 49–57.

80. Beatson J, Taryan S. Predisposition to depression: the role of attachment. Aust NZ J Psychiatry 2003; 37: 219–225.

81. Post RM. Do the epilepsies, pain syndromes, and affective disorders share common kindling-like mechanisms? Epilepsy Res 2002; 50: 203–219.

82. Post RM, Ballenger JC, Uhde TW, et al. Kindling and drug sensitization: implications for the progressive development of psychopathology and treatment with carbamazepine. In: Sandler M, ed. Psychopharmacology of Anticonvulsants. Oxford: Oxford University Press, 1982: 27.

83. Post RM, Weiss SRB. Sensitization and kindling phenomena in mood, anxiety, and obsessive–compulsive disorders: the role of serotonergic mechanisms in illness progression. Biol Psychiatry 1998; 44: 193–206.

84. Post RM, Weiss SR. Convergences in course of illness and treatments of the epilepsies and recurrent affective disorders. Clin EEG Neurosci 2004; 35: 14–24.

85. Savage DD, Reigel CE, Jobe PC. Angular bundle kindling is accelerated in rats with a genetic predisposition to acoustic stimulus-induced seizures. Brain Res 1986; 376: 412–415.

86. Savage DD, Reigel CE, Jobe PC. The development of kindled seizures is accelerated in the genetically epilepsy-prone rat. Life Sci 1986; 39: 879–886.

87. McIntyre DC, Kelly ME, Dufresne C. FAST and SLOW amygdala kindling rat strains: comparison of amygdala, hippocampal, piriform and perirhinal cortex kindling. Epilepsy Res 1999; 35: 197–209.

88. Racine RJ, Steingart M, McIntyre DC. Development of kindling-prone and kindling-resistant rats: selective breeding and electrophysiological studies. Epilepsy Res 1999; 35: 183–195.

89. Porsolt RD, Anton G, Blavet N, Jalfre M. Behavioural despair in rats: a new model sensitive to antidepressant treatments. Eur J Pharmacol 1978; 47: 379–391.

90. Detke MJ, Lucki I. Detection of serotonergic and noradrenergic antidepressants in the rat forced swimming test: the effects of water depth. Behav Brain Res 1996; 73: 43–46.

91. Weber RJ, Dailey JW, Jobe PC. A genetic model of depression. Neuroimmunomodulation 1998; 5: 61–120.

92. Willner P, Sampson D, Papp M, et al. Animal Models of Anhedonia. Anxiety Depression and Mania. New York: Karger, 1991: 71.

93. Harris RB, Zhou J, Youngblood BD, et al. Failure to change exploration or saccharin preference in rats exposed to chronic mild stress. Physiol Behav 1997; 63: 91–100.

94. Volkers AC, Tulen JH, van den Broek WW, et al. Motor activity and autonomic cardiac functioning in major depressive disorder. J Affect Disord 2003; 76: 23–30.

95. Wolff EA III, Putnam FW, Post RM. Motor activity and affective illness. The relationship of amplitude and temporal distribution to changes in affective state. Arch Gen Psychiatry 1985; 42: 288–294.

96. Wehr TA, Muscettola G, Goodwin FK. Urinary 3-methoxy-4-hydroxyphenylglycol circadian rhythm. Early timing (phase-advance) in manic-depressives compared with normal subjects. Arch Gen Psychiatry 1980; 37: 257–263.

97. van Londen L, Kerkhof GA, van den BF, et al. Plasma arginine vasopressin and motor activity in major depression. Biol Psychiatry 1998; 43: 196–204.

98. Yirmiya R. Endotoxin produces a depressive-like episode in rats. Brain Res 1996; 711: 163–174.

99. Mills SA, Savage DD. Evidence of hypothyroidism in the genetically epilepsy-prone rat. Epilepsy Res 1988; 2: 102–110.

100. Razani-Boroujerdi S, Savage DD. Evidence of transient neonatal hypothyroidism in the moderate seizure genetically epilepsy-prone (GEPR-3) rat. Society for Neuroscience Abstracts 1990; 16: 781.

101. Nathan KI, Musselman DL, Schatzberg AF, Nemeroff CB. Biology of mood disorders. In: Schatzberg AF, Nemeroff CB, eds. The American Psychiatric Press Textbook of Psychopharmacology. Washington, DC: American Psychiatric Press, 1995: 439.

102. Jefferson JW, Greist JH. Mood disorders. In: Hales RE, Yudofsky S, Talbott JA, eds. The American Psychiatric Press Textbook of Psychiatry. Washington, DC: American Psychiatric Press, 1994: 465.

103. Holsboer F. Neuroendocrinology of mood disorders. In: Bloom FE, Kupfer DJ, eds. Psychopharmacology: The Fourth Generation of Progress. New York: Raven Press, 1995: 957.

104. Hickie I, Bennett B, Mitchell P, et al. Clinical and subclinical hypothyroidism in patients with chronic and treatment-resistant depression. Aust NZ J Psychiatry 1996; 30: 246–252.

105. Whybrow PC, Prange AJ Jr. A hypothesis of thyroid–catecholamine-receptor interaction. Its relevance to affective illness. Arch Gen Psychiatry 1981; 38: 106–113.

106. Jackson IM. The thyroid axis and depression. Thyroid 1998; 8: 951–956.

107. Charney DS, Miller HL, Licinio J, Salomon R. Treatment of depression. In: Schatzberg AF, Nemeroff CB, eds. The American Psychiatric Press Textbook of Psychopharmacology. Washington, DC: American Psychiatric Press, 1995: 575.

108. Joffe RT. The use of thyroid supplements to augment antidepressant medication. J Clin Psychiatry 1998; 59: 26–29.

109. Hendrick V, Altshuler L, Whybrow P. Psychoneuroendocrinology of mood disorders. The hypothalamic–pituitary–thryoid axis. Psychiatr Clin North Am 1998; 21: 277–292.

110. Robertson MM, Coppen A, Trimble MR. The dexamethasone suppression test in medicated epileptic patients. Biol Psychiatry 1986; 21: 225–228.

111. Mills SA, Reigel CE, Jobe PC, Savage DD. Serum growth hormone and body weight are decreased in developing genetically epilepsy-prone rats. Society for Neuroscience Abstracts 1987; 13: 943.

112. Sakkas PN, Soldatos CR, Bergiannaki JD, et al. Growth hormone secretion during sleep in male depressed patients. Prog Neuropsychopharmacol Biol Psychiatry 1998; 22: 467–483.

113. Contreras F, Navarro MA, Menchon JM, et al. Growth hormone response to growth hormone releasing hormone in non-delusional and delusional depression and healthy controls. Psychol Med 1996; 26: 301–308.

114. Carr JA, Ortiz KA, Paxton LL, et al. Alterations in spleen norepinephrine and lymphocyte [3H]dihydroalprenolol binding site number in genetically epilepsy-prone rats. Brain Behav Immun 1993; 7: 113–120.

115. Rowland RR, Tokuda S, Weiss GK, et al. Evidence of immunosuppression in the genetically epilepsy-prone rat. Life Sci 1991; 48: 1821–1826.

116. Ahn JJ, Jobe PC, Stewart JJ, Weber RJ. Altered immune system in genetically

epilepsy-prone rats. Society for Neuroscience Abstracts 1994; 20(1): 104.

117. Ahn JJ, O'Donnell MS, Polonus RM, et al. Analysis of immune cells populations and activitation markers in genetically epilepsy-prone rats (GEPRs). Psychoneuroimmunology Research Society Abstracts B11. 1994, November 16–20.

118. O'Donnell MS, Ahn JJ, Polonus RM, et al. Characterization of immune system function in genetically epilepsy-prone rats (GEPRs). Psychoneuroimmunology Research Society Abstracts, B12. 1994.

119. Wang X, Shen G, Sun B, Su N. Studies on lymphocyte subpopulations and Nk cell activities in epileptic patients. J Tongji Med Univ 1989; 1: 25–28.

120. Barinaga M. Antibodies linked to rare epilepsy. Science 1995; 268: 362–363.

121. Bartolomei F, Boucraut J, Barrie M, et al. Cryptogenic partial epilepsies with anti-GM1 antibodies: a new form of immune-mediated epilepsy? Epilepsia 1996; 37: 922–926.

122. Eeg-Olofsson O, Prchal JF, Andermann F. Abnormalities of T-lymphocyte subsets in epileptic patients. Acta Neurol Scand 1985; 72: 140–144.

123. Farrell MA, Droogan O, Secor DL, et al. Chronic encephalitis associated with epilepsy: immunohistochemical and ultrastructural studies. Acta Neuropathol (Berl) 1995; 89: 313–321.

124. Rogers SW, Andrews PI, Gahring LC, et al. Autoantibodies to glutamate receptor GluR3 in Rasmussen's encephalitis. Science 1994; 265: 648–651.

125. Xie XK, Tang LO. Observation on anti-brain antibody in serum of 110 epileptics. Chin Med J 1990; 103: 71–75.

126. Leonard BE, Song C. Changes in the immune system in rodent models of depression. Int J Neuropsychopharmacol 2002; 5: 345–356.

127. Leonard BE, Song C. Changes in the immune–endocrine interrelationships in anxiety and depression. Stress Med 1997; 13: 217–227.

128. Maes M, Stevens W, DeClerck L, et al. Immune disorders in depression: higher T helper/T suppressor–cytotoxic cell ratio. Acta Psychiatr Scand 1992; 86: 423–431.

129. Kronfol Z, Silva J Jr, Greden J, et al. Impaired lymphocyte function in depressive illness. Life Sci 1983; 33: 241–247.

130. Kronfol Z. Immune dysregulation in major depression: a critical review of existing evidence. Int J Neuropsychopharmacol 2002; 5: 333–343.

131. Anderson JL. The immune system and major depression. Adv Neuroimmunol 1996; 6: 119–129.

132. Anisman H, Hayley S, Turrin N, Merali Z. Cytokines as a stressor: implications for depressive illness. Int J Neuropsychopharmacol 2002; 5: 357–373.

133. Leonard B. Stress, depression and the activation of the immune system. World J Biol Psychiatry 2000; 1: 17–25.

134. Petitto JM, Folds JD, Ozer H, et al. Abnormal diurnal variation in circulating natural killer cell phenotypes and cytotoxic activity in major depression. Am J Psychiatry 1992; 149: 694–696.

135. Schleifer SJ, Keller SE, Siris SG, et al. Depression and immunity. Lymphocyte function in ambulatory depressed patients, hospitalized schizophrenic patients, and patients hospitalized for herniorrhaphy. Arch Gen Psychiatry 1985; 42: 129–133.

136. Solomon GF, Moos RH. Emotions, immunity, and disease. Arch Gen Psychiatry 1964; 11: 657–674.

137. Stein M, Miller AH, Trestman RL. Depression and the immune system. In: Ader R, Felten DL, Cohen N, eds. Psychoneuroimmunology, 2nd edn. San Diego, CA: Academic Press, 1991: 897.

138. Asberg M, Traskman L, Thoren P. 5-HIAA in the cerebrospinal fluid. A biochemical suicide predictor? Arch Gen Psychiatry 1976; 33: 1193–1197.

139. Brown GL, Ebert MH, Goyer PF, et al. Aggression, suicide, and serotonin: relationships to CSF amine metabolites. Am J Psychiatry 1982; 139: 741–746.

140. Brown GL, Linnoila MI. CSF serotonin metabolite (5-HIAA) studies in depression, impulsivity, and violence. J Clin Psychiatry 1990; 51 (Suppl): 31–41.

141. Roy A, De Jong J, Linnoila M. Cerebrospinal fluid monoamine metabolites and suicidal behavior in depressed patients. A 5-year follow-up study. Arch Gen Psychiatry 1989; 46: 609–612.

142. Cheetham SC, Crompton MR, Czudek C, et al. Serotonin concentrations and turnover in brains of depressed suicides. Brain Res 1989; 502: 332–340.

143. Stanley M, Virgilio J, Gershon S. Tritiated imipramine binding sites are decreased in the frontal cortex of suicides. Science 1982; 216: 1337–1339.

144. Perry EK, Marshall EF, Blessed G, et al. Decreased imipramine binding in the brains of patients with depressive illness. Br J Psychiatry 1983; 142: 188–192.

145. Leake A, Fairbairn AF, McKeith IG, Ferrier IN. Studies on the serotonin uptake binding site in major depressive disorder and control post-mortem brain: neurochemical and clinical correlates. Psychiatry Res 1991; 39: 155–165.

146. Briley MS, Langer SZ, Raisman R, et al. Tritiated imipramine binding sites are decreased in platelets of untreated depressed patients. Science 1980; 209: 303–305.

147. Langer SZ, Zarifian E, Briley M, et al. High-affinity binding of 3H-imipramine in brain and platelets and its relevance to the biochemistry of affective disorders. Life Sci 1981; 29: 211–220.

148. Langer SZ, Galzin AM. Studies on the serotonin transporter in platelets. Experientia 1988; 44: 127–130.

149. Nemeroff CB, Knight DL, Krishnan RR, et al. Marked reduction in the number of platelet-tritiated imipramine binding sites in geriatric depression. Arch Gen Psychiatry 1988; 45: 919–923.

150. Malison RT, Price LH, Berman R, et al. Reduced brain serotonin transporter availability in major depression as measured by [123I]-2 beta-carbomethoxy-3 beta-(4-iodophenyl)tropane and single photon emission computed tomography. Biol Psychiatry 1998; 44: 1090–1098.

151. Ogilvie AD, Harmar AJ. Association between the serotonin transporter gene and affective disorder: the evidence so far. Mol Med 1997; 3: 90–93.

152. Ogilvie AD, Battersby S, Bubb VJ, et al. Polymorphism in serotonin transporter gene associated with susceptibility to major depression. Lancet 1996; 347: 731–733.

153. Lowther S, Katona CL, Crompton MR, Horton RW. 5-HT1D and 5-HT1E/1F binding sites in depressed suicides: increased 5-HT1D binding in globus pallidus but not cortex. Mol Psychiatry 1997; 2: 314–321.

154. Cheetham SC, Crompton MR, Katona CL, Horton RW. Brain 5-HT1 binding sites in

depressed suicides. Psychopharmacology 1990; 102: 544–548.

155. Stockmeier CA, Shapiro LA, Dilley GE, et al. Increase in serotonin-1A autoreceptors in the midbrain of suicide victims with major depression–postmortem evidence for decreased serotonin activity. J Neurosci 1998; 18: 7394–7401.

156. Callado LF, Meana JJ, Grijalba B, et al. Selective increase of alpha 2A-adrenoceptor agonist binding sites in brains of depressed suicide victims. J Neurochem 1998; 70: 1114–1123.

157. Meana JJ, Barturen F, Garcia-Sevilla JA. Alpha 2-adrenoceptors in the brain of suicide victims: increased receptor density associated with major depression. Biol Psychiatry 1992; 31: 471–490.

158. Meana JJ, Garcia-Sevilla JA. Increased alpha 2-adrenoceptor density in the frontal cortex of depressed suicide victims. J Neural Transm 1987; 70: 377–381.

159. Ordway GA, Widdowson PS, Smith KS, Halaris A. Agonist binding to alpha 2-adrenoceptors is elevated in the locus coeruleus from victims of suicide. J Neurochem 1994; 63: 617–624.

160. Klimek V, Stockmeier C, Overholser J, et al. Reduced levels of norepinephrine transporters in the locus coeruleus in major depression. J Neurosci 1997; 17: 8451–8458.

161. Matussek N, Ackenheil M, Hippius H, et al. Effect of clonidine on growth hormone release in psychiatric patients and controls. Psychiatry Res 1980; 2: 25–36.

162. Siever LJ, Uhde TW, Jimerson DC, et al. Differential inhibitory noradrenergic responses to clonidine in 25 depressed patients and 25 normal control subjects. Am J Psychiatry 1984; 141: 733–741.

163. Siever LJ, Uhde TW, Silberman EK, et al. Growth hormone response to clonidine as a probe of noradrenergic receptor responsiveness in affective disorder patients and controls. Psychiatry Res 1982; 6: 171–183.

164. Siever LJ, Trestman RL, Coccaro EF, et al. The growth hormone response to clonidine in acute and remitted depressed male patients. Neuropsychopharmacology 1992; 6: 165–177.

165. Ordway GA, Schenk J, Stockmeier CA, et al. Elevated agonist binding to alpha2-adrenoceptors in the locus coeruleus in major depression. Biol Psychiatry 2003; 53: 315–323.

166. Pranzatelli MR. Putative neurotransmitter abnormalities in infantile spasms: cerebrospinal fluid neurochemistry and drug effects. J Child Neurol 1994; 9: 119–129.

167. Silverstein F, Johnston MV. Cerebrospinal fluid monoamine metabolites in patients with infantile spasms. Neurology 1984; 34: 102–105.

168. Yamamoto H, Egawa B, Horiguchi K, et al. Changes in CSF tryptophan metabolite levels in infantile spasms. No To Hattatsu 1992; 24: 530–535.

169. Yamamoto H. Studies on CSF tryptophan metabolism in infantile spasms. Pediatr Neurol 1991; 7: 411–414.

170. Giroud M, Dumas R, Dauvergne M, et al. 5-Hydroxyindoleacetic acid and homovanillic acid in cerebrospinal fluid of children with febrile convulsions. Epilepsia 1990; 31: 178–181.

171. Verdot L, Garreau B, Barthelemy C, et al. Immunoreactivity of sera to a peptide derived from the serotonin 5-ht1a receptor in a group of children with developmental disorders: possible role in non-autistic epilepsy. Int J Mol Med 1998; 1: 185–189.

172. Toczek MT, Carson RE, Lang L, et al. PET imaging of 5-HT1A receptor binding in patients with temporal lobe epilepsy. Neurology 2003; 60: 749–756.

173. Zoghbi HY, Percy AK, Glaze DG, et al. Reduction of biogenic amine levels in the Rett syndrome. N Engl J Med 1985; 313: 921–924.

174. Langlais PJ, Wardlow ML, Yamamoto H. Changes in CSF neurotransmitters in infantile spasms. Pediatr Neurol 1991;7:440–445.

175. Briere R, Sherwin AL, Robitaille Y, et al. Alpha-1 adrenoceptors are decreased in human epileptic foci. Ann Neurol 1986; 19: 26–30.

176. Sherwin A, Robitaille Y, Quesney L, et al. Noradrenergic abnormaities in human cortical seizure foci. In: Engel J, ed. Fundamental Mechanisms of Human Brain Function. New York: Raven Press, 1987: 249.

177. Pacia SV, Doyle WK, Broderick PA. Biogenic amines in the human neocortex in patients with neocortical and mesial temporal lobe epilepsy: identification with in situ microvoltammetry. Brain Res 2001; 899: 106–111.

178. Merlet I, Ostrowsky K, Costes N, et al. 5-HT1A receptor binding and intracerebral activity in temporal lobe epilepsy: an [18F]MPPF-PET study. Brain 2004; 127: 900–913.

179. Theodore WH. Epilepsy and depression: imaging potential common factors. Clin EEG Neurosci 2004; 35: 38–45.

180. Clough RW, Peterson BR, Steenbergen JL, et al. Neurite extension of developing noradrenergic neurons is impaired in genetically epilepsy-prone rats (GEPR-3s): an in vitro study on locus coeruleus. Epilepsy Res 1998; 29: 135–146.

181. Ryu JR, Jobe PC, Milbrandt JC, et al. Morphological deficits in noradrenergic neurons in GEPR-9s stem from abnormalities in both the locus coeruleus and its target tissues. Exp Neurol 1999; 156: 84–91.

182. Yan QS, Jobe PC, Dailey JW. Thalamic deficiency in norepinephrine release detected via intracerebral microdialysis: a synaptic determinant of seizure predisposition in the genetically epilepsy-prone rat. Epilepsy Res 1993; 14: 229–236.

183. Dailey JW, Mishra PK, Ko KH, et al. Serotonergic abnormalities in the central nervous system of seizure-naive genetically epilepsy-prone rats. Life Sci 1992; 50: 319–326.

184. Owens MJ, Ballenger CA, Knight DL, Nemeroff CB. Platelet 5-hydroxytryptamine (5-HT) transporter and 5-HT2A receptor binding after chronic hypercorticosteronemia, (+/-)-1-(2,5-dimethoxy-4-iodophenyl)-2-aminopropane administration or neurotoxin-induced depletion of central nervous system 5-HT in the rat. J Pharmacol Exp Ther 1996; 278: 1040–1049.

185. Chien YH, Chiang SC, Huang A, et al. Treatment and outcome of Taiwanese patients with 6-pyruvoyltetrahydropterin synthase gene mutations. J Inherit Metab Dis 2001; 24: 815–823.

186. Horwath E, Johnson J, Klerman GL, Weissman MM. Depressive symptoms as relative and attributable risk factors for first-onset major depression. Arch Gen Psychiatry 1992; 49: 817–823.

187. Long TD, Kathol RG. Critical review of data supporting affective disorder caused by nonpsychotropic medication. Ann Clin Psychiatry 1993; 5: 259–270.

188. Kovacs DA, Zoll JG. Seizure inhibition by median raphe nucleus stimulation in rat. Brain Res 1974; 70: 165–169.

189. Buterbaugh GG. Effect of drugs modifying central serotonergic function on the response of extensor and nonextensor rats to maximal electroshock. Life Sci 1978; 23: 2393–2404.

190. Loscher W, Czuczwar SJ. Evaluation of the 5-hydroxytryptamine receptor agonist 8-hydroxy- 2-(di-n-propylamino)tetralin in different rodent models of epilepsy. Neurosci Lett 1985; 60: 201–206.

191. Prendiville S, Gale K. Anticonvulsant effect of fluoxetine on focally evoked limbic motor seizures in rats. Epilepsia 1993; 34: 381–384.

192. Meldrum BS, Balzamo E, Wada JA, Vuillon-Cacciuttolo G. Effects of l-tryptophan, l-3, 4, dihydroxyphenylalanine and tranylcypromine on the electroencephalogram and on photically induced epilepsy in the baboon, Papio papio. Physiol Behav 1972; 9: 615–621.

193. Wada JA, Balzamo E, Meldrum BS, Naquet R. Behavioral and electrographic effects of l-5-hydroxytryptophan and d,l-parachlorophenyl-alanine on epileptic Senegalese baboon (Papio papio). Electroencephalogr Clin Neurophysiol 1972; 33: 520–526.

194. Lam RW, Tam EM, Grewal A, Yatham LN. Effects of alpha-methyl-para-tyrosine-induced catecholamine depletion in patients with seasonal affective disorder in summer remission. Neuropsychopharmacology 2001; 25: S97–S101.

195. Miller HL, Delgado PL, Salomon RM, et al. Clinical and biochemical effects of catecholamine depletion on antidepressant-induced remission of depression. Arch Gen Psychiatry 1996; 53: 117–128.

196. Miller HL, Delgado PL, Salomon RM, et al. Effects of alpha-methyl-para-tyrosine (AMPT) in drug-free depressed patients. Neuropsychopharmacology 1996; 14: 151–157.

197. Neumeister A, Turner EH, Matthews JR, et al. Effects of tryptophan depletion vs catecholamine depletion in patients with seasonal affective disorder in remission with light therapy. Arch Gen Psychiatry 1998; 55: 524–530.

198. Jobe PC, Picchioni AL, Chin L. Effect of lithium carbonate and alphamethyl-p-tyrosine on audiogenic seizure intensity [Letter]. J Pharm Pharmacol 1973; 25: 830–831.

199. Anand A, Darnell A, Miller HL, et al. Effect of catecholamine depletion on lithium-induced long-term remission of bipolar disorder. Biol Psychiatry 1999; 45: 972–978.

200. Salomon RM, Miller HL, Delgado PL, Charney D. The use of tryptophan depletion to evaluate central serotonin function in depression and other neuropsychiatric disorders. Int Clin Psychopharmacol 1993; 8: 41–46.

201. Reilly JG, McTavish SF, Young AH. Rapid depletion of plasma tryptophan: a review of studies and experimental methodology. J Psychopharmacol 1997; 11: 381–392.

202. Young SN, Pihl RO, Benkelfat C, et al. The effect of low brain serotonin on mood and aggression in humans. Influence of baseline mood and genetic factors. In: Filippini GA, Costa CVL, Bertazzo A, eds. Recent Advances in Tryptophan Research. Tryptophan and Serotonin Pathways. New York: Plenum Press, 1996: 45.

203. Jobe PC, Picchioni AL, Chin L. Role of brain 5-hydroxytryptamine in audiogenic seizure in the rat. Life Sci 1973; 13: 1–13.

204. Wang C, Mishra PK, Dailey JW, et al. Noradrenergic terminal fields as determinants of seizure predisposition in GEPR-3s: a neuroanatomic assessment with intracerebral microinjections of 6-hydroxydopamine. Epilepsy Res 1994; 18: 1–9.

205. Statnick MA, Maring-Smith ML, Clough RW, et al. Effect of 5,7-dihydroxytryptamine on audiogenic seizures in genetically epilepsy-prone rats. Life Sci 1996; 69: 1763–1771.

206. Clough R, Statnick M, Maringsmith M, et al. Fetal raphe transplants reduce seizure severity in serotonin-depleted GEPRS. NeuroReport 1996; 8: 341–346.

207. Clough RW, Browning RA, Maring ML, et al. Effects of intraventricular locus coeruleus transplants on seizure severity in genetically epilepsy-prone rats following depletion of brain norepinephrine. J Neural Transplant Plast 1994; 5: 65–79.

208. Shaw K, Turner J, Del Mar C. Are tryptophan and 5-hydroxytryptophan effective treatments for depression? A meta-analysis. Aust NZ J Psychiatry 2002; 36: 488–491.

209. Ko KH, Dailey JW, Jobe PC. Effect of increments in norepinephrine concentrations on seizure intensity in the genetically epilepsy-prone rat. J Pharmacol Exp Ther 1982; 222: 662–669.

210. Jobe PC, Woods TW, Dailey JW. Proconvulsant and anticonvulsant effects of tricyclic antidepressants in genetically epilepsy-prone rats. Epilepsy International Symposium Abstracts 1983: 229.

211. Dailey JW, Jobe PC. Anticonvulsant drugs and the genetically epilepsy-prone rat. Fed Proc 1985; 44: 2640–2644.

212. Dailey JW, Yan QS, Mishra PK, et al. Effects of fluoxetine on convulsions and on brain serotonin as detected by microdialysis in genetically epilepsy-prone rats. J Pharmacol Exp Ther 1992; 260: 533–540.

213. Dailey JW, Yan QS, Adams-Curtis LE, et al. Neurochemical correlates of antiepileptic drugs in the genetically epilepsy-prone rat (GEPR). Life Sci 1996; 58: 259–266.

214. Yan QS, Jobe PC, Dailey JW. Evidence that a serotonergic mechanism is involved in the anticonvulsant effect of fluoxetine in genetically epilepsy-prone rats. Eur J Pharmacol 1993; 252: 105–112.

215. Yan QS, Jobe PC, Dailey JW. Further evidence of anticonvulsant role for 5-hydroxytryptamine in genetically epilepsy-prone rats. Br J Pharmacol 1995; 115: 1314–1318.

216. Yan QS, Jobe PC, Cheong JH, et al. Role of serotonin in the anticonvulsant effect of fluoxetine in genetically epilepsy-prone rats. Naunyn Schmiedebergs Arch Pharmacol 1994; 350: 149–152.

217. Browning RA, Statnick MA, Dailey JW, Jobe PC. Effect of intranigral fluoxetine on audiogenic seizures in genetically epilepsy-prone rats. Society for Neuroscience Abstracts 1993; 19: 1468.

218. Browning RA, Wood AV, Merrill MA, et al. Enhancement of the anticonvulsant effect of fluoxetine following blockade of 5-ht1a receptors. Eur J Pharmacol 1997; 336: 1–6.

219. Dailey JW, Naritoku DK. Antidepressants and seizures: clinical anecdotes overshadow neuroscience. Biochem Pharmacol 1996; 52: 1323–1329.

220. Meldrum BS, Anlezark GM, Adam HK, Greenwood DT. Anticonvulsant and proconvulsant properties of viloxazine hydrochloride: pharmacological and pharmacokinetic studies in rodents and the epileptic baboon. Psychopharmacology (Berl) 1982; 76: 212–217.

221. Lange SC, Julien RM, Fowler GW. Biphasic effects of imipramine in experimental models of epilepsy. Epilepsia 1976; 17: 183–195.

222. Polc P, Schneeberger J, Haefely W. Effects of several centrally active drugs on the sleep–wakefulness cycle of cats. Neuropharmacology 1979; 18: 259–267.

223. Yanagita T, Wakasa Y, Kiyohara H. Drug-dependence potential of viloxazine hydrochloride tested in rhesus monkeys. Pharmacol Biochem Behav 1980; 12: 155–161.

224. Buge A, Poisson M, Rancurel G, et al. Un cas rebelle de petit mal. Action de la clomipramine. Nouv Presse Med 1974; 3: 373–375.

225. Favale E, Rubino V, Mainardi P, et al. Anticonvulsant effect of fluoxetine in humans. Neurology 1995; 45: 1926–1927.

226. Fromm GH, Rosen JA, Amores CY. A clinical and experimental investigation of the effect of imipramine on epilepsy. Epilepsia 1971; 12: 282–283.

227. Fromm GH, Wessel HB, Glass JD, et al. Imipramine in absence and myoclonic-astatic seizures. Neurology 1978; 28: 953–957.

228. Fromm GH, Amores CY, Thies W. Imipramine in epilepsy. Arch Neurol 1972; 27: 198–204.

229. Ojemann LM, Friel PN, Trejo WJ, Dudley DL. Effect of doxepin on seizure frequency in depressed epileptic patients. Neurology 1983; 33: 646–648.

230. Ojemann LM, Baugh-Bookman C, Dudley DL. Effect of psychotropic medications on seizure control in patients with epilepsy. Neurology 1987; 37: 1525–1527.

231. Pineda MR, Russell SC. The use of a tricyclic antidepressant in epilepsy. Dis Nerv Syst 1974; 35: 322–323.

232. Favale E, Audenino D, Cocito L, Albano C. The anticonvulsant effect of citalopram as an indirect evidence of serotonergic impairment in human epileptogenesis. Seizure 2003; 12: 316–318.

233. Specchio LM, Iudice A, Specchio N, et al. Citalopram as treatment of depression in patients with epilepsy. Clin Neuropharmacol 2004; 27: 133–136.

234. Gigli GL, Diomedi M, Troisi A, et al. Lack of potentiation of anticonvulsant effect by fluoxetine in drug-resistant epilepsy. Seizure 1996; 3: 221–224.

235. Hovorka J, Herman E, Nemcova I, I. Treatment of interictal depression with citalopram in patients with epilepsy. Epilepsy Behav 2000; 1: 444–447.

236. Jobe PC, Picchioni AL, Chin L. Role of brain norepinephrine in audiogenic seizure in the rat. J Pharmacol Exp Ther 1973; 184: 1–10.

237. Yan QS, Dailey JW, Steenbergen JL, Jobe PC. Anticonvulsant effect of enhancement of noradrenergic transmission in the superior colliculus in genetically epilepsy-prone rats (GEPRs) – a microinjection study. Brain Res 1998; 780: 199–209.

238. Yan QS, Jobe PC, Dailey JW. Noradrenergic mechanisms for the anticonvulsant effects of desipramine and yohimbine in genetically epilepsy-prone rats – studies with microdialysis. Brain Res 1993; 610: 24–31.

239. Preskorn SH, Fast GA. Tricyclic antidepressant-induced seizures and plasma drug concentration. J Clin Psychiatry 1992; 53: 160–162.

240. Jobe PC, Mishra PK, Dailey JW. Genetically epilepsy-prone rats: actions of antiepileptic drugs and monoaminergic neurotransmitters. In: Faingold C, Fromm GH, eds. Drugs for the Control of Epilepsy: Actions on

Neuronal Networks Involved in Seizure Disorders. Boca Raton, FL: CRC Press, 1992: 253.

241. Hauser WA. Incidence and prevalence. In: Engel J Jr, Pedley TA, eds. Epilepsy: A Comprehensive Textbook, vol 1. Philadelphia: Lippincott-Raven, 1997: 47.

242. Mitchell PB, Malhi GS. The expanding pharmacopoeia for bipolar disorder. Annu Rev Med 2002; 53: 173–188.

243. Yan QS, Mishra PK, Burger RL, et al. Evidence that carbamazepine and antiepilepsirine may produce a component of their anticonvulsant effects by activating serotonergic neurons in genetically epilepsy-prone rats. J Pharmacol Exp Ther 1992; 261: 652–659.

244. Dailey JW, Reith MEA, Yan QS, et al. Anticonvulsant doses of carbamazepine increase hippocampal extracellular serotonin in genetically epilepsy-prone rats: dose response relationships. Neurosci Lett 1997; 227: 13–16.

245. Dailey JW, Reith ME, Steidley KR, et al. Carbamazepine-induced release of serotonin from rat hippocampus in vitro. Epilepsia 1998; 39: 1054–1063.

246. Dailey JW, Reith ME, Yan QS, et al. Carbamazepine increases extracellular serotonin concentration: lack of antagonism by tetrodotoxin or zero Ca2+. Eur J Pharmacol 1997; 328: 153–162.

247. Southam E, Kirkby D, Higgins GA, Hagan RM. Lamotrigine inhibits monoamine uptake in vitro and modulates 5-hydroxytryptamine uptake in rats. Eur J Pharmacol 1998; 358: 19–24.

248. Dailey JW, Seo D, Yan QS, et al. The anticonvulsant effect of the broad spectrum anticonvulsant loreclezole may be mediated in part by serotonin in rats: a microdialysis study. Neurosci Lett 1994; 178: 179–183.

249. Whitton PS, Fowler LJ. The effect of valproic acid on 5-hydroxytryptamine and 5-hydroxyindoleacetic acid concentration in hippocampal dialysates in vivo. Eur J Pharmacol 1991; 200: 167–169.

250. Kaneko S, Okada M, Hirano T, et al. Carbamazepine and zonisamide increase extracellular dopamine and serotonin levels in vivo, and carbamazepine does not antagonize adenosine effect in vitro: mechanisms of blockade of seizure spread. Jpn J Psychiatry Neurol 1993; 47: 371–373.

251. Browning RA, Lanker ML, Faingold CL. Injections of noradrenergic and GABAergic agonists into the inferior colliculus: effects on audiogenic seizures in genetically epilepsy-prone rats. Epilepsy Res 1989; 4: 119–125.

252. Chakravarty DN, Faingold CL. Increased responsiveness and failure of habituation in neurons of the external nucleus of inferior colliculus associated with audiogenic seizures of the genetically epilepsy-prone rat. Exp Neurol 1996; 141: 280–286.

253. Doretto MC, Burger RL, Mishra PK, et al. A microdialysis study of amino acid concentrations in the extracellular fluid of the substantia nigra of freely behaving GEPR-9s – relationship to seizure predisposition. Epilepsy Res 1994; 17: 157–165.

254. Faingold CL, Randall ME, Anderson CA. Blockade of GABA uptake with tiagabine inhibits audiogenic seizures and reduces neuronal firing in the inferior colliculus of the genetically epilepsy-prone rat. Exp Neurol 1994; 126: 225–232.

255. Faingold CL, Gehlbach G, Caspary DM. Decreased effectiveness of GABA-mediated inhibition in the inferior colliculus of the

genetically epilepsy-prone rat. Exp Neurol 1986; 93: 145–159.

256. Faingold CL. The genetically epilepsy-prone rat. Gen Pharmacol 1988; 19: 331–338.

257. Lasley SM, Yan QS. Diminished potassium-stimulated GABA release in vivo in genetically epilepsy-prone rats. Neurosci Lett 1994; 175: 145–148.

258. Lasley SM. Roles of neurotransmitter amino acids in seizure severity and experience in the genetically epilepsy-prone rat. Brain Res 1991; 560: 63–70.

259. Molnar LR, Fleming WW, Taylor DA. Alterations in neuronal gamma-aminobutyric acid(A) receptor responsiveness in genetic models of seizure susceptibility with different expression patterns. J Pharmacol Exp Ther 2000; 295: 1258–1266.

260. Ribak CE, Lauterborn JC, Navetta MS, Gall CM. The inferior colliculus of GEPRs contains greater numbers of cells that express glutamate decarboxylase (GAD67) mRNA. Epilepsy Res 1993; 14: 105–113.

261. Roberts RC, Ribak CE. Anatomical changes of the GABAergic system in the inferior colliculus of the genetically epilepsy-prone rat. Life Sci 1986; 39: 789–798.

262. Roberts RC, Ribak CE, Oertel WH. Increased numbers of gabaergic neurons occur in the inferior colliculus of an audiogenic model of genetic epilepsy. Brain Res 1985; 361: 324–338.

263. Bernard C, Cossart R, Hirsch JC, et al. What is GABAergic inhibition? How is it modified in epilepsy? Epilepsia 2000; 41: S90–S95.

264. Bradford HF. Glutamate, GABA and epilepsy. Prog Neurobiol 1995; 47: 477–511.

265. De Deyn PP, Marescau B, Macdonald RL. Epilepsy and the GABA-hypothesis a brief review and some examples. Acta Neurol Belg 1990; 90: 65–81.

266. di Michele F, Verdecchia M, Dorofeeva M, et al. GABA(A) receptor active steroids are altered in epilepsy patients with tuberous sclerosis. J Neurol Neurosurg Psychiatry 2003; 74: 667–670.

267. Jaeken J. Genetic disorders of gamma-aminobutyric acid, glycine, and serine as causes of epilepsy. J Child Neurol 2002; 17: 3S84–3S87.

268. Aldenkamp AP, Vermeulen J, Mulder OG, et al. Gamma-vinyl GABA (vigabatrin) and mood disturbances. Epilepsia 1994; 35: 999–1004.

269. Bjork JM, Moeller FG, Kramer GL, et al. Plasma GABA levels correlate with aggressiveness in relatives of patients with unipolar depressive disorder. Psychiatry Res 2001; 101: 131–136.

270. Papadimitriou GN, Dikeos DG, Karadima G, et al. Association between the GABA(A) receptor alpha5 subunit gene locus (GABRA5) and bipolar affective disorder. Am J Med Genet 1998; 81: 73–80.

6

Depression in stroke: cause and consequence of brain dysfunction

Andres M Kanner and John J Barry

INTRODUCTION

There is no topic that is more fitting for a book on *Depression and Brain Dysfunction* than that of depression and stroke, as depression is not only a relatively frequent complication of stroke but has been identified in recent studies as a risk factor for stroke occurrence, poor recovery of cognitive and neurological functions and higher mortality risk[1]. Indeed, recent data clearly demonstrate a bidirectional relationship between stroke and depression. The purpose of this chapter is to review in some detail both sides of this relationship. First, we examine the impact of stroke on the risk of depression occurrence, for which we review the available data on the prevalence, clinical manifestations, potential pathogenic mechanisms and treatment of post-stroke depression (PSD). Second, we review the data suggesting that depression may be a risk factor for stroke, an obstacle to recovery of neurological functions and a higher mortality risk.

Bidirectional relationships between mood and neurological disorders are not restricted to stroke and depression, but can be identified also between depression and epilepsy and depression and Parkinson's disease (PD)[2–4]. Not only are patients with epilepsy and PD at higher risk of suffering from depression, but patients with depression have a four–six-fold higher risk of developing epilepsy and a 2.2–fold higher risk of developing PD.

THE IMPACT OF STROKE ON DEPRESSION

Prevalence of post-stroke depression

Depressive disorders are a common complication of stroke. Several cross-sectional studies have identified PSD in 30–50% of patients following cerebral infarction[1,5–7]. Huff *et al.* found that the prevalence of PSD peaks 3–6 months after stroke[8]. Prevalence rates vary, depending on the patient population studied. In a recent review article, Robinson[1] calculated a pooled prevalence rate of all types of depression of 31.8% (range 30–44%) from four community-based studies[7,9–11]. The prevalence rates from studies carried out in acute hospitals ranged from 25 to 47%[6,12–14], while in rehabilitation hospitals prevalence rates ranged from 35 to 72%[5,15–18].

PSD can present as major depression or meet criteria of minor depression. In community-based studies, the prevalence rates of major depression ranged from 11 to 15%, while those of minor depression ranged from 8 to 12%[7,9–11]. In acute hospital-based studies the prevalence of major depression ranged from 10 to 27% and minor depression between 11 and 20%[6,12–14], while in rehabilitation-based studies the prevalence rates of major depression ranged from 10 to 40% and of minor depression from 21 to 44%[5,15–18].

Clinical manifestations of post-stroke depression

Two major categories of depressive disorders have been described in patients with stroke. The most frequently recognized is PSD that, as stated above, can present as a major depression, dysthymia or minor depression. The second type, referred to as vascular depression, was described by several investigators as a late-onset (after the age of 65) depressive disorder identified in patients who may have had overt or silent stroke(s) or subcortical bilateral white matter ischemic disease. We will describe the two types of depression separately.

Post-stroke depression

In general, the clinical manifestations of PSD are similar to those of idiosyncratic late-onset depression, with the caveat that psychomotor retardation may be more frequently identified among patients with PSD. In fact, Lipsey *et al.* found that the presence of slowness/psychomotor retardation in PSD patients were the only differentiating symptoms from idiopathic depressed patients, who, in turn, reported more anhedonia and more difficulty with concentration[19]. The severity of PSD correlates with that of impairment of activities of daily living (ADL), during both its acute and its chronic phases. In a study of 142 stroke patients followed over a 24-month period, Federoff *et al.* found that patients with PSD were younger than those who never experienced it[6]. Gainotti *et al.* have also suggested that patients with PSD are more likely to present catastrophic reactions, hyperemotionalism and diurnal mood variation than patients with idiopathic depression, although these findings have not been confirmed by other investigators[20].

A cautionary note is in order: vegetative symptoms (i.e. changes in sleep, appetite, sexual drive) and fatigue are common symptoms in depressive and (acute or chronic) neurological disorders, including stroke. Thus, clinicians must interpret with great caution the diagnostic significance of these symptoms in the absence of overt symptoms of depression (e.g. anhedonia, crying bouts, hopelessness, helplessness). On the other hand, the presence of a depressive disorder can enhance the severity of vegetative symptoms. In fact, in the study by Federoff *et al.* cited above, vegetative symptoms consisting of disturbances of sleep, libido and level of energy were significantly more frequent among depressed than non-depressed patients at initial evaluation, and at 3, 6, 12 and 24 months[6]. This observation bears significant implications with respect to the way we interpret self-rating scales, as a false-positive diagnosis of depression may result when relying solely on diagnostic instruments that have a relatively large number of items identifying vegetative symptoms, such as the Beck Depression Inventory and the Hamilton Depression Rating Scale[21,22].

Several studies have shown that the course of PSD can be rather lengthy. For example, symptoms of major depression identified in 27% of patients who had suffered a stroke persisted for approximately 1 year, while symptoms of minor depression identified in 20% of patients lasted for more than 2 years[16,23]. In a recent study of 100 patients followed for 18 months, Berg *et al.* identified symptoms of PSD at follow-up visits: 46% of patients were depressed during the first 2 months while only 12% of patients experienced their first symptoms 12 months after their stroke[24]. Among the patients with early-onset PSD, symptoms of depression had persisted 12 and 18 months later.

Duration of PSD symptoms appears to depend on the vascular territory of the stroke. Starkstein *et al.*, for example, found that PSD in patients with a stroke in the territory of the middle cerebral artery (MCA) was longer in duration than when the stroke involved the posterior circulation. At 6 months of follow-up, 82% of patients with MCA stroke continued to be symptomatic, whereas 20% of those with posterior circulation strokes did not. At 12 and 24 months, none of the latter exhibited symptoms, but 62% of the former did[25].

Relationship of post-stroke depression to cognitive and motor deficits

Two studies have demonstrated a significant correlation between severity of disability and depression[26,27], but this association has been inconsistent in studies in which the post-stroke follow-up evaluation time was longer[28]. These data suggest that disability may cause a reactive depressive process in the early stages after stroke, but probably does not mediate the development of PSD in the long term. In a review of the literature, Robinson and Szetela concluded that the severity of impairment in ADL had a limited impact on PSD[29].

With respect to the relationship between aphasia and PSD, Robinson and Benson observed a significantly higher prevalence among patients with non-fluent aphasia but not among patients with fluent aphasia[30].

Vascular depression

Alexopoulos *et al.* have described the existence of a clinically defined depressive disorder consisting of late-onset symptoms of depression in patients with clinical and/or neuroradiological evidence of diffuse bilateral white-matter small vessel disease who present chronic cerebrovascular risk factors such as hypertension, diabetes, carotid stenosis, atrial fibrillation and hyperlipidemia[31]. These patients may or may not have experienced an overt stroke or may have reported only a transient ischemic attack. In a recently published study, Mast *et al.* compared the clinical characteristics of PSD and vascular depression in 670 geriatric rehabilitation patients who were grouped into those with overt stroke, those with cerebrovascular risk factors but without overt stroke, and those without stroke or cerebrovascular risk factors. There was no difference in the rate of depression among patients with stroke (36.4%) and cerebrovascular risk factors (35.2%). The frequency of depression increased significantly with the burden of risk factors, but this was not observed among patients with overt stroke[32].

Clinically, vascular depression appears after the age of 65; in patients with a history of depression, however, there is a change in the course of their symptoms. Patients with vascular depression are more likely to present neuropsychological disturbances with impairment of executive functions, a greater tendency to psychomotor retardation, poor insight and a less frequent family history of depression. We must emphasize, however, that this form of depression has not been universally recognized or accepted.

Potential pathogenic factors that mediate the occurrence and severity of post-stroke depression

The various factors investigated in the occurrence of PSD have included location of the stroke, size of the lesion, timing of the onset of depression relative to the occurrence of stroke and size of the ventricles. Some investigators have also suggested that the same cardiovascular factors that increase the risk of stroke also increase the risk of depression[1].

In a community-based, prospective epidemiological study, Whyte *et al.* investigated the role of cerebrovascular risk factors (hypertension, diabetes, atherosclerotic heart disease, tabacism), number of impaired ADL and psychiatric history, in their ability to predict symptoms of depression 2 years after a stroke among 1134 subjects aged 65 or older living in a rural community in southern Pennsylvania[33]. The presence of symptoms of depression was defined as the identification of at least five symptoms on the modified Center for Epidemiological Studies Depression scale (CES-D). Subjects with a history of stroke were six times more likely to report symptoms of depression 2 years later. This association was independent of functional disability, age, sex, education and previous depressive symptoms and any of the cardiovascular risk factors. The epidemiological nature of this study suggests that the data are representative of a generalized (and not hospital-based) population. Of note is their finding of an absent association between level of disability and symptoms of depression at 2 years.

The relationship between location of stroke and PSD has been the most extensively debated by multiple groups of investigators. The available data reveal contradictory findings among the different studies, however. Singh *et al.* addressed this question in a review of data from 12 published studies and found no relationship between location of stroke and PSD in six studies, a higher likelihood of PSD in right sided-stroke in two studies and the reverse in four studies[12]. The relationship between location of stroke and PSD was first suggested in 1984 by Robinson *et al.*, who reported a higher occurrence of left hemispheric stroke among patients with PSD, more often involving left frontal dorsolateral cortical regions[34]. They also found that strokes in left (but not right) basal ganglia were significantly associated with PSD. Robinson and Szetela had also observed an inverse relationship between the severity of PSD and proximity of the stroke to the frontal pole[29]. These findings have been replicated by other authors in only left hemispheric stroke[5], while others found the same relationship among patients with left- and right-sided stroke[35,36]. However, some authors have not found a relationship between location of stroke and PSD[27].

After completing a review of the PSD literature, Cummings and Mega concluded that the relationship between location of stroke and PSD was tied to timing of the onset of depression symptoms after the stroke[37]. Thus, in the first 10 days following a stroke, there is a higher frequency of left- than right-hemisphere lesions, while this relationship disappears if the onset of symptoms takes place 3 months after the stroke. Furthermore, these authors reported that, when symptoms of depression appear more than 1 year after the stroke, right-sided lesions are more frequent.

The presence of cortical atrophy and enlarged ventricles has been suggested as a potential risk factor for PSD. Starkstein *et al.* compared the magnitude of subcortical atrophy in brain computed tomography (CT) studies obtained immediately after the stroke[38]. Patients with PSD had a significantly greater degree of atrophy than those without depression. Finally, small subcortical lesions of the left hemisphere are associated with a higher frequency of depression than right-sided subcortical lesions[25]. In terms of duration of depression, patients with subcortical, cerebellar or brain-stem lesions have briefer depressions than patients with MCA infarctions[26].

THE OTHER SIDE OF THE COIN: THE IMPACT OF DEPRESSION ON THE RISK OF STROKE OCCURRENCE, THE RECOVERY OF NEUROLOGICAL IMPAIRMENTS AND MORTALITY

When addressing this important question, we have to examine whether depression has a direct or indirect impact on the development of stroke. In other words, does depression increase the risk of stroke, *per se*, or does it have an impact on the development of vascular disease, in general? If so, is this process mediated by a direct injury to the vessels (e.g. through an inflammatory reaction) or by increasing the risk factors that facilitate the development of vascular disease such as diabetes, obesity, smoking and hypertension?

Does depression increase the risk of stroke occurrence?

Five studies have investigated this question in large cohorts ranging from 1703 to 6675 subjects[39–43]. Four studies found that depression increased the risk of developing stroke, after controlling for other risk factors of stroke[39,40,42,43]. Of note is the study by Larson *et al.*[40]; these investigators followed 1703 subjects for a period of 13 years. The diagnosis of depression was established with the Diagnostic Interview Schedule. Patients with a depressive disorder or depressive symptoms had a 2.67 (95% CI 1.08–6.63) relative risk of developing a stroke, after controlling for vascular risk factors (hypertension, diabetes, hyperlipidemia, heart disease and use of tobacco). In a separate study, May *et al.* followed 2201 men aged 45–59 years for 14 years[39]. Patients with scores of > 5 on the General Hospital Questionnaire, indicative of significant symptoms of depression, had a 3.36 (95% CI 1.29–8.71) relative risk of developing a fatal stroke.

Does depression increase the risk and/or severity of vascular disease?

In a recently published epidemiological longitudinal study of 93 676 postmenopausal women followed for an average of 4.1 years, Wassertheil-Smoller *et al.* investigated the impact of symptoms of depression on risks of cardiovascular disease[44]. Symptoms of depression were identified with a short form of the CES-D, and a history of depression was established with two items from the Diagnostic Interview Schedule. Overall, 15.8% of women reported current depressed mood and 12.3% had a history of depressed mood. Depression was significantly related to risk factors of cardiovascular disease and a history of cardiovascular morbidity. Among women with no history of cardiovascular disease, depression was an independent predictor of cardiovascular disease and all-cause mortality, while among those with a history of cardiovascular disease, depression was significantly associated with stroke.

Five other studies found that depression predicted a worse prognosis of coronary artery disease and/or cardiac mortality[45–52].

Depression can contribute to the development of vascular disease through four additional mechanisms: activation of the hypothalamic–pituitary–adrenocortical (HPA) axis, increased sympathoadrenal hyperactivity, vascular inflammation and facilitation of hypercoagulability. The activation of the hypothalamic–pituitary–adrenal axis causes elevation of cortisol, which promotes the development of atherosclerosis and causes sympathoadrenal hyperactivity. This results in increased secretion of catecholamines, which in turn leads to vasoconstriction, platelet activation and an elevated heart rate[52].

The vascular inflammatory reaction has been attributed to elevated plasma levels of inflammatory markers such as C-reactive protein (CRP), and pro-inflammatory cytokines, such as interleukin (IL)-1, and IL-6 all of which have been found to be elevated in patients with depression with and without vascular disease[53–57]. These cytokines are known to be secreted in the presence of increased stress, and Leonard has suggested that the increased levels of inflammatory molecules may be related to chronic psychological stress[56,57].

Conversely, the increase of these cytokines has been associated with the development of vascular depression through damage of the vascular endothelium in vessels in the central nervous system (CNS), although no relation has been established to date between the presence of cardiovascular disease and vascular depression.

The suspected role of depression in the development of a hypercoagulable state has been related to the HPA axis and sympathoadrenal hyperactivity, as both are known to stimulate the coagulation process. Indeed, hypercortisolism can cause elevation of factor VIII and Von Willebrand factor, and a decrease in fibrinolytic activity, while elevation in catecholamines is associated with an increase in both coagulation and fibrinolysis[58]. In addition, abnormal platelet function in untreated depressed patients has also been implicated in the hypercoagulability state. Depressed patients have been found to have platelet reactivity of up to 40% greater than in controls[59]. In addition, abnormal platelet function has been associated with abnormalities in platelets' serotonin receptors[60]. In fact, the exposure to selective serotonin reuptake inhibitors (SSRIs) has been shown to normalize platelet activation. In a study of 126 patients with cardiovascular disease, patients exposed to an SSRI had lower platelet activation by several measures than those who never took an SSRI[61,62]. Musselaman *et al.* also found that platelet activation was reduced to levels comparable to those in controls after 6 weeks of treatment with paroxetine[63].

Does depression increase the risk and/or severity of hypertension?

A review of the available literature yields mixed results on the impact of depression on hypertension. For example, in a prospective study of 508 adults followed for a 4-year period, Shinn *et al.* failed to find any association between depression and hypertension[64]. Two other prospective studies in which adults were followed for 20 and five years, respectively, revealed that patients with symptoms of depression at baseline were more likely to develop hypertension at follow-up[65,66].

Does depression increase the risk of cardiac arrhythmias?

Cardiac arrhythmias are associated with a higher mortality risk and are also a risk factor for stroke. Decreased heart rate variability has been blamed for the development of cardiac arrhythmias and sudden death, and has been attributed to increased sympathetic or decreased parasympathetic activities. Patients with depression have been found to have a higher risk of ventricular tachycardia[67]. In a study of 700 post-myocardial infarction patients, those with depression had significantly decreased heart rate variability relative to controls[68]. In a separate study, the presence of symptoms of depression was associated with a decrease of 30% of baroreflex sensitivity[69], while other studies revealed higher QT variability among depressed patients than among controls[70,71].

Does depression increase the risk and/or severity of diabetes?

It is well known that patients with psychiatric disorders have several risk factors for developing type-2 diabetes, including obesity and lack of exercise. Recent data, however, have revealed that depression may be an independent risk factor for the development of type-2 diabetes after controlling for confounding factors such as race, age, gender, socioeconomic status, level of education and body weight[72]. Furthermore, once patients develop diabetes, the presence of depression may be associated with worse glycemic controls, related to increased difficulty to adhere to a proper diet and worse compliance with hypoglycemic agents.

The depression-related biological mechanisms responsible for the greater risk of developing diabetes include an increased release of counter-regulatory hormones (glucocorticoids, growth hormone) and catecholamines, which counteract the hypoglycemic action of insulin, by raising levels of glucagon. These hormonal changes may be responsible, in turn, for the development of insulin resistance.

Clearly, these data provide evidence of the negative impact of depression on the development of stroke through various pathogenic mechanisms (Figure 6.1).

IMPACT OF POST-STROKE DEPRESSION ON THE QUALITY OF LIFE OF STROKE PATIENTS

The presence of PSD has an impact on the quality of life of post-stroke patients at three levels: (1) recovery of cognitive impairments; (2) recovery of ADL; and (3) mortality risks.

As stated above, a significant association has been established between the occurrence of

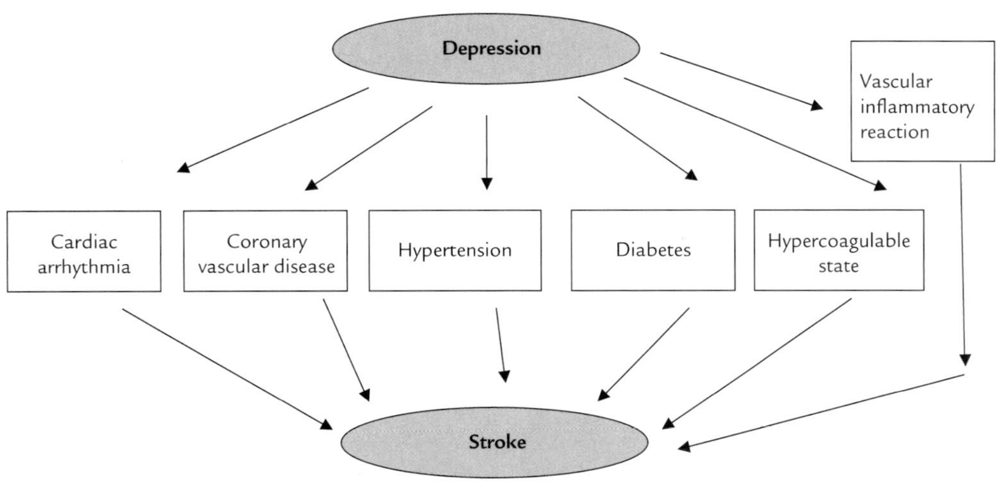

Figure 6.1 Potential pathogenic mechanisms mediating the increased risk of stroke in depressed patients

PSD and cognitive impairments. Starkstein *et al.* demonstrated that patients with major PSD had significantly more cognitive deficits than non-depressed patients with similar location and size of strokes in the left hemisphere. This association, however, was not demonstrated in right hemisphere strokes[38]. Robinson *et al.* have also shown that the presence of major PSD was associated with greater cognitive impairment 2 years after a stroke in a follow-up study of 140 patients[13].

The presence of PSD also has a negative impact on the recovery of ADL. Parikh *et al.*, for example, found that in-hospital PSD was the most important variable that predicted poor recovery in ADL over a 2-year period[73]. Of note, the score of in-hospital ADL was not associated with the 2-year recovery[73].

Of greater concern is the relationship between the presence of PSD and the higher mortality risk following stroke. Wade *et al.* showed that, among 976 patients followed for 1 year, those with PSD had 50% higher mortality than those without[11].

The impact of treatment on cognitive and ADL recovery has been investigated in several studies. Kimura *et al.* have shown that responders to treatment with antidepressant medication had higher scores in the Mini-Mental State Examination compared to non-responders[74]. Other double-blind controlled studies of citalopram and nortriptyline failed to show improvement in cognitive deficits, even in the presence of improvement of depression[19,75]. Clearly, more studies are necessary to clarify this point.

Chemerinski *et al.* found that successful treatment of PSD with nortriptyline was significantly associated with recovery in ADL[76,77]. Gainotti *et al.* compared the recovery of motor deficits and disability in 49 patients with PSD; 25 patients were treated with antidepressant medication and 24 were not. Patients who were not treated had a worse recovery than those treated and those without depression[78]. In addition, recovery from depression was greater in the treated patients.

Robinson *et al.* found that patients treated with fluoxetine or nortriptyline had an increased survival probability at 6 years (61%) compared to that of patients given placebo (34%)[75]. Jorge *et al.* also found that treatment with antidepressant medication was an independent predictor of increased survival[79].

Treatment of post-stroke depression

Several double-blind controlled studies have been carried out to assess the efficacy of SSRIs and tricyclic antidepressant drugs (TCA) in PSD. Fruehwald *et al.* recently compared the efficacy of 20 mg/day of fluoxetine to placebo in 54 patients with PSD[80]. Patients on both arms improved during the initial 4 weeks and no significant differences were found between the two treatments. However, by 12 weeks, patients on fluoxetine continued to improve while those on placebo had experienced recurrence of their symptoms of depression.

Robinson *et al.* compared fluoxetine at 40 mg/day, nortriptyline at 100 mg/day and placebo in 56 patients with PSD[75]. Patients on placebo and fluoxetine did not differ in their response to therapy, while those on nortriptyline showed a significant improvement. On the other hand, Wiast *et al.* found a superior response to fluoxetine at 10 mg/day than to placebo in 31 patients with PSD[81].

Lipsey *et al.* had demonstrated efficacy of nortriptyline in a double-blind, placebo-controlled study carried out in 34 patients, 14 of whom were on the TCA and 20 on placebo[19]. Patients on nortriptyline showed a significantly greater improvement in their depression after 6 weeks of therapy relative to those on placebo.

These data show that the treatment of PSD is effective in most studies using SSRIs and with the TCA nortriptyline. Successful treatment of PSD is associated with a better recovery of ADL and possibly of cognitive disturbances, and a lower mortality risk.

Prevention of PSD with antidepressant medication has been shown in two controlled studies. In one study, 137 patients were randomized to sertraline at doses of 50–150 mg/day and placebo within the first month after their stroke, and were treated for 1 year[82]. Among the treated patients, 8.3% developed a depressive disorder compared to 22.8% of untreated patients ($p = 0.037$). In a separate study, 48 non-depressed patients were randomized to receive fluoxetine, nortriptyline or placebo for a 3-month period[83]. Patients on placebo were more likely to develop a depressive episode than those treated with the antidepressant drugs ($p = 0.036$).

CONCLUDING REMARKS

The data reviewed in this chapter clearly indicate a bidirectional relationship between depression and stroke. Indeed, not only is stroke a risk factor for depression, but depression is a risk factor for stroke. Furthermore, the presence of PSD is associated with a worse recovery from neurological disabilities. The negative impact of PSD on stroke and its outcome as well as on the risk factors of co-morbid medical conditions (e.g. diabetes, coronary artery disease) that are commonly associated with stroke oblige us to identify its existence at its early stages and institute the proper pharmacological and supportive treatment. Sadly, depression remains under-recognized and undertreated by both clinicians and investigators of the treatment of stroke. Indeed, the presence of PSD is yet to be included in the variables associated with outcome in most of the randomized studies of stroke therapies. It is time that neurologists recognize depression as an expression of brain disease and not as a mere reactive process to an adverse situation.

REFERENCES

1. Robinson RG. Poststroke depression: Prevalence, diagnosis, treatment and disease progression. Biol Psychiatry 2003; 54: 376–387.
2. Hesdorffer DC, Hauser WA, Annegers JF, et al. Major depression is a risk factor for seizures in older adults. Ann Neurol 2000; 47: 246–249.
3. Nilsson FM, Kessing LV, Bowlig TG. Increased risk of developing Parkinson's disease for patients with major affective disorder: a register study. Acta Psychiatr Scand 2001; 104: 380–386.
4. Leentgens AFG, Van Der Akker M, Metsemakers JFM, et al. Higher incidence of depression preceding the onset of Parkinson's disease: a register study. Mov Disord 2003; 18: 414–418.
5. Eastwood MR, Rifat SL, Nobbs H, Ruderman J. Mood disorder following cerebrovascular accident. Br J Psychiatry 1989; 154: 195–200.
6. Fedoroff JP, Starkstein SE, Parikh RM, et al. Are depressive symptoms non-specific in patients with acute stroke? Am J Psychiatry 1991; 148: 1172–1176.
7. Burvill PW, Johnson GA, Jamrozik KD, et al. Prevalence of depression after stroke: the

Perth Community Stroke Study. Br J Psychiatry 1995; 166: 320–327.

8. Huff W, Steckel R, Sistzer M. Poststroke depression: risk factors and effects on the course of the stroke. Nervenarzt 2003; 74: 104–114.

9. House A, Dennis M, Mogridge L, et al. Mood disorders in the year after first stroke. Br J Psychiatry 1991; 158: 83–92.

10. Kotila M, Numminen H, Waltimo O, Kaste M. Depression after stroke. Results of the FINNSTROKE study. Stroke 1998; 29: 368–372.

11. Wade DT, Legh-Smith J, Hewer RA. Depressed mood after stroke, a community study of its frequency. Br J Psychiatry 1987; 151: 200–205.

12. Singh A, Black SE, Herrmann N, et al. Functional and neuroanatomic correlations in poststroke depression. The Sunnybrook Stroke Study. Stroke 2000; 31: 637–644.

13. Robinson RG, Starr LB, Kubos KL, Price TR. A two year longitudinal study of post-stroke mood disorders: findings during the initial evaluation. Stroke 1983; 14: 736–744.

14. Ebrahim S, Barer D, Nouri F. Affective illness after stroke. Br J Psychiatry 1987; 151: 52–56.

15. Folstein MF, Maiberger R, McHugh PR. Mood disorder as a specific complication of stroke. J Neurol Neurosurg Psychiatry 1977; 40: 1018–1020.

16. Morris PLP, Robinson RG, Raphael B. Prevalence and course of depressive disorders in hospitalized stroke patients. Int J Psychiatr Med 1990; 20: 349–364.

17. Schubert DSP, Taylor C, Lee S, et al. Physical consequences of depression in the stroke patient. Gen Hosp Psychiatry 1992; 14: 69–76.

18. Schwartz JA, Speed NM, Brunberg JA, et al. Depression in stroke rehabilitation. Biol Psychiatry 1993; 33: 694–699.

19. Lipsey JR, Robinson RG, Pearlson GD, et al. Nortriptyline treatment of post-stroke depression. A double-blind study. Lancet 1984; 1: 297–300.

20. Gainotti G, Azzoni A, Marra C. Frequency, phenomenology and anatomical–clinical correlates of major post-stroke depression. Br J Psychiatry 1999; 175: 163–167.

21. Hamilton MA. A rating scale for depression. J Neurol Neurosurg Psychiatry 1960; 23: 56–62.

22. Beck AT, Ward CH, Mendelson M, et al. An inventory for measuring depression. Arch Gen Psychiatry 1961; 4: 561–571.

23. Robinson RG, Price TR. Post-stroke depressive disorders: a follow-up study of 103 outpatients. Stroke 1982; 13: 635–641.

24. Berg A, Palomaki H, Letitihalmes M, et al. Post stroke depression: an 18-month follow-up. Stroke 2003; 34: 138–143.

25. Starkstein SE, Robinson RG, Berther ML, Price TR. Depressive disorders following posterior circulation as compared with middle cerebral artery infarcts. Brain 1988; 11: 375–387.

26. Starkstein SE, Moran TH, Bowersox JA, Robinson RG. Behavioral abnormalities induced by frontal cortical and nucleus accumbens lesions. Brain Res 1988; 473: 74–80.

27. Bolla-Wilson K, Robinson RG, Starkstein SE, et al. Lateralization of dementia of depression in stroke patients. Am J Psychiatry 1989; 146: 627–634.

28. Downhill J, Robinson RG. Longitudinal assessment of depression and cognitive impairment following stroke. J Nerv Ment Dis 1994; 182: 425–431.

29. Robinson RG, Szetela B. Mood changes following left hemisphere brain injury. Ann Neurol 1981; 91: 447–453.

30. Robinson RG, Benson DF. Depression in aphasic patients: frequency, severity and clinical pathological correlations. Brain Lang 1981; 14: 282–291.

31. Alexopoulos GS, Meyers BS, Young RC, et al. Vascular depression hypothesis. Arch Gen Psychiatry 1997; 54: 915–922.

32. Mast BT, MacNeill SE, Lichtenberg PA. Poststroke and clinically-defined vascular depression in geriatric rehabilitation patients. Am J Geriatr Psychiatry 2004; 12: 84–92.

33. Whyte EM, Mulsant BH, Vanderbilt J, et al. Depression after stroke: a prospective epidemiologic study. J Am Geriatr Soc 2004; 52: 774–778.

34. Robinson RG, Kubos KL, Starr LB, Price TR. Mood changes in stroke patients: Relation to lesion location. Brain 1984; 107: 81–93.

35. Sinyor D, Amato P, Kaloupek P. Post-stroke depression: relationship to functional impairment, coping strategies, and rehabilitation outcome. Stroke 1986; 17: 112–117.

36. House A, Dennis M, Warlow C, et al. The relationship between intellectual impairment and mood disorder in the first year after stroke. Psychol Med 1990; 20: 805–814.

37. Cummings JL, Mega JL. Disturbances of mood and affect: cerebrovascular disease. In: Cumming JL, Mega JL, eds. Neuropsychiatry and Behavioral Neuroscience.New York: Oxford University Press, 2003.

38. Starkstein SE, Robinson RG, Price TR. Comparison of patients with and without post-stroke major depression matched for age and location of lesion. Arch Gen Psychiatry 1988; 45: 247–252.

39. May M, McCarron P, Stansfeld S, et al. Does psychological distress predict the risk of ischemic stroke and transient ischemic attack? The Caerphilly Study. Stroke 2002; 33: 7–12.

40. Larson SL, Owens PL, Ford D, Eaton W. Depressive disorder, dysthymia, and risk of stroke. Thirteen-year follow-up from the Baltimore Epidemiological Catchment Area Study. Stroke 2001; 32: 1979–1983.

41. Colantonio A, Kasi SV, Ostfeld AM. Depressive symptoms and other psychosocial factors as predictors of stroke in the elderly. Am J Epidemiol 1992; 136: 884–894.

42. Everson SA, Roberts RE, Goldberg DE, Kaplan GA. Depressive symptoms and increased risk of stroke mortality over a 29-year period. Arch Intern Med 1998; 158: 1133–1138.

43. Jonas BS, Mussolino ME. Symptoms of depression as a prospective risk factor for stroke. Psychosom Med 2000; 62: 463–471.

44. Wassertheil-Smoller S, Applegate WB, Berge K, et al. Change in depression as a precursor of cardiovascular events. SHEP cooperative research group. Arch Intern Med 1996; 156: 553–561.

45. Lesperance F, Frasure-Smith N, Juneau M, Theroux P. Depression and 1-year prognosis in unstable angina. Arch Intern Med 2000; 160: 1354–1360.

46. Lesperance F, Frasure-Smith N, Talajic M. Major depression before and after myocardial infarction: its nature and consequences. Psychosom Med 1996; 58: 99–110.

47. Carney RM, Rich MW, Freedland KE, et al. Major depressive disorder predicts cardiac events in patients with coronary artery disease. Psychosom Med 1988; 50: 627–633.

48. Carney RM, Rich MW, Tevelde A, et al. Major depressive disorder in coronary artery disease. Am J Cardiol 1987; 60: 1273–1275.

49. Frasure-Smith N, Lesperance F, Talajic M. Depression following myocardial infarction. Impact on 6 month survival. JAMA 1993; 270: 1819–1825.

50. Frasure-Smith N, Lesperance F, Talajic M. Depression and 18-month prognosis after myocardial infarction. Circulation 1995; 91: 999–1005.

51. Connerney I, Shapiro PA, McLaughlin JS, et al. Relationship between depression and coronary artery bypass surgery and 12 month outcome. Lancet 2001; 358: 1766–1771.

52. Troxler RG, Sprague EA, Albanese RA, et al. The association of elevated plasma cortisol and early atherosclerosis as demonstrated by coronary angiography. Atherosclerosis 1977; 26: 151–162.

53. Mulvihill NT, Foley JB. Inflammation in acute coronary syndromes. Heart 2002; 87: 201–204.

54. Koenig W. Inflammation and coronary heart disease: an overview. Cardiol Rev 2001; 9: 31–35.

55. Thompson SG, Kienast J, Pyke SD, et al. Hemostatic factors and the risk of myocardial infarction or sudden death in patients with angina pectoris. European Concerted Action on Thrombosis and Disabilities Angina Pectoris Study Group. N Engl J Med 1995; 332: 635–641.

56. Leonard BE. Evidence for a biochemical lesion in depression. J Clin Psychiatry 2000; 61 (Suppl 6): 12–17.

57. Leonard BE. The immune system, depression and the action of antidepressants. Prog Neuropsychopharmacol Biol Psychiatry 2001; 25: 767–780.

58. Musselman DL, Tomer A, Manatunga AK, et al. Exaggerated platelet reactivity in major depression. Am J Psychiatry 1996; 153: 1313–1317.

59. Kuijpers PM, Hamulyak K, Strik JJ, et al. Beta-thromboglobulin and platelet factor 4 levels in post-myocardial infarction patients with major depression. Psychiatry Res 2002; 109: 207–210.

60. Mendelson SD. The current status of the platelet 5-HT(2A) receptor in depression. J Affect Disord 2002; 57: 13–24.

61. Serebruany VL, O'Connor CM, Gurbel PA. Effect of selective serotonin reuptake inhibitors on platelets in patients with coronary artery disease. Am J Cardiol 2001; 87: 1398–1400.

62. Serebruany VL, Gurbel PA, O'Connor CM. Platelet inhibition by sertraline and N-demethylsertraline: a possible missing link between depression, coronary events, and mortality benefits of selective serotonin reuptake inhibitors. Pharmacol Res 2001; 43: 453–462.

63. Musselman DL, Marzec UM, Manatunga A, et al. Platelet reactivity in depressed patients treated with paroxetine: preliminary findings. Arch Gen Psychiatry 2000; 57: 875–882.

64. Shinn EH, Poston WS, Kimball KT, et al. Blood pressure and symptoms of depression and anxiety: a prospective study. Am J Hypertens 2001; 14: 660–664.

65. Jonas BS, Lando JF. Negative affect as a prospective risk factor for hypertension. Psychosom Med 2000; 62: 188–196.

66. Davidson K, Jonas BS, Dixon KE, Markovitz JH. Do depression symptoms predict early hypertension incidence in

young adults in the CARDIA study? Coronary Artery Risk Development in Young Adults. Arch Intern Med 2000; 160: 1495–1500.

67. Carney RM, Freedland KE, Rich MW, et al. Ventricular tachycardia and psychiatric depression in patients with coronary artery disease. Am J Med 1993; 95: 23–28.

68. Carney RM, Blumenthal JA, Stein PK, et al. Depression, heart rate variability, and acute myocardial infarction. Circulation 2001; 104: 2024–2028.

69. Watkins LL, Grossman P. Association of depressive symptoms with reduced baroreflex cardiac control in coronary artery disease. Am Heart J 1999; 137: 453–457.

70. Yeragani VK, Pohl R, Jampala VC, et al. Increased QT variability in patients with panic disorder and depression. Psychiatry Res 2000; 93: 225–235.

71. Nahshoni E, Aizenberg D, Strasberg B, et al. QT dispersion in the surface electrocardiogram in elderly patients with major depression. J Affect Disord 2000; 60: 197–200.

72. Eaton WW, Armenian H, Gallo J, et al. Depression and risk for onset of type II diabetes: a prospective population-based study. Diabetes Care 1996; 22: 1097–1102.

73. Parikh RM, Robinson RG, Lipsey JR, et al. The impact of post-stroke depression on recovery in activities of daily living over two year follow-up. Arch Neurol 1990; 47: 785–789.

74. Kimura M, Robinson RG, Kosier T. Treatment of cognitive impairment after poststroke depression. Stroke 2000; 31: 1482–1486.

75. Robinson RG, Schultz SK, Castillo C, et al. Nortriptyline versus fluoxetine in the treatment of depression and in short term recovery after stroke: a placebo controlled, double-blind study. Am J Psychiatry 2000; 157: 351–359.

76. Chemerinski E, Robinson RG, Arndt S, Kosier JT. The effect of remission of post-stroke depression on activities of daily living in a double-blind randomized treatment study. J Nerv Ment Dis 2001; 189: 421–425.

77. Chemerinski E, Robinson RG, Kosier JT. Improved recovery in activities of daily living associated with remission of post-stroke depression. Stroke 2001; 32: 113–117.

78. Gainotti G, Antonucci G, Marra C, Paolucci S. Relation between depression after stroke, antidepressant therapy and functional recovery. J Neurol Neurosurg Psychiatry 2001; 71: 258–261.

79. Jorge RE, Robinson RG, Arndt S, Starkstein S. Mortality and poststroke depression: a placebo-controlled trial of antidepressants. Am J Psychiatry 2003; 160: 1823–1829.

80. Fruehwald S, Gatterbauer E, Rehak P, Baumhacke U. Early fluoxetine treatment of poststroke depression: a three month double-blind-placebo-controlled study with an open-label long-term follow-up. J Neurol 2003; 250: 347–351.

81. Wiast L, Petit H, Joseph PA, et al. Fluoxetine in early post-stroke depression: a double-blind-placebo-controlled study. Stroke 2000; 31: 1829–1832.

82. Rasmussen A, Lunde M, Poulsen DL, et al. A double-blind-placebo-controlled study of sertraline in the prevention of depression in stroke patients. Psychosomatics 2003; 44: 216–222.

83. Narushima K, Kosier JT, Robinson RG. Preventing of poststroke depression. A 12-week double-blind randomized treatment trial with 21 month follow-up. J Nerv Ment Dis 2002; 190: 296–303.

7

Depression and epilepsy

Frank G Gilliam

Depression and epilepsy share many clinical and neurobiological characteristics[1–3]. Both are paroxysmal disorders that tend to be chronic and recurring. Depression has been noted for many decades to have an increased prevalence in persons with epilepsy, and numerous retrospective studies have evaluated predictive factors for depression in epilepsy[4–6]. The literature on co-morbid epilepsy and depression suggests a pattern of interactions that potentially offers unique insights into factors related to depression in the general population. Environmental[7], personality[8] and neurobiological influences[9–14] appear to play a role in the increased rates of depression in epilepsy, but the relative importance of each factor has not been elucidated. Epilepsy may serve as a catalyst or amplifier of predisposition to depression. For example, social isolation and vocational disability due to epilepsy may serve as 'life stressors' for the initiation of a major depressive episode. Perhaps more importantly, the brain injury or dysfunction present in common forms of epilepsy, such as mesial temporal sclerosis, traumatic contusion, or cortical malformations, frequently occur in anatomical locations included in the limbic network[15]. Defects in certain neurotransmitter systems, especially serotonin and norepinephrine, could plausibly influence both mood and seizure susceptibility. Genetic determinants of cerebral structure and function also have strong potential to adversely affect mood regulation and seizure threshold.

This chapter reviews the available data on the epidemiological, health outcomes and neurobiological relationships of depression and epilepsy. Results of the analysis of these aspects of the strong associations of depression and epilepsy suggest that their coexistence may serve as a research model to investigate and better understand either disorder.

EPIDEMIOLOGY OF DEPRESSION AND EPILEPSY

Most published studies have shown increased rates of depression in persons with epilepsy. Depression ranges in prevalence from 20 to 55% in patients with recurrent seizures and from 3 to 9% in patients with well-controlled seizures, based on multiple epidemiological studies. In a consecutive series of epilepsy patients admitted to a video/electroencephalogram (EEG) seizure monitoring unit, Boylan *et al.*[16] found that 50% were depressed, 19% had recent suicidal ideation, and only 17% were being treated with an antidepressant medication. Mendez *et al.*[17] used the Hamilton Depression Rating Scale in 175 consecutive patients in an outpatient epilepsy clinic, finding that 55% met the criteria for depression. A subsequent study using the same instrument in a similar sample of 96 patients replicated the finding of 50% of patients meeting DSM-IV criteria for major depression[18].

Several community-based studies have also demonstrated increased rates of significant symptoms of depression in epilepsy patients. Using the Hospital Anxiety and Depression Scale in a large survey, Jacoby *et al.*[19] observed that 21% of 168 patients with recurrent seizures were depressed. Approximately 9% of patients without a seizure during the past year endorsed increased symptoms of depression. O'Donoghue *et al.*[20] used the same scale to demonstrate that, of a group of 155 patients identified through two large primary care practices in the UK, 33% with recurrent seizures and 6% of those in remission had probable depression. Edeh and Toone[21] found that 22% of 88 epilepsy patients identified from general practices in the UK endorsed significant symptoms of depression using the Clinical Interview Schedule.

More recently Ettinger *et al.*[22] used the household panel maintained by the National Family Opinion to study depression and quality of life in persons with epilepsy, asthma and healthy controls. The response rate for the survey was approximately 50% in each group, with a total of 1532 responses. Patients with epilepsy were significantly more likely to score in the depressed range on the Center for Epidemiological Studies-Depression Scale (CES-D) (37%), than were those with asthma (28%), or healthy subjects (12%). Although nearly half of the group with epilepsy had not had a seizure during the past year, the mean scores on the short form-36 scales for role limitations, emotional well-being and social well-being were significantly worse than those of the asthma group. None of the scales were lower for asthma.

Beghi *et al.*[23] compared depression severity in epilepsy to type I diabetes mellitus and community controls. Fifty-five patients with idiopathic or cryptogenic epilepsy were compared to age- and sex-matched subjects with type I diabetes or persons donating blood in the local medical clinic. Epilepsy subjects with any structural brain abnormality were excluded from the study, and only 37% reported a seizure within the past year, suggesting that the sample represented the less severe aspect of the spectrum of epilepsy. Although reliable and valid measures of frustration, aggressiveness and apathy were not different between groups, subjects with depression had significantly higher mean Beck Depression Inventory Scores. Thirty-four per cent of epilepsy patients scored in the depressed range compared to 27% of type I diabetics and 7% of blood donors. Although these studies have methodological considerations, including potential selection bias and varying diagnostic techniques, depression is consistently found to be at least 5–10 times more prevalent in samples with uncontrolled epilepsy than in the general population.

Suicide has one of the highest standardized mortality rates (SMR) of all causes of death in persons with epilepsy. Robertson reviewed 17 studies pertaining to mortality in epilepsy, summarizing that the proportion of deaths due to suicide was ten times greater in epilepsy patients than in the general population[24]. Rafnsson *et al.*[25] recently reported the results of a population-based incidence cohort study from Iceland in which suicide had the highest SMR (5.8) of all causes of death. A Swedish study of cause-specific mortality among 9000 previously hospitalized patients with epilepsy found an SMR of 3.5[26]. Jones *et al.* reported that 12% of patients attending five outpatient epilepsy clinics reported recent suicidal ideation[27]. The findings from these studies emphasize the importance of depression in epilepsy and the need for effective screening and treatment of depressive disorders.

ASSOCIATION OF HEALTH STATUS WITH DEPRESSION IN EPILEPSY

Perrine *et al.*[28] evaluated 257 epilepsy patients to determine the relationship of neuropsychological function to health-related quality of life (HRQOL). The independent neuropsychological variables included mood, verbal memory, psychomotor function, visuospatial function, language and cognitive inhibition. The mood factor had the highest correlations with scales of the Quality of Life in Epilepsy Inventory (QOLIE)-89 ($r=-0.20$ to $r=-0.73$ with the 17 subscales) and was the strongest predictor of quality of life in regression analyses. Forty-six per cent of the variance in overall quality of life was explained by the mood factor ($p<0.0001$).

Lehrner *et al.*[29] investigated 56 consecutive patients with temporal lobe epilepsy confirmed by video/EEG monitoring. Reliable and valid instruments developed for native German speakers assessed HRQOL and depression. Depression was the single strongest predictor for each domain of HRQOL. The significant association of depression with HRQOL persisted after controlling for seizure frequency, seizure severity and other psychosocial variables, as shown in Table 7.1. This study offers cross-cultural validation of the importance of mood for health outcomes in epilepsy.

Gilliam *et al.*[30] also reported the strong correlation of severity of mood dysfunction with subjective health status. In a consecutive sample of 125 patients who had undergone temporal resection more than 1 year earlier for refractory complex partial seizures, the Profile of Mood States (POMS) total score had the highest correlation coefficient with subjective health status, including role function, compared to other clinical endpoints, as shown in Table 7.2. Using stepwise regression to evaluate the independent association of each clinical variable with self-reported health, mood status remained the most important predictor of each composite scale score of the POMS (Table 7.3). The POMS total score explained over a quarter of the variance in the Role Function and Physical Function composite scales.

In 195 consecutive patients from outpatient epilepsy clinics, Beck Depression Inventory (BDI) scores were found to have a much higher

Table 7.1 Results of stepwise regression analyses comparing demographic, clinical and neuropsychological variables to health-related quality of life (HRQOL) domains. Adapted from reference 29

HRQOL scale (dependent variable)	*Significant independent variables in the model*	r^2
Physical well-being	Depression	0.27
Activity	Depression	0.21
Relations and family	Depression	0.11
Emotion/mood	1. Depression 2. ADL–cultural scale	0.38
Independence	Depression	0.33
Coping/control	1. Depression 2. Seizure frequency	0.38

ADL, activities of daily living; all r^2 values significant at $p<0.001$

correlation with QOLIE-89 total scores than monthly seizure rates[31] (Figure 7.1).

The data were replicated by Johnson *et al.*[14] in a subsequent study of 87 patients with temporal lobe epilepsy. They also showed that 35% of the variance in quality of life was explained by the severity of depression symptoms, whereas epilepsy-related factors explained less than

Table 7.2 Univariate Pearson correlation coefficients for ESI-55 Composite Scale Scores with clinical variables within the postoperative group ($n = 125$). From reference 30

	Mental health	*Physical health*	*Role function*
Profile of Mood States summary score†	-0.76★★	-0.53★★	-0.53★★
Employment	0.27★★	0.37★★	0.41★★
Driving	0.29★★	0.38★★	0.35★★
AED-free	0.21★	0.28★★	0.30★★
Seizure-free	0.19	0.19★	0.26★★
Age†	0.08	0.02	-0.01
Gender	0.02	-0.02	-0.07
Postoperative performance IQ†	0.11	0.18	0.14
Postoperative verbal IQ†	0.10	0.15	0.15

ESI, Epilepsy Surgery Inventory; AED, antiepileptic drug. ★Correlation is significant at $p < 0.05$ level (2-tailed); ★★correlation is significant at $p < 0.01$ level (2-tailed); †continuous variables evaluated with the Pearson r correlation; all other clinical variables are dichotomous and were evaluated with the Pearson point-biserial correlation

Table 7.3 Final stepwise regression models for the association of Epilepsy Surgery Inventory (ESI)-55 composite scales with conventional clinical variables ($n = 125$). From reference 27

ESI-55 composite scales (dependent variables)	*Significant independent variables in the model*	r^2
Role function	1. Mood Status	0.28
	2. Employment	0.37
	3. Medication use	0.40
Mental health	1. Mood status	0.57
	2. Driving	0.59
Physical health	1. Mood status	0.27
	2. Employment	0.34
	3. Driving	0.37

All r^2 values in each model are cumulative and significant at a level of $p < 0.0001$. Additional independent variables in the analysis included seizure-free status, age, gender, and verbal and performance IQ, but were not significantly associated with ESI-55 composite scale scores

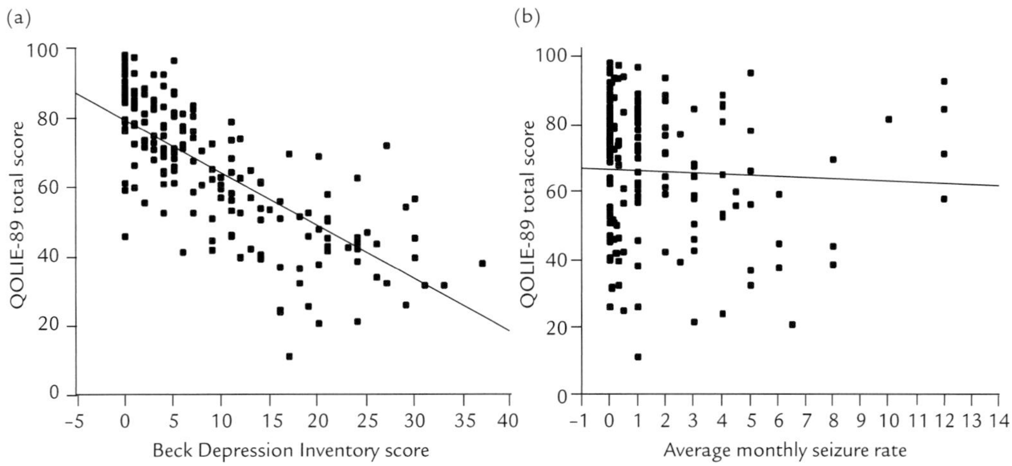

Figure 7.1 Scatterplot of Quality Of Life In Epilepsy (QOLIE-89) total score with (a) severity of depression symptoms ($r = -0.49$, $p = 0.001$); and (b) average monthly seizure rates ($r = 0.01$, $p = 0.93$) ($n = 195$). From reference 31, with permission

20%. Cramer *et al.*[32] found similarly strong correlation of severity of depression with subjective health status in a postal survey of over 500 persons with epilepsy. Additional analysis[33] demonstrated that health-care utilization was significantly increased in epilepsy patients with depression compared to a non-depressed epilepsy sample; depressed patients had more than twice as many non-psychiatric office visits and emergency room visits.

PERSONALITY AND ENVIRONMENTAL FACTORS IN DEPRESSION IN EPILEPSY

The psychosocial ramifications of epilepsy have been recognized for many decades[34,35], and the long-term disability is clearly characterized even for patients in extended remission[36]. Few studies, however, have systematically evaluated the potential relationship between epilepsy-related disability and depression. Hermann *et al.*[8] examined the relevance of the theory of learned helplessness and also attributional (pessimistic, fatalistic attitude) style to depression in epilepsy. Utilizing the BDI, the CES-D and the Optimism/Pessimism Scale, they demonstrated that a negative or pessimistic attributional style was significantly associated with increased self-reported depression and remained significant when the effects of several confounding variables were controlled (age, age at onset, laterality of temporal lobe epilepsy (TLE), sex and method variance). They concluded that attributional style was a component of the genesis of depression in some patients with epilepsy, as it is in the general population. In another study, Hermann and Wyler found an association with certain psychosocial factors such as locus of control and depression scale scores[7].

Schmitz *et al.*[37] compared clinical and psychosocial variables in 25 patients with epilepsy and depression to these variables in 50 otherwise healthy epilepsy patients. Although the mean age was higher in patients with depression and epilepsy, no differences were found between any psychosocial variables. Based on their findings, the authors questioned the 'simple hypothesis' that social and vocational problems could predict depression in epilepsy. The results of these studies emphasize the complexity of the etiologies of depression in epilepsy, and also the potential exacerbating factors of specific personality traits and possibly psychosocial problems in patients with recurrent seizures.

NEUROIMAGING STUDIES OF DEPRESSION IN EPILEPSY

Adverse psychosocial outcomes are an intuitively compelling explanation for the high prevalence of depression in epilepsy, but emerging evidence suggests that brain dysfunction in regions related to mood regulation may be more important. The recognition of depression as a clinical expression of underlying brain dysfunction has important implications for the care of patients with epilepsy, as well as for future research[15]. As Engel *et al.*[13] stated nearly two decades ago, 'although arguments favoring the occurrence of epilepsy-induced disorders of behavior must not be presented in a way that adds to the stigmata associated with epilepsy, it is not in the best interest of epileptic patients to deny this possible relationship and overlook an opportunity to prevent or treat a major cause of disability.'

More than 10 years ago Bromfield *et al.*[38] evaluated ^{18}F-fluorodeoxyglucose–positron emission tomography (FDG-PET) abnormalities and BDI scores in nine brain regions of 23 patients, and found a significant correlation of higher depression scores with hypometabolism in the inferior frontal regions. It is noteworthy that similar observations have been described in patients with depression and no history of neurological disorder[39]. Victoroff *et al.*[40] studied 53 epilepsy patients with FDG-PET and structured psychiatric interviews, finding a significant association of left temporal lateralization and severity of hypometabolism with a history of major depressive disorders. However, severity of hypometabolism in the right temporal lobes also correlated with a history of depression.

Quiske *et al.*[41] examined the association of magnetic resonance imaging (MRI)-defined mesial temporal sclerosis with BDI scores in 60 patients with temporal lobe epilepsy. Mean depression scores were significantly higher in patients with mesial temporal sclerosis (MTS), independent of the lateralization of MTS. The investigators described depression as 'a good indicator of MTS, but not vice versa'. Figure 7.2 shows an example of MRI atrophy and signal changes in the left hippocampus of a patient with recurrent depression and temporal lobe epilepsy.

An additional study of the association of severity of depression symptoms, clinical variables and FDG-PET evaluated 62 consecutive patients from outpatient clinics at Washington University[15]; each had previously had a PET scan as part of a presurgical evaluation. Figure 7.3 presents the FDG-PET scan from a subject in this series with right temporal lobe and basal ganglia hypometabolism[15]. The mean age, complex partial and generalized tonic–clonic seizure rate, number of current antiepileptic drugs, and Adverse Events Profile (AEP) scores were similar between the patients who had an abnormal PET scan ($n = 55$) compared to the group with a normal PET scan ($n = 7$), as shown in Table 7.4. However, the median and mean BDI scores were higher in the abnormal

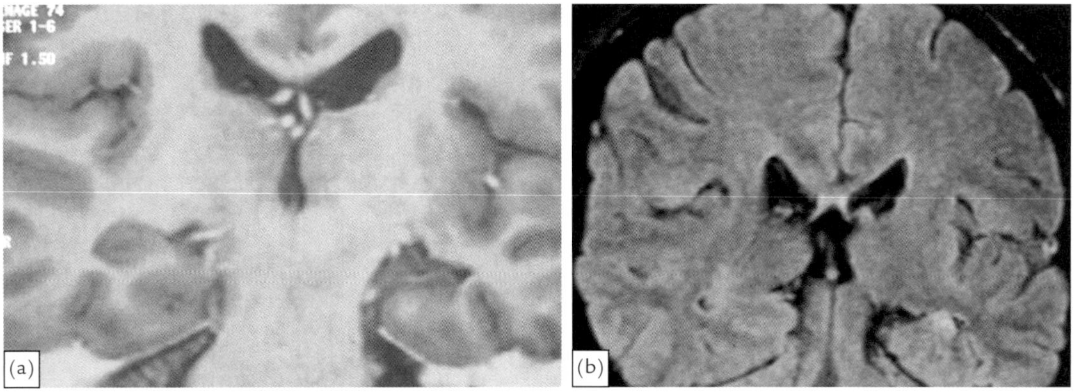

Figure 7.2 Coronal magnetic resonance image showing left hippocampal atrophy and signal changes in (a) inversion recovery, and (b) fluid-attenuated inversion recovery (FLAIR) sequences

PET group (Mann–Whiney *U*, $p<0.01$) as shown in the box-and-whisker plot in Figure 7.4. Fifty-one (82%) of the PET scans showed abnormalities limited to the temporal lobes (maximal hypometabolism; 23 left versus 28 right).

NEUROBIOLOGICAL SIMILARITIES OF DEPRESSION AND EPILEPSY

Both depression and epilepsy have been difficult to study with traditional neurobiological research techniques. The subjective nature of mood and complex associations with specific aspects of behavior limit the potential of molecular research techniques to explain the neurobiology of depression. Similarly, the complexity of epileptogenesis and the multiple components of the seizure threshold in the human brain have limited the usefulness of single neurotransmitter systems or animal models to explicate epilepsy. Despite these inherent restrictions, recent research has made advances in our understanding of the neuronal mechanisms of depression and epilepsy, and has yielded insights into aspects of their overlapping pathophysiology[10,42].

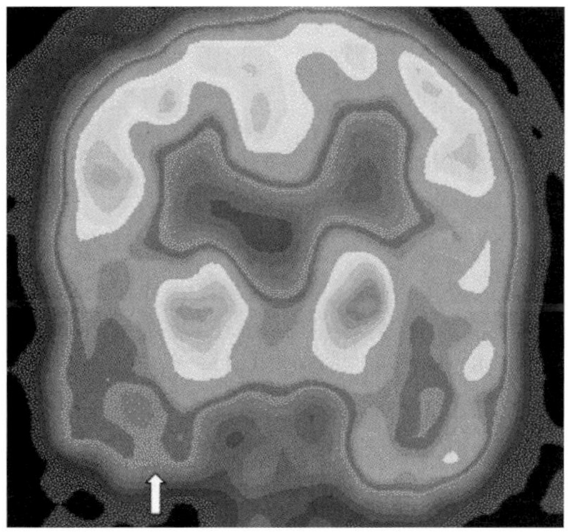

Figure 7.3 ^{18}F-fluorodeoxyglucose–positron emission tomography image of right temporal (white arrow) and right basal ganglia hypometabolism in a patient with refractory right temporal lobe epilepsy. From reference 15, with permission. (See also color image on p. xxii)

Substantial evidence supports the important role of serotonin (5-hydroxytryptamine, 5-HT)

Table 7.4 Comparison of clinical and mood assessments in 62 patients with refractory localization-related epilepsy divided by findings from ^{18}F-fluorodeoxyglucose–positron emission tomography (FDG-PET). From reference 15

	Abnormal FDG-PET (n = 55; mean and SD)	*Normal FDG-PET (n = 7; mean and SD)*
Age	35.9 (11.7)	31.2 (8.6)
Monthly complex partial and GTC seizures	8.7 (14.6)	10.2 (14.3)
Number of AEDs	2.2 (0.85)	1.7 (1.0)
Adverse Events Profile score	39.8 (10.8)	33.5 (10.7)
Beck Depression Inventory score*	9.7 (8.5)	1.8 (2.6)

GTC, generalized tonic–clonic; AED, antiepileptic drug; *Mann–Whitney U $p<0.01$, no other significant between-group differences

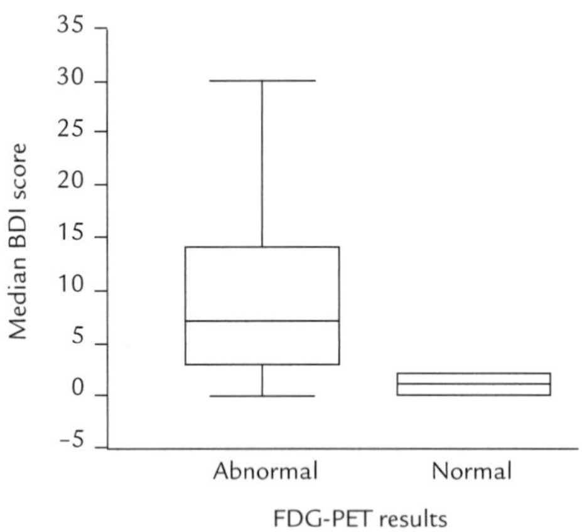

Figure 7.4 Box-and-whisker plots of median and quartile Beck Depression Inventory scores from epilepsy patients presented in Table 7.4. From reference 15. Mann–Whitney U, $p<0.01$

system dysfunction in depression, as reviewed in Chapter 2. Recent cellular and animal experiments indicate a potential role for monoamine neurotransmitters in epileptogenesis[2,10,43]. Areas of the brain involved in common epilepsy syndromes, such as the hippocampus and prefrontal regions, have a high density of 5-HT_{1A} receptors. Activation of 5-HT_{1A} receptors results in membrane hyperpolarization and inhibition of firing, mediated by an increase in K^+ conductance. These observations are indirect evidence for a clinically relevant connection between the underlying neurobiology of both depression and epilepsy[44].

Additional studies have provided evidence that serotonergic neurotransmission modulates some types of epilepsy and experimentally induced seizures[42,45]. Examples include pentylenetetrazole, neocortical and hippocampal kindling, and injections of bicuculline into the area tempestas[2]. The severity of audiogenic seizures was also inversely related to brain 5-HT_{1A} levels in genetically epilepsy-prone rats (GEPRs)[46]. Drugs that increased extracellular 5-HT_{1A} decreased convulsion intensity. Also, 5-HT_{1A} receptor sites are decreased in the hippocampus of GEPRs compared to non-epileptic control rats.

Lu and Gean[47] recently demonstrated several interesting influences of serotonin on epileptiform activity in rat hippocampal CA1 neurons. Excitatory postsynaptic potentials were reversibly depressed by serotonin in a concentration-dependent manner (Figure 7.5). Epileptiform activity induced by picrotoxin, a $GABA_A$ receptor antagonist, was also dose-dependently suppressed by serotonin. Fluoxetine, a selective serotonin reuptake inhibitor (SSRI), potently inhibited epileptiform activity, and a $5\text{-}HT_{1A}$ inhibitor blocked this effect. The anticonvulsant effect of carbamazepine can be blocked by 5-HT-depleting drugs.

Savic *et al.*[48] recently investigated $5\text{-}HT_{1A}$ binding in limbic regions in human temporal lobe epilepsy. Fourteen patients received (Carbonyl-^{11}C)WAY-100 635 to determine $5\text{-}HT_{1A}$ mesial temporal and frontal regions, as well as the raphe nuclei. In 14 patients the mean $5\text{-}HT_{1A}$ binding potential was significantly reduced compared to that in normal controls in the epileptogenic and contralateral hippocampi, the raphe nuclei, and the ipsilateral amygdala, anterior cingulate, insular and lateral temporal neocortex. However, the analyses did not appear to adjust for multiple comparisons. Only the anterior cingulate region had a significant correlation with the depression rating scale scores, but the study did not appear to be powered to identify less than a very strong correlation as significant.

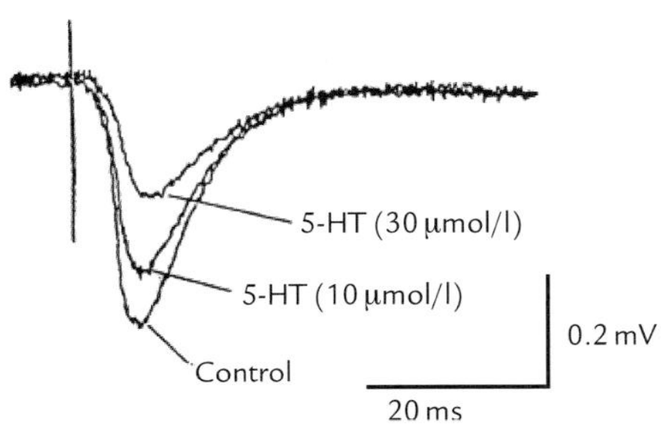

Figure 7.5 Serotonin ($5\text{-}HT_{1A}$) depresses excitatory postsynaptic potentials in a concentration-dependent manner. Control, CsCl-filled electrode. From reference 47, with permission

Recent information on noradrenergic deficits also suggests a potential linkage between depression and epilepsy[2,10,42]. Studies of both humans and GEPRs indicate deficiencies of noradrenergic receptors and norepinephrine concentrations. Noradrenergic deficits in GEPRs result from deficient arborization of neurons arising from the locus coeruleus, as well as excessive suppression of stimulated norepinephrine release in the terminal fields and lack of postsynaptic compensatory up-regulation. Storage vesicle inactivators such as reserpine and tetrabenazine exacerbate seizures in GEPRs, epileptic baboons and other mammals, as the norepinephrine false transmitter α-methyl-*m*-tyrosine does in GEPRs.

DEPRESSION AND EPILEPSY SURGERY

As early as 1989 Hermann and Wyler[7] reported that depression symptoms improved as a group effect after anterior temporal lobectomy (ATL) for refractory temporal lobe epilepsy. Additional analyses indicated that depression declined significantly only in patients rendered completely seizure-free by ATL, whereas patients who were significantly improved (≥ 75% reduction in seizure frequency) but continued to experience some seizure activity showed no changes in self-reported depression. Finally, a wide variety of neurological and seizure-related variables were shown to be unrelated to depression.

Multiple studies have investigated the association of depression with the laterality of the epileptogenic zone, and with the effects of surgery related to laterality[49–51]. Several series have demonstrated more prominent symptoms of depression in patients with left hemisphere epilepsy, whereas at least two studies have found greater depression in right temporal lobe epilepsy[51,52]. As previously mentioned, epilepsy surgery appears to improve negative mood symptoms as a group effect, but with individual variablility[27,53–56]. However, most studies have found that more patients have a worsening of mood status after right hemisphere surgery than left-sided surgery. No definite conclusions can be drawn from these data, but the overall results suggest that right temporal injury associated with epilepsy may have a higher degree of mood dysfunction than left temporal injury[51].

Symptoms of fear during the aura have been associated with amygdala atrophy in temporal lobe epilepsy[57]. Kohler *et al.*[58] found that fear auras were not associated with increased rates of interictal depression or anxiety compared to patients with non-fear auras or no aura. Interestingly, mood and anxiety disorders persisted after surgery at a higher rate in the ictal fear group than the other two groups. Conditioned fear, forced normalization and kindling were offered by the authors as possible explanations for the persistence of mood and anxiety disorders. These observations underscore the complexity of mood regulation in the brain, and the complex effects of limbic system dysfunction and subsequent surgical lesions.

DIAGNOSIS OF DEPRESSION IN EPILEPSY

The diagnosis of depression presents a clinical challenge due to the confounding influences of the symptoms of certain epilepsy syndromes and antiepileptic medications. For example, temporal lobe epilepsy is frequently associated with memory complaints, poor concentration and sleep disturbance. Antiepileptic drugs may cause sedation and fatigue in a substantial proportion of patients[59,60]. These confounding factors may decrease the accuracy of screening instruments for depression that include vegetative and somatic symptoms[61].

Systematic screening appears to increase the rate of detection of depression by 10–47%[62]. Although few studies have evaluated the utility of screening instruments in epilepsy, Karzmark *et al.*[63] suggested that instruments with less somatic content in the question battery may be more accurate for the diagnosis of depression in persons with epilepsy.

Several studies have indicated that most patients with epilepsy are not evaluated or treated for depression by their physician. Furthermore, the standard of care for the treatment of depression in epilepsy is not well characterized in the literature. The few available studies indicate that depression usually goes untreated[5,6]. Ojemann *et al.*[64] investigated an adult clinic population of 2000 at a tertiary epilepsy center caring for complicated cases, finding that less than 8% were on antidepressant medications. If an estimated 30% of patients were depressed, then only 27% of the depressed group was receiving treatment.

A subsequent study evaluated a representative sample of neurologists in the USA in order to determine their practice patterns and also their perspective on relevant factors that might affect their screening for depression in epilepsy patients[15]. Through a brief email questionnaire, 135 neurologists randomly selected from the American Academy of Neurology directory were asked to answer the following two questions: (1) Do you routinely screen epilepsy patients for depression in your outpatient clinics? and (2) If a randomized controlled trial demonstrated that the treatment of depression improved compliance and health-related quality of life in epilepsy patients, would you systematically screen for depression in your outpatient clinic? Fifty-three (79% of responders who see epilepsy patients) reported that they did not screen for depression in their clinic, but that they would if an appropriate trial demonstrated that treatment improved compliance and health outcomes such as quality of life.

Although appropriate interpretation of the results of available depression screening tools is necessary for patients with epilepsy, the risk of missing depression appears to outweigh any potential problems with available instruments[15,65,66]. Standard screening instruments should be used cautiously until validation and reliability studies are completed in samples of persons with epilepsy, but they provide the opportunity to identify patients at risk for depression who might otherwise have gone undiagnosed.

TREATMENT OF DEPRESSION IN EPILEPSY

Reasons for inadequate diagnosis and treatment include lack of appreciation for the impact of depression on quality of life[12,67], fear of seizure exacerbation by antidepressants[68,69] and belief that antidepressants are ineffective in epilepsy patients[6]. Seizures were first reported in patients taking antidepressants in 1959, soon after the release of imipramine in the UK[70]. Although numerous cases of isolated seizures after initiation of tricyclic and SSRI antidepressants have been reported[5,68,69], the only prospective trial of antidepressants in epilepsy patients found no difference in seizure rates between placebo and tricyclic agents[71]. This study included 42 patients in three treatment arms, which did not provide adequate power to detect any less than a very large effect. Ojemann *et al.*[64] found, in a retrospective 2-month study of the effect of doxepin on seizure frequency in 19 patients with epilepsy, that 15 patients experienced a reduction in seizures, and only two patients experienced increased seizures. Kanner *et al.*[72] reported that only one of 100 patients started on sertraline had significant increase in seizures that could be attributed to the drug. Pisani *et al.*[69] concluded, in a recent comprehensive review, that 'the use of antidepressants in patients with epilepsy still raises uncertainties because of the wide-spread conviction that this class of drugs facilitates seizures.' As summarized in a recent review by Devinsky[65], the risk of worsened seizures by SSRIs does not warrant restriction of treatment for depression in persons with epilepsy based on available data.

Very few studies have systematically assessed the efficacy of antidepressants in persons with epilepsy. Several potential problems could plausibly affect the efficacy of antidepressants. Considering the common syndromes of temporal lobe epilepsy and frontal lobe epilepsy that involve brain regions with dense serotonin receptors that receive projections from the raphe nuclei, it is possible that priming the system by blocking serotonin reuptake would not have the expected effect on mood[15]. It is noteworthy that the only published double-blind, randomized, placebo-controlled trial of antidepressants in epilepsy did not demonstrate efficacy[71].

In a retrospective analysis of 75 patients with temporal lobe epilepsy and major depression, no difference was found in efficacy of three antidepressants (mirtazapine, citalopram and reboxetine)[73]. Between 30 and 50% of subjects had a greater than 50% reduction in Hamilton Rating Scale for Depression scores between baseline and 20-week assessment, but only 20% were considered in remission. Approximately half of the patients exited the study prior to 20 weeks of treatment. Most of the dropouts were due to adverse events or loss to follow-up. No patient had an increase in seizures based on clinical judgment. Efficacy of citalopram[74] and sertraline[72] has been suggested by uncontrolled, unmasked studies. Preliminary data suggest that psychotherapeutic techniques may be effective for depression associated with epilepsy[75], but no controlled trials have been published. Further investigation is clearly indicated for both pharmacological and interpersonal therapies for depression in epilepsy.

REFERENCES

1. Engel J Jr, Rocha LL. Interictal behavioral disturbances: a search for molecular substrates. Epilepsy Res Suppl 1992; 9: 341–349; discussion 350.
2. Kanner AM, Balabanov A. Depression and epilepsy: how closely related are they? Neurology 2002; 58 (8 Suppl 5): S27–S39.
3. Tsopelas ND, Saintfort R, Fricchione GL. The relationship of psychiatric illnesses and seizures. Curr Psychiatry Rep 2001; 3: 235–242.
4. Hermann BP, Seidenberg M, Bell B. Psychiatric comorbidity in chronic epilepsy: identification, consequences, and treatment of major depression. Epilepsia 2000; 41 (Suppl 2): S31–S41.
5. Lambert MV, Robertson MM. Depression in epilepsy: etiology, phenomenology, and treatment. Epilepsia 1999; 40 (Suppl 10): S21–S47.
6. Kanner AM, Nieto JC. Depressive disorders in epilepsy. Neurology 1999; 53 (5 Suppl 2): S26–S32.
7. Hermann BP, Wyler AR. Depression, locus of control, and the effects of epilepsy surgery [published erratum appears in Epilepsia 1990; 31: 825]. Epilepsia 1989; 30: 332–338.
8. Hermann BP, Trenerry MR, Colligan RC. Learned helplessness, attributional style, and depression in epilepsy. Bozeman Epilepsy Surgery Consortium. Epilepsia 1996; 37: 680–686.
9. Jobe PC. Common pathogenic mechanisms between depression and epilepsy: an experimental perspective. Epilepsy Behav 2003 (Suppl 3): 14–24.
10. Hecimovic H, Goldstein JD, Sheline YI, Gilliam FG. Mechanisms of depression in epilepsy from a clinical perspective. Epilepsy Behav 2003; 4 (Suppl 3): 25–30.
11. Hermann BP, Seidenberg M, Haltiner A, Wyler AR. Mood state in unilateral temporal lobe epilepsy. Biol Psychiatry 1991; 30: 1205–1218.
12. Wiegartz P, Seidenberg M, Woodard A, et al. Co-morbid psychiatric disorder in chronic epilepsy: recognition and etiology of depression. Neurology 1999; 53 (5 Suppl 2): S3–S8.
13. Engel J Jr, Caldecott-Hazard S, Bandler R. Neurobiology of behavior: anatomic and physiological implications related to epilepsy. Epilepsia 1986; 27 (Suppl 2): S3–S13.
14. Johnson EK, Jones JE, Seidenberg M, Hermann BP. The relative impact of anxiety, depression, and clinical seizure features on health-related quality of life in epilepsy. Epilepsia 2004; 45: 544–550.

15. Gilliam FG, Santos J, Vahle V, et al. Depression in epilepsy: ignoring clinical expression of neuronal network dysfunction? Epilepsia 2004; 45: 28–33.

16. Boylan LS, Flint LA, Labovitz DL, et al. Depression but not seizure frequency predicts quality of life in treatment-resistant epilepsy. Neurology 2004; 62: 258–261.

17. Mendez MF, Cummings JL, Benson DF. Depression in epilepsy. Significance and phenomenology. Arch Neurol 1986; 43: 766–770.

18. Indaco A, Carrieri PB, Nappi C, et al. Interictal depression in epilepsy. Epilepsy Res 1992; 12: 45–50.

19. Jacoby A, Baker GA, Steen N, et al. The clinical course of epilepsy and its psychosocial correlates: findings from a U.K. Community study. Epilepsia 1996; 37: 148–161.

20. O'Donoghue MF, Goodridge DM, Redhead K, et al. Assessing the psychosocial consequences of epilepsy: a community-based study. Br J Gen Pract 1999; 49: 211–214.

21. Edeh J, Toone B. Relationship between interictal psychopathology and the type of epilepsy. Results of a survey in general practice. Br J Psychiatry 1987; 151: 95–101.

22. Ettinger A, Reed M, Cramer J. Depression and comorbidity in community-based patients with epilepsy or asthma. Neurology 2004; 63: 1008–1014.

23. Beghi E, Spagnoli P, Airoldi L, et al. Emotional and affective disturbances in patients with epilepsy. Epilepsy Behav 2002; 3: 255–261.

24. Robertson MM. Suicide, parasuicide, and epilepsy. In: Engel J, Pedley TA, eds. Epilepsy: a Comprehensive Textbook. Philadelphia: Lippincott-Raven, 1997.

25. Rafnsson V, Olafsson E, Hauser WA, Gudmundsson G. Cause-specific mortality in adults with unprovoked seizures. A population-based incidence cohort study. Neuroepidemiology 2001; 20: 232–236.

26. Nilsson L, Tomson T, Farahmand BY, et al. Cause-specific mortality in epilepsy: a cohort study of more than 9,000 patients once hospitalized for epilepsy. Epilepsia 1997; 38: 1062–1068.

27. Jones JE, Hermann BP, Barry JJ, et al. Rates and risk factors for suicide, suicidal ideation, and suicide attempts in chronic epilepsy. Epilepsy Behav 2003; 4 (Suppl 3): 31–38.

28. Perrine K, Hermann BP, Meador KJ, et al. The relationship of neuropsychological functioning to quality of life in epilepsy [see comments]. Arch Neurol 1995; 52: 997–1003.

29. Lehrner J, Kalchmayr R, Serles W, et al. Health-related quality of life (HRQOL), activity of daily living (ADL) and depressive mood disorder in temporal lobe epilepsy patients. Seizure 1999; 8: 88–92.

30. Gilliam F, Kuzniecky R, Meador K, et al. Patient-oriented outcome assessment after temporal lobectomy for refractory epilepsy [see comments]. Neurology 1999; 53: 687–694.

31. Gilliam F. Optimizing health outcomes in active epilepsy. Neurology 2002; 58 (Suppl 5): S9–S19.

32. Cramer JA, Blum D, Reed M, Fanning K. The influence of comorbid depression on quality of life for people with epilepsy. Epilepsy Behav 2003; 4: 515–521.

33. Cramer JA, Blum D, Fanning K, Reed M. The impact of comorbid depression on

health resource utilization in a community sample of people with epilepsy. Epilepsy Behav 2004; 5: 337–342.

34. Collings JA. Psychosocial well-being and epilepsy: an empirical study. Epilepsia 1990; 31: 418–426.

35. Taylor DC. Mental state and temporal lobe epilepsy. A correlative account of 100 patients treated surgically. Epilepsia 1972; 13: 727–765.

36. Sillanpaa M, Jalava M, Kaleva O, Shinnar S. Long-term prognosis of seizures with onset in childhood [see comments]. N Engl J Med 1998; 338: 1715–1722.

37. Schmitz EB, Robertson MM, Trimble MR. Depression and schizophrenia in epilepsy: social and biological risk factors. Epilepsy Res 1999; 35: 59–68.

38. Bromfield EB, Altshuler L, Leiderman DB, et al. Cerebral metabolism and depression in patients with complex partial seizures [published erratum appears in Arch Neurol 1992; 49: 976]. Arch Neurol 1992; 49: 617–623.

39. Drevets WC, Price JL, Simpson JRJ, et al. Subgenual prefrontal cortex abnormalities in mood disorders. Nature 1997; 386: 824–827.

40. Victoroff JI, Benson F, Grafton ST, et al. Depression in complex partial seizures. Electroencephalography and cerebral metabolic correlates. Arch Neurol 1994; 51: 155–163.

41. Quiske A, Helmstaedter C, Lux S, Elger CE. Depression in patients with temporal lobe epilepsy is related to mesial temporal sclerosis. Epilepsy Res 2000; 39: 121–125.

42. Jobe PC. Common pathogenic mechanisms between depression and epilepsy: an experimental perspective. Epilepsy Behav 2003; 4 (Suppl 3): 14–24.

43. Chugani DC, Chugani HT. Does serotonin have trophic effects in temporal lobe epilepsy? Neurology 2003; 60: 736–737.

44. Theodore WH. Does serotonin play a role in epilepsy? Epilepsy Curr 2003; 3: 173–177.

45. Post RM. Neurobiology of seizures and behavioral abnormalities. Epilepsia 2004; 45 (Suppl 2): 5–14.

46. Dailey JW, Yan QS, Adams-Curtis LE, et al. Neurochemical correlates of antiepileptic drugs in the genetically epilepsy-prone rat (GEPR). Life Sci 1996; 58: 259–266.

47. Lu KT, Gean PW. Endogenous serotonin inhibits epileptiform activity in rat hippocampal CA1 neurons via 5-hydroxytryptamine1A receptor activation. Neuroscience 1998; 86: 729–737.

48. Savic I, Lindstrom P, Gulyas B, et al. Limbic reductions of 5-HT1A receptor binding in human temporal lobe epilepsy. Neurology 2004; 62: 1343–1351.

49. Glosser G, Zwil AS, Glosser DS, et al. Psychiatric aspects of temporal lobe epilepsy before and after anterior temporal lobectomy. J Neurol Neurosurg Psychiatry 2000; 68: 53–58.

50. Paradiso S, Hermann BP, Blumer D, et al. Impact of depressed mood on neuropsychological status in temporal lobe epilepsy. J Neurol Neurosurg Psychiatry 2001; 70: 180–185.

51. Quigg M, Broshek DK, Heidal-Schiltz S, et al. Depression in intractable partial epilepsy varies by laterality of focus and surgery. Epilepsia 2003; 44: 419–424.

52. Kohler C, Norstrand JA, Baltuch G, et al. Depression in temporal lobe epilepsy before

epilepsy surgery. Epilepsia 1999; 40: 336–340.

53. Altshuler L, Rausch R, Delrahim S, et al. Temporal lobe epilepsy, temporal lobectomy, and major depression. J Neuropsychiatry Clin Neurosci 1999; 11: 436–443.

54. Derry PA, Rose KJ, McLachlan RS. Moderators of the effect of preoperative emotional adjustment on postoperative depression after surgery for temporal lobe epilepsy. Epilepsia 2000; 41: 177–185.

55. Naylor AS, Rogvi-Hansen B, Kessing L, Kruse-Larsen C. Psychiatric morbidity after surgery for epilepsy: short-term follow up of patients undergoing amygdalohippocampectomy. J Neurol Neurosurg Psychiatry 1994; 57: 1375–1381.

56. Wrench J, Wilson SJ, Bladin PF. Mood disturbance before and after seizure surgery: a comparison of temporal and extratemporal resections. Epilepsia 2004; 45: 534–543.

57. Cendes F, Andermann F, Gloor P, et al. Relationship between atrophy of the amygdala and ictal fear in temporal lobe epilepsy. Brain 1994; 117: 739–746.

58. Kohler CG, Carran MA, Bilker W, et al. Association of fear auras with mood and anxiety disorders after temporal lobectomy. Epilepsia 2001; 42: 674–681.

59. Mattson RH, Cramer JA, Collins JF, et al. Comparison of carbamazepine, phenobarbital, phenytoin, and primidone in partial and secondarily generalized tonic–clonic seizures. N Engl J Med 1985; 313: 145–151.

60. Baker GA, Jacoby A, Buck D, et al. Quality of life of people with epilepsy: a European study. Epilepsia 1997; 38: 353–362.

61. Gilliam F, Kanner A. Treatment of depressive disorders in epilepsy patients. Epilepsy Behav 2002; 3: S2–S9.

62. Pignone MP, Gaynes BN, Rushton JL, et al. Screening for depression in adults: a summary of the evidence for the U.S. Preventive Services Task Force. Ann Intern Med 2002; 136: 765–776.

63. Karzmark P, Zeifert P, Barry J. Measurement of depression in epilepsy. Epilepsy Behav 2001; 2: 124–128.

64. Ojemann LM, Baugh-Bookman C, Dudley DL. Effect of psychotropic medications on seizure control in patients with epilepsy. Neurology 1987; 37: 1525–1527.

65. Devinsky O. Therapy for neurobehavioral disorders in epilepsy. Epilepsia 2004; 45 (Suppl 2): 34–40.

66. Kanner AM. Depression in epilepsy: prevalence, clinical semiology, pathogenic mechanisms, and treatment. Biol Psychiatry 2003; 54: 388–398.

67. Whalley D, McKenna SP. Measuring quality of life in patients with depression or anxiety. Pharmacoeconomics 1995; 8: 305–315.

68. Settle EC Jr. Antidepressant drugs: disturbing and potentially dangerous adverse effects. J Clin Psychiatry 1998; 59 (Suppl 16): 25–30; discussion 40–22.

69. Pisani F, Spina E, Oteri G. Antidepressant drugs and seizure susceptibility: from in vitro data to clinical practice. Epilepsia 1999; 40 (Suppl 10): S48–S56.

70. Brooke G, Weatherly JRC. Imipramine. Lancet 1959; 2: 568–569.

71. Robertson MM, Trimble MR. The treatment of depression in patients with epilepsy. A double-blind trial. J Affect Disord 1985; 9: 127–136.

72. Kanner AM, Kozak AM, Frey M. The use of sertraline in patients with epilepsy: is it safe? Epilepsy Behav 2000; 1: 100–105.

73. Kuhn K-U, Quednow BB, Thiel M, et al. Antidepressive treatment in patients with temporal lobe epilepsy and major depression: a prospective study with three different antidepressants. Epilepsy Behav 2003; 4: 674–679.

74. Hovorka J, Herman E, Nemcova I. Treatment of interictal depression with citalopram in patients with epilepsy. Epilepsy Behav 2000; 1: 444–447.

75. Gillham RA. Refractory epilepsy: an evaluation of psychological methods in outpatient management. Epilepsia 1990; 31: 427–432.

8

Atypical presentations of depression in epilepsy: are they the expression of seizure-related brain dysfunction?

Andres M Kanner

INTRODUCTION

The Diagnostic and Statistical Manual of Psychiatric Disorders, 4th edition (DSM-IV) has subdivided depressive disorders into those that meet criteria for a major depressive disorder (MDD), dysthymic disorder (DD), minor depression (MD) and depressive disorder not otherwise specified (DDNOS) or secondary to a medical condition or substance[1]. The difference between MDD and DD is based largely on severity, persistence and chronicity. According to DSM-IV criteria, symptoms in both disorders may consist of depressed mood, anhedonia, weight loss or gain, insomnia or hypersomnia, psychomotor agitation or retardation, fatigue, worthlessness and guilt, decreased concentration ability and recurrent thoughts of death. In patients with a MDD, at least 2 weeks of either a depressed mood or anhedonia must accompany four of these symptoms. In contrast, DD is a more chronic but less severe disorder with symptoms persistent for more days than not for at least 2 years. MD encompasses at least two but less than five of the depressive symptoms of MDD, noted above.

Patients with epilepsy can experience depressive disorders that meet DSM-IV diagnostic criteria of MDD, DD and MD, yet a significant percentage of depressive disorders do not meet any of these criteria and end up with a classification of atypical depression not otherwise specified. For example, in a study of 101 patients with epilepsy who were suffering from a depressive disorder that required a psychiatric evaluation, Mendez and colleagues found 25 patients with depression that met DSM-III-R criteria of MDD; in 50 (49.5%) patients, the depressive disorder was classified as a DDNOS[2]. In a study of 97 consecutive patients with epilepsy and a depressive disorder that required pharmacotherapy carried out at the Rush Epilepsy Center in Chicago, 28 met DSM-IV criteria for MDD but the remaining 69 patients did not meet criteria for other DSM-IV categories[3]. They reported symptoms of depression and irritability with a waxing and waning course and interspersed symptom-free periods. The clinical picture of these patients' depressive disorder resembled a DD, but failed to meet diagnostic criteria. We therefore termed these patients' disorder as dysthymic-like disorder of epilepsy (DLDE; see below). Blumer and Altshuler have suggested that atypical forms of depression are relatively frequent in patients with epilepsy, accounting for up to one-third of all patients with mood disorders[4].

The failure of depressive disorders to meet DSM or International Classification of Diseases (ICD) diagnostic criteria among patients with epilepsy should not come as a surprise, as the proposed diagnostic criteria in these classifications were based on the clinical manifestations of depressive disorders of patients *without* epilepsy. It is likely that the 'atypical' clinical expression of depression in epilepsy may be accounted for by their variable manifestations according to the temporal relationship between depressive symptoms and seizure occurrence (i.e. presenting as ictal, preictal and postictal symptomatology)[5,6]. In addition, iatrogenic processes related to either pharmacological or surgical therapies of seizure disorders may often present with atypical features[3,7,8]. Finally, epilepsy-induced chronic electrophysiological and neurochemical changes in limbic structures must result in a symptomatology that differs from that of patients with idiopathic mood disorders in whom these changes are absent.

Whether the atypical clinical manifestations of depression in epilepsy are specific to these patients or may occur in other neurological and psychiatric disorders is yet to be established. In this chapter, we review the clinical characteristics of these atypical expressions of depressive disorders in epilepsy and their treatment.

DEPRESSION AS AN EXPRESSION OF A PARAICTAL PROCESS

As stated above, depressive disorders in epilepsy can be classified according to their temporal relationship between the onset of psychiatric symptomatology and seizure occurrence into three categories: (1) *ictal* (the depressive symptoms are a clinical manifestation of the seizure); (2) *peri-ictal* (symptoms precede and/or follow the seizure occurrence); and (3) interictal (symptoms that occur independently of seizure occurrence). We consider *preictal and postictal* symptoms to be an expression of a paraictal process.

Peri-ictal depressive episodes

Preictal depressive episodes

There are very few reports in the literature of this form of depressive disorder, which typically presents as a dysphoric mood preceding seizures. At times, prodromal symptoms may extend for hours or even 1–2 days prior to the onset of a seizure. Blanchet and Frommer carried out a systematic analysis of this type of phenomenon and studied mood changes during 56 days in 27 patients who were asked to rate their mood on a daily basis[9]. Mood ratings pointed to a dysphoric state 3 days prior to a seizure in 22 patients. This change in mood was more accentuated during the 24 hours preceding the seizure. In children, these dysphoric moods took the form of irritability, poor frustration tolerance, restlessness and overt motor hyperactivity as well as aggressive behavior, and were relatively commonly reported by parents.

Postictal depressive episodes

Postictal symptoms of depression have been recognized for a long time but have been poorly studied. In a recently published study, we investigated the prevalence and type of postictal psychiatric symptoms in 100 consecutive patients with poorly controlled partial seizure disorders who were off all psychotropic drugs[6]. Patients were given a 42-item questionnaire designed to identify psychiatric symptoms occurring in the postictal period, defined as the 72 hours that followed recovery of consciousness from the last seizure. We estimated the average duration of each symptom, investigated the presence of these symptoms during the interictal period and compared their severity, when present during the two periods. We included a symptom in our analyses only if it occurred following more than 50% of seizures, so as to include 'habitual' postictal psychiatric symptomatology. We coded neurovegetative symptoms (changes in sleep, appetite and sexual drive) and fatigue as separate symptom categories to avert false-positive increases in the prevalence of postictal symptoms of depression, anxiety and hypomania, as they are primary symptoms of mood and anxiety disorders but are also common postictal symptoms.

We identified a mean of 4.8 ± 2.4 postictal symptoms of depression (range 2–9; median 5) in 43 (43%) of patients. A median duration of 24 hours was reported in all but one of the symptoms (range 0.1–240 hours). Thirteen of these 43 patients reported a cluster of at least seven symptoms, and 18 patients reported six symptoms that lasted 24 hours or longer and mimicked symptoms of a major depressive episode, with exception of the time-frame criterion. Of note, postictal symptoms of depression always occurred in combination with other postictal psychiatric symptoms, particularly with postictal symptoms of anxiety, irritability and postictal neurovegetative symptoms. For example, seven of the 100 patients who reported postictal psychotic symptoms also experienced postictal symptoms of depression and anxiety. Clearly, postictal depressive episodes have a pleomorphic presentation.

Thirteen patients reported habitual postictal suicidal ideation; eight experienced passive and active suicidal thoughts, while only five reported passive suicidal ideation. No patient ever acted on these symptoms. Ten of these 13 patients (77%) had a history of either major depression or bipolar disorder and this association was highly significant.

Among the 43 patients with postictal symptoms of depression, 25 had a history of a mood disorder and 11 of an anxiety disorder. In fact, there were a significantly greater number of postictal symptoms of depression in the presence of a history of depression and anxiety disorders. Furthermore, there was a significant association between the occurrence of postictal symptoms of depression and prior psychiatric hospitalizations, primarily with the occurrence of symptoms related to postictal suicidal ideation. It is worth noting that some reports have suggested that adverse life events may be contributory to the occurrence of postictal depression, but this point is yet to be confirmed. Also, symptoms of depression can outlast the ictus for up to 2 weeks and, at times, have led patients to suicide.

Postictal symptoms of anxiety were identified among 34 of the 43 patients with postictal symptoms of depression. Accordingly, it is fitting to include our findings of postictal symptoms of anxiety when discussing postictal depressive episodes. We identified a mean of 2 ± 1 postictal symptoms of anxiety (range 1–5; median 2) in 45 patients. Fifteen patients (33%) experienced a cluster of four symptoms that lasted at least 24 hours. A history of anxiety disorder was identified in 15 patients (33%). There was a significant association between a history of anxiety disorder and the report of postictal worrying and panicky feelings. In addition, there were a significantly greater number of postictal symptoms of anxiety in the presence of a history of anxiety and depressive disorders.

Interictal symptoms may be impacted in the ictal event and their severity may worsen postictally. In the same study[6], we had identified interictal symptoms of depression and anxiety that worsened in severity or were not affected during the postictal period. We referred to the former symptoms as interictal symptoms with a postictal exacerbation (ISPE). Symptoms in which the severity did not differ during interictal and postictal periods were coded as interictal symptoms. We identified ISPE of depression and anxiety in 36 patients; in 19, all recorded symptoms were coded as ISPE, while the other 17 patients reported symptoms coded as both ISPE and interictal symptoms.

Among the 36 patients with ISPE, 30 (83%) also experienced separate or *de novo* postictal symptoms of depression or anxiety. In fact, there was a significant association between the occurrence of ISPE and *de novo* postictal symptoms of depression and anxiety.

Clearly, postictal symptoms of depression are relatively frequent among patients with poorly controlled epilepsy and must account in great part for the poor quality of life of these patients and for the patients' perceived chronic dysphoric states. In our opinion, these symptoms contribute to the atypical presentation of the semiology of depression in patients with epilepsy.

The relatively high prevalence of pre- and postictal symptoms of depression identified in patients with persistent seizures raises the question of their role in 'shaping' the psychiatric symptomatology in intractable partial epilepsy. To the best of our knowledge, in none of the published studies aimed at identifying psychiatric disorders according to the DSM criteria did investigators ever discriminate between psychiatric symptoms with an interictal, preictal or postictal occurrence. Had we attempted to formulate a psychiatric diagnosis according to DSM-IV criteria, most of our patients would have ended with a diagnosis of 'atypical depression (or anxiety) not otherwise specified'. As stated above, several authors have classified a significant percentage of depression in epilepsy in this category, or alluded to its atypical presentation (see also below).

Ictal depressive episodes

Ictal depressive episodes are the clinical expression of simple partial seizures in which the symptoms of depression consist of its sole (or predominant) semiology. It has been estimated that psychiatric symptoms occur in 25% of 'auras'; 15% of these involve affect or mood changes[10–12]. For example, ictal depression ranked second after anxiety/fear as the most common type of ictal affect in one study. At times, mood changes represent the only expression of simple partial seizures, and consequently, it may be difficult to recognize them as epileptic phenomena. They are typically brief and stereotypical, they occur out of context and they are associated with other ictal phenomena. The most frequent symptoms include feelings of anhedonia, guilt and suicidal ideation. More typically, however, ictal symptoms of depression are followed by alteration of consciousness as the ictus evolves from a simple to a complex partial seizure.

DEPRESSIVE EPISODES AS AN EXPRESSION OF AN IATROGENIC PROCESS

Pharmacologially induced iatrogenic depression

Every antiepileptic drug (AED), including those with positive psychotropic properties (e.g. carbamazepine, valproic acid, gabapentine and lamotrigine) can cause symptoms of depression in patients with epilepsy, some to a greater degree than others[5]. For example, phenobarbital can cause depressive episodes that at times may be associated with the occurrence of suicidal ideation and of suicidal and parasuicidal behavior. Primidone, tiagabine, vigabatrin, felbamate, topiramate and levetiracetam are other AEDs known to cause symptoms of depression[13–18]. The AEDs with mood-stabilizing properties, such as carbamazepine and valproic acid, can occasionally cause depressive episodes, albeit with a significantly lower frequency than other AEDs.

There are limited data on the clinical characteristics of iatrogenic depressive disorders. In a study carried out at our epilepsy center, we identified 27 patients with an iatrogenic depressive disorder among 97 patients with a depressive disorder severe enough to warrant pharmacotherapy[3]. The depressive episodes were caused primarily by the addition of primidone and vigabatrine, which had resulted in seizure freedom, or best seizure control ever, so that patients elected to treat the depressive episode rather than discontinue the AED. Of the 27 patients, 21 were on a polytherapy regimen. Only three patients had a clinical picture of depression that met the criteria of MDD. On the other hand, in 24 patients, the depressive disorder failed to meet any DSM criteria, and resembled a dysthymic disorder, but did not meet the actual DSM criteria of this disorder. In 17 of these patients, irritability was one of the most prominent symptoms and as was anhedonia in the other seven patients. Patients were treated with the selective serotonin reuptake inhibitor sertraline with a mean dose of 105 ± 50 mg/day which they took for a mean period of 8 ± 7.9 months. Complete symptom remission occurred in 43% of patients.

Depression as an expression of an iatrogenic process caused by surgical treatment of epilepsy

Surgery has become an established therapy for intractable epilepsy, above all temporal lobe epilepsy (TLE). With the significant surge in the number of surgical resections in the past three decades, there have been an increasing number of reports of depressive disorders following an anterotemporal lobectomy[7,8]. It is not unusual to see 'mood lability' with significant symptoms of anxiety within the initial 6 weeks after surgery. Often these symptoms subside, but, in up to 30% of patients, overt symptoms of depression become apparent within the first 6 months. Characteristically, symptoms of depression vary in severity from mild to very severe, including suicidal attempts, and the clinical manifestations may present as MDD or DD, or may fail to meet any DSM criteria. In fact, Blumer *et al.* described post-surgical depressive disorders as intermittent and pleomorphic with respect to their symptomatology and suggested that they be termed dysphoric disorders with prominent depressive mood[8].

In a recently completed study of 93 patients who underwent a temporal lobectomy at the Rush Epilepsy Center, 34 experienced a post-surgical depressive episode (Kanner *et al.*, submitted). In 14 patients, the depressive episode was the first ever, while the remaining 20 experienced a recurrence or exacerbation in severity of a pre-surgical depressive disorder that required reintroduction or dose adjustment of antidepressant medication. In fact, having a history of depression was a strong predictor of a post-surgical depressive episode. Furthermore, reaching a post-surgical state free of disabling seizures (no seizures, or only auras) was not significantly associated with the occurrence of post-surgical depression. Treatment with sertraline resulted in remission of symptoms in 22 of the 34 patients. However, 11 patients developed a chronic depressive disorder that failed to respond to multiple trials of psychotropic drugs.

INTERICTAL DEPRESSION WITH AN ATYPICAL PRESENTATION

In this category, we include those interictal depressive disorders that fail to meet any of the DSM diagnostic criteria. They probably are the expression of the impact of chronic epileptic activity-related electrophysiological and neurochemical changes in limbic structures. These forms of depressive disorder often include interictal and peri-ictal symptoms.

Blumer and Altshuler have been among the strongest proponents of an atypical presentation of interictal depression in a significant proportion of patients with epilepsy[4]. They based their description on the observations originally made by Kraepelin in the beginning of the 20th century in institutionalized patients with epilepsy in whom he identified an intermittent and pleomorphic dysphoric disorder as the most common psychiatric disorder associated with epilepsy[18]. Kraepelin described dysphoric symptoms during the preictal and postictal periods, but also during interictal periods. Blumer and Altshuler coined the term 'interictal dysphoric disorder' (IDD) to refer to this form of depression in epilepsy[4].

According to Blumer and Altshuler, IDD is a chronic disorder consisting of a pleomorphic presentation of symptoms that may last from a few hours to several days, to be interrupted by symptom-free periods of variable duration. They suggested that IDD consists of the following eight affective–somatoform symptoms: irritability, depressive moods, anergia, insomnia, pains, anxiety, phobic fears and euphoric moods. They concluded that the presence of three symptoms is associated with significant disability. Interictal dysphoric disorders tend to develop 2 years or longer following the onset of epilepsy. They suggested that the use of a low dose of antidepressant medication was effective in the management of this form of depressive disorder and advocated a combination of a tricyclic antidepressant such as amytriptyline or imipramine with a selective serotonin reuptake inhibitor. No controlled study has been carried out on the efficacy of this regimen in IDD, and the observations of Blumer and Altshuler were based solely on open trials. Furthermore, the specificity of this form of depression to epilepsy has yet to be established in controlled studies.

In a recently completed study we were able to confirm the pleomorphic nature of interictal depressive episodes in a study of 205 consecutive epilepsy patients from five epilepsy centers[19]. The presence of a DSM-IV (axis I) diagnosis was identified in all patients with the Structured Clinical Interview for DSM-IV diagnosis (SCID) and the Mini-International psychiatric interview (MINI). To identify the existence of the pleomorphic presentation of symptoms in depressive disorders in epilepsy, we elaborated a 46-item instrument, based in part on the observations of Blumer and Altshuler. The instrument, called 'Mood and Anxiety Symptoms in Epilepsy Inventory' (MASEI), is a self-rating questionnaire that includes symptoms from the following eight domains: depression, anxiety, irritability, self-consciousness, physical symptoms, disturbances in socialization, suicidal ideation and increased energy symptoms. Each item is rated by the patient (and a companion) on a 4-point Lickert scale (1–4) and was completed twice, 2 weeks apart, as the items inquire about the presence of symptoms for the previous 2 weeks. MASEI had a high internal consistency with a Cronbach's alpha of 0.96. Among the 205 patients, 67 met a DSM-IV criterion of mood and/or anxiety disorders: 19 patients had MDD, four had minor depression or DD, 27 had an anxiety disorder and 17 a mixed MDD and anxiety disorder. Six patients were excluded because of incomplete data. However among the 132 patients who were free of any DSM-IV diagnosis according to the SCID and MINI, 52 were symptomatic on the MASEI and the clinical profile was reminiscent of a dysphoric disorder. Indeed, more than 90% of these 52 patients reported symptoms of depression, anxiety and irritability, 51% reported symptoms of self-consciousness, 67% increased energy symptoms (concurrent with symptoms of depression) and 71% physical symptoms. There was a high correlation in the occurrence of symptoms of depression, anxiety and irritability, while symptoms of self-consciousness correlated significantly with symptoms of depression and anxiety.

We investigated the impact of the 'typical' depressive and anxiety disorders (three groups of patients with MDD, anxiety disorders and mixed MDD/anxiety disorder), and the group of 52 patients with dysphoric-like episodes on their quality of life, assessed with the Quality of Life in Epilepsy Inventory (QOLIE-89). Patients with MDD and mixed MDD and anxiety disorder had the worst scores on the QOLIE-89, while patients with anxiety disorders and patients symptomatic only on the MASEI did not differ in their QOLIE-89 scores, but scored significantly worse than asymptomatic patients on the SCID, MINI and MASEI. Thus, 'sub-syndromic' presentations of mood disorders appear to have a significant negative impact on the quality of life of patients with epilepsy, comparable in magnitude to that yielded by (DSM-IV) anxiety disorders.

These data confirm the pleomorphic presentation of symptoms among patients with epilepsy who suffer from typical mood disorders (according to DSM-IV criteria) as well as with atypical presentation. Clearly, the diagnostic instruments developed to identify psychiatric symptomatology in non-epilepsy patients will fail to identify a significant proportion of symptomatic patients in whom the atypical mood disorder is having a significant impact on their quality of life.

In a separate study, we identified the presence of an interictal mood disorder in patients with chronic epilepsy who failed to meet any DSM-IV criteria among a group of 97 patients with an interictal depressive disorder severe enough to warrant pharmacotherapy with antidepressant medication[3]. Indeed, of these 97 patients, 28 met DSM-IV criteria for MDD, while the depressive symptoms of the remaining 69 patients did not meet the criteria of any of the DSM-IV-recognized categories of affective disorders. These 69 patients presented a pleomorphic clinical picture consisting of anhedonia, with or without hopelessness, fatigue, anxiety, irritability and poor frustration tolerance, and mood lability with bouts of crying. Changes in appetite and sleep patterns, and problems with concentration were also reported by some patients. In 33 of these 69 patients (34% of total), the predominant and most disabling symptom was anhedonia, while in the remaining 36 patients irritability and poor frustration tolerance were the most disabling symptoms. Most symptoms presented with a waxing and waning course, with repeated interspersed symptom-free periods of one to several days' duration. Nevertheless, symptomatology was severe enough in all patients to disrupt their activities, interpersonal relations and overall quality of life and to make them seek treatment. Since these clinical phenomena had the strongest similarity with DD, we referred to them as 'dysthymic-like disorder of epilepsy' (DLDE). We then classified them into two categories – DLDE-A and DLDE-I – to reflect whether anhedonia or irritability was the predominant symptom. Patients with DLDE-A and DLDE-I differed in two additional ways. A history of MDD was elicited with a significantly greater frequency among patients with DLDE-A. On the other hand, patients with DLDE-I were more likely to suffer from an iatrogenic depressive disorder than patients with DLDE-A and MDD.

Fifty-seven patients had a psychiatric disorder of more than 1 year's duration, while in the remaining 40 it had been present for less than 4 months. Despite the long duration of their depressive disorder, only 21 (37%) of the 57 patients had requested or had been offered treatment within 6 months of the onset of their symptoms. Failure to identify the need for therapy did not significantly differ among patients with chronic MDD and chronic DLDE. Fifty-four patients of the 97 patients treated (55.6%) with sertraline experienced a complete resolution of their identified symptoms with. In our opinion, our DLDE and some of Blumer's IDD are probably the same type of event.

REMAINING QUESTIONS

The classic forms of mood disorder in epilepsy remain under-recognized and undertreated. It is not surprising that the atypical presentations of depression have been investigated even less. In our opinion, these atypical forms may be the expression of the 'brain dysfunction' associated with the seizure disorder (and/or its treatment) in the development of psychiatric symptomatology. It is yet to be established whether the clinical manifestations of the atypical presentations of mood disorders described in this chapter are specific to epilepsy or can be identified in other neurological and psychiatric disorders.

REFERENCES

1. Diagnostic and Statistical Manual of Mental Disorders, 4th edn. Washington, DC: American Psychiatric Press, 1994.

2. Mendez MF, Cummings J, Benson D, et al. Depression in epilepsy. Significance and phenomenology. Arch Neurol 1986; 43: 766–770.

3. Kanner AM, Kozak AM, Frey M. The use of sertraline in patients with epilepsy: is it safe? Epilepsy Behav 2000; 1: 100–105.

4. Blumer D, Altshuler LL. Affective disorders. In: Engel J, Pedley TA, eds. Epilepsy: a Comprehensive Textbook, vol II. Philadelphia: Lippincott-Raven, 1998: 2083–2099.

5. Kanner AM, Balabanov A. Depression in epilepsy: how closely related are these two disorders? Neurology 2002; 58 (Suppl 5): S27–S39.

6. Kanner AM, Soto A, Gross-Kanner H. Prevalence and clinical characteristics of postictal psychiatric symptoms in partial epilepsy. Neurology 2004; 62: 708–713.

7. Savard G, Andermann LF, Reutens D, Andermann F. Epilepsy, surgical treatment and postoperative psychiatric complications: a re-evaluation of the evidence. In: Trimble MR, Schmitz B, eds. Forced Normalization and Alternative Psychosis of

Epilepsy. Petersfield, UK: Wrightson Biomedical Publishing, 1998: 179–192.

8. Blumer D, Wakhlu S, Davies K. Psychiatric outcome of temporal lobectomy for epilepsy: incidence and treatment of psychiatric complications. Epilepsia 1998; 39: 478–486.

9. Blanchet P, Frommer GP. Mood change preceding epileptic seizures. J Nerv Ment Dis 1986; 174: 471–476.

10. Daly D. Ictal affect. Am J Psychiatry 1958; 115: 97–108.

11. Williams D. The structure of emotions reflected in epileptic experiences. Brain 1956; 79: 29–67.

12. Weil A. Depressive reactions associated with temporal lobe uncinate seizures. J Nerv Ment Dis 1955; 121: 505–510.

13. Brent D, Crumrine P, Varma R. Phenobarbital treatment and major depressive disorder in children with epilepsy. Pediatrics 1987; 80: 909–917.

14. Ferrari N, Barabas G, Matthews W. Psychological and behavioral disturbance among epileptic children treated with barbiturate anticonvulsants. Am J Psychol 1983; 140: 112–113.

15. Ring H, Reynolds E. Vigabatrin and behavior disturbance. Lancet 1990; 335: 970.

16. McConnell H, Duffy J, Cress K. Behavioral effects of felbamate. J Neuropsych Clin Neurosci 1994; 6: 323.

17. Kanner AM, Wuu J, Faught E, et al. A past psychiatric history may be a risk factor for topiramate-related psychiatric and cognitive adverse events. Epilepsy Behav 2003; 4: 548–552.

18. Kraepelin E. Psychiatrie, vol 3. Leipzig: Johann Ambrosius Barth, 1923.

19. Kanner AM, Wuu J, Barry J, et al. Atypical depressive episodes in epilepsy: a study of their clinical characteristics and impact on quality of life. Neurology 2004; 62 (Suppl 5): A249.

9

Depression in Parkinson's disease

Kevin J Black and Avni Pandya

INTRODUCTION

The scope of the problem

In 1817 the British physician James Parkinson reported his observations of patients (and passersby!) with 'the shaking palsy'[1]. The disease, which has since the late 1800s been known eponymously[2], is a chronic, progressive disorder of later life characterized by a variable mix of abnormal movements including most prominently tremor at rest, involuntary muscle resistance to passive movement (rigidity), diminished amplitude and slowing of voluntary movement (bradykinesia), impaired righting reflexes (postural instability), and reduction in the overall quantity of movement (akinesia)[3]. In the past half century, physician scientists discovered the primary pathological attribute of the disease: a loss of dopamine-producing neurons in the midbrain[4,5]. The motor signs of the disease are most closely associated with death of substantia nigra pars compacta neurons that project to the posterior putamen. However, the disease is still diagnosed clinically, and accuracy of diagnosis is only moderately good compared to eventual autopsy diagnosis[6].

Parkinson's disease (PD) is a common disorder. Of any million people in the general population, about 3000 have PD and about 170 will develop PD in the next year[7,8]. Among people 60 or older, 1 in 100 have PD. As the modal age of onset is about 60 years, the prevalence is likely to rise as the population continues to age.

Although the diagnosis of the disease is based on motor signs, other features are also common. These include other neurological signs (e.g. dystonia), autonomic symptoms (e.g. orthostasis, constipation), sensory symptoms, akathisia and numerous psychiatric symptoms. In fact, only a small minority of PD patients have purely motor symptoms[9]. Psychiatric manifestations occur in a large majority of PD patients and may include depression, anxiety, delusions, hallucinations, dementia, rapid eye movement (REM) sleep behavior disorder and personality change[10]. These are no mere annoyances. The most common reason for nursing home admission in PD is psychosis[11–13], and major depression is the strongest indicator of poor quality of life in PD patients, far surpassing severity of motor disability, and accounting for over half of the variance in quality-of-life measures[14–16]. Similarly, depression may affect activities of daily living (ADL) at least as much as motor signs of PD[16,17]. Depressed patients also exhibit a significantly greater cognitive decline and faster progression of parkinsonism compared to non-depressed patients or those with non-syndromal depressive symptoms[14,17–20].

Obviously patients are not expected to rejoice at having PD, but occasional frustration and sadness is very different from a most-of-the-day, most-days-of-the-week, full depressive syndrome. In fact, most patients with PD never develop major depression.

The challenge posed to clinicians is recognition, diagnosis and treatment of depression in PD. Substantial debate continues among experts as to the basic aspects of definition and diagnosis. Many more PD patients have disabling depression than receive effective treatment[17,21,22], and there are almost no methodologically sound data on which to base treatment. In this chapter we attempt to summarize what is known about the clinical features, epidemiology, etiology, pathophysiology, diagnosis and treatment of depression in PD. Valuable prior reviews of the subject include the comprehensive summary by Cummings[23] and more recent articles by Brown and Jahanshahi[24], Mayberg and Solomon[19], Slaughter *et al.*[25] and McDonald *et al.*[26].

Definitions

Disability and other important prognostic variables have been associated with depression in PD, and it is likely that patients with more severe, prolonged and syndromal depression may benefit more from treatment. Thus, careful diagnosis is felt to be important.

Unfortunately, 'depression' is a poor choice of terminology, as it is used variably in common speech to refer to a symptom or a syndrome, mood changes whether mild or severe, and sadness that is either brief or prolonged. In this review we touch only briefly on depression the symptom – i.e. mild, transient and limited sadness – even though this may be the most common 'depressive' problem in PD. We use 'depressive symptoms' when discussing noncategorical assessment of mood and 'major depressive episode' (MDE) to refer to the full syndrome. This somewhat awkward nomenclature results from the way that DSM-IV-TR treats illness that may result directly from a known pathological lesion.

For diagnosis, DSM-IV-TR requires two clinical judgments as to causation: one at the symptom level and one at the syndrome level. Each depressive *symptom* can count towards a major depressive episode only if it is not better explained by the neurological condition irrespective of other depressive features. An MDE (*syndrome*) is diagnosed either as 'major depression' (or bipolar depression, etc.) or as 'mood disorder due to Parkinson's disease' depending on the clinician's judgment in each patient. The symptom judgment requires substantial clinical expertise, discussed below, and current knowledge does not allow a definitive syndrome judgment for individual patients.

Dysthymia is the presence of chronic depressive symptoms (for at least 2 years). Research criteria for minor depression are essentially those for major depression, except that both dysthymia and minor depression are diagnosed in people with fewer than the five depressive criterion symptoms required for diagnosing an MDE. Minor depression is more frequent than major depression, with 8–20% of older community residents displaying symptoms[27].

One point that is often forgotten but that may prove to be important is that the three main DSM-IV-TR depressive diagnoses – MDE, dysthymia and minor depressive disorder – all require depression to be present 'most of the day, more days than not', for at least several weeks[28]. This rule may be important in excluding from diagnosis the normal and nearly ubiquitous experience in PD patients of frequent but brief frustration or sadness. However, it may also exclude from clinical attention patients with clinically important symptoms. 'Subsyndromal depression' is discussed below.

These nosological issues may seem abstruse but affect the interpretation of many of the research studies discussed below. The current nosology is probably imperfect, yet already some diagnostic distinctions are known to have prognostic value, and they may prove important in treatment choice.

CLINICAL DESCRIPTION

Major depressive episodes

Symptom profile

Major depressive episodes in PD are by definition similar to episodes of major depression in the normal, non-PD population. However, there appear to be some differences[29]. Although these details are debated[24], depressed patients with PD have low rates of self-reproach[30]. Suicide is not common in clinical samples of depressed PD patients[23], although one epidemiological study did not find it less common than in the general population[31]. Depression in PD often begins in late life, whereas MDD most often begins before the age of 40. Anxiety is a prominent and frequent concomitant of major depression in PD, to a degree greater than that seen in primary major depression[32–34]. Depressed PD patients use coping skills less well than non-depressed PD patients[35] (Table 9.1).

Numerous authors have agonized over certain features of depression and PD that appear similar and might be confused. Some similarities or overlaps do occur, and Table 9.2 lists some of these points. However, we believe that the overlap is less problematic than it may look at face value, and that many superficial similarities can be resolved with careful clinical observation. This point is discussed under the section on Diagnosis, below, but Table 9.2 describes some clinical features of depression and how they may differ from signs of PD that occur in the absence of depression.

Table 9.1 Major depressive episodes in Parkinson's disease: differences from non-parkinsonian depression

Guilt and self-reproach are infrequent
Anxiety is more common
There is a low suicide rate
Depression often begins in late life
Men and women are nearly equally affected
Mood-congruent psychosis is infrequent
There is small diurnal variation
Mania is rare (excluding advanced, treated Parkinson's disease or dose-related mood fluctuations)

In patients with earlier-onset parkinsonism, depression is more strongly associated with cognitive impairment and disease duration than in patients with older-onset parkinsonism, but is less associated with disability[36]. Early-onset patients are also more likely to have problematic dose-related mood fluctuations[37].

Co-morbidity

Depression in PD is associated with greater cognitive impairment and prefrontal dysfunction, as seen in tasks of verbal fluency, short-term memory and concept formation[38,39]. Definite dementia is more common in PD patients with major depression, as is levodopa-induced psychosis[17,40–43].

Other depressive disorders in PD

Researchers increasingly are focusing on the burden of illness related to depressive symptoms

Table 9.2 Distinguishing symptoms or signs of major depression from those of uncomplicated Parkinson's disease (PD)

Symptom	*Distinguishing characteristics in (non-PD) major depression*	*Distinguishing characteristics in uncomplicated PD*	*Notes*
Fatigue	'Mental fatigue'[207]	'Physical fatigue'[207]	Fatigue is common in PD with or without depression
Sleep disorder	Early-morning awakening and inability to return to sleep, often with prominent worry	REM sleep behavior disorder; pain; dystonia; akathisia; restless legs syndrome	Insomnia is common in PD[210], but depression explains much of this[211]. Early morning awakening did not discriminate sad versus euthymic PD patients in one study[155], but this needs replication
Loss of interests	Consummatory anhedonia (does not enjoy activities that the patient can actually perform); loss of interest in previously enjoyed activities that are not substantially affected by motor disability	Decreased activities purely because of disability; enjoys activities not compromised by motor disability (such as family visits or watching sports)	Apathy (loss of motivation, loss of wish to do things) can occur independently of depression in PD and is associated with executive dysfunction and disability[161]
Motor retardation or agitation	Markedly increased latency of speech onset; decreased initiation of complex movement ('just sits there'); often with downcast eyes; agitation not attributable to akathisia	Diminished amplitude of volitional movements; diminished frequency of simple movements such as blinking or shifting position; latency of response greater with self-generated than with externally cued actions	
Circumscribed facial expression	Smiles little (i.e. rare expressions of happiness); looks sad rather than neutral	Little smiles (i.e. usual frequency of happy expressions but smiles are diminished in amplitude), also called facial masking	
Low appetite	Decreased interest in food and decreased enjoyment of food	Abnormal or diminished taste of food; may not like having external dietary restrictions	

continued

Table 9.2 *continued*

Symptom	*Distinguishing characteristics in (non-PD) major depression*	*Distinguishing characteristics in uncomplicated PD*	*Notes*
Low self-esteem or guilt	Out of proportion; excessive guilt; may be delusional in intensity	Regrets placing a burden on caretakers	
Impaired mental concentration	Giving up on cognitive tasks; distracted by worries	Some difficulty with high-attention and 'frontal lobe' tasks, even when not sad	Cognition may fluctuate with levodopa dosing

that do not meet the criteria for an MDE. In decreasing order of severity, the current nosology includes dysthymia and minor depression.

Dysthymia and minor depression appear to be similar to the same disorders in non-PD populations, although some of the same differences apply that were noted for an MDE in PD. Starkstein and colleagues[14] reported data on minor depression in PD. Patients with an MDE showed a significantly greater decline in cognition, ADL and disability than did patients with minor depression.

Depressive symptoms that are associated with distress or social dysfunction, but do not meet the criteria for any of the diagnoses just discussed, have been termed subsyndromal depression[44]. In the non-PD population, subsyndromal depression is common, especially in the physically ill elderly[45,46]. One operational definition was: any two or more simultaneous symptoms of depression, present for at least 2 weeks, associated with social dysfunction, in the absence of major depression, minor depression, or dysthymia[46]. In a representative general population sample, this was about as prevalent as the three DSM-III mood disorders. Twice as many women as men were affected, and the diagnosis was associated with disability and welfare benefits. The most common symptoms were insomnia and feeling tired (> 40% each), followed by recurrent thoughts of death, trouble concentrating, weight gain, slowed thinking and hypersomnia. This symptom profile does not correspond closely to that seen in PD patients with more obvious depression, but may describe those with a similar severity of depressive symptoms. Similar results were obtained in geriatric patients from an outpatient medical practice, using an alternative definition based on symptom severity[47].

Mood fluctuations in treated Parkinson's disease

There exists a specific mood syndrome that is unique to PD. With prolonged treatment and progression of disease, individual doses of antiparkinsonian drugs confer benefit for shorter and shorter periods of time. Commonly patients and physicians refer to 'on' periods, when the medication is providing motor benefit, and 'off' periods, when the parkinsonian symptoms return (Figure 9.1). These motor fluctuations are often accompanied by fluctuations of non-motor central nervous system (CNS) symptoms, including pain, autonomic dysfunction, cognitive problems, or mood changes. Some degree of sadness with moments

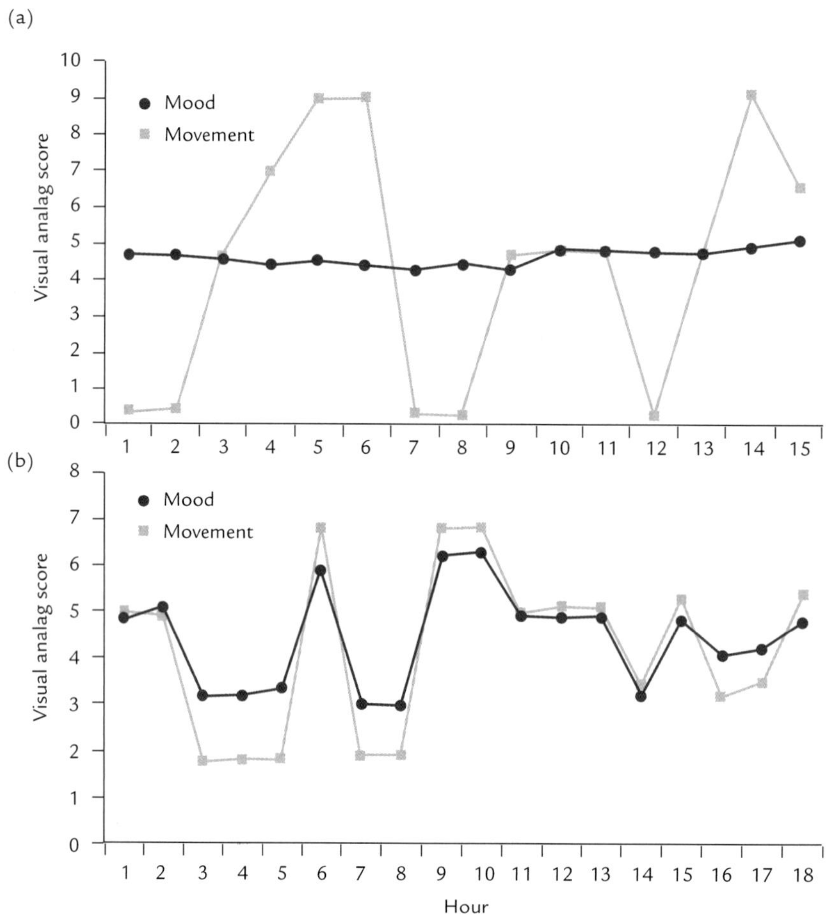

Figure 9.1 Ultradian fluctuations of mood occur in some patients with Parkinson's disease. In both panels, up means better mood or movement and down means worse function. (a) Many patients have marked fluctuation in mobility but little change in mood during the course of a day. (b) Some patients have off-period depression. Reprinted with permission from *J Neuropsychiatry Clin Neurosci* copyright 2001, American Psychiatric Association (reference 54)

of increased disability is expected, and in fact mild mood fluctuations occur in almost all PD patients with motor fluctuations[48–53]. These fluctuations are not always tightly correlated with levodopa dosing or motor status[50,54]. However, a minority of patients develop severe depression, anxiety or mania, usually with a more predictable relationship to levodopa dosing[48–52,55–62]. This occurs more in PD than in equally disabled patients with rheumatoid arthritis[63]. Patients consider mood fluctuations more disabling than their motor deficits, and caregivers consider them more stressful[64,65]. The most common clinically meaningful mood fluctuations are, in decreasing order, off-period anxiety, off-period depressive symptoms and on-period manic symptoms[37]. Patients with clinically significant mood fluctuations tend to have severe PD, with early onset, long duration of illness and extremely high rates of psychiatric co-morbidity including dementia, prior (non-fluctuating) MDEs and drug-induced psychosis[37].

NATURAL HISTORY

Timing of onset of depression versus onset of motor symptoms of Parkinson's disease

Depression is much more common in people with PD than in healthy people of the same age. Surprisingly, this excess of depression in PD begins *before* the onset of parkinsonism. Clinical experience first provided this observation; in fact, the early presence of major depression was proposed as a diagnostic feature of PD[66]. Case–control studies confirmed this impression[67–69]. More recently, several independent epidemiological studies have confirmed the temporal precedence of depression, using a 'pseudo-prospective' study design (i.e. using health registries maintained for decades)[70–73]. The odds ratios are in the range of 2–5 and the results are highly significant.

Various studies differ somewhat in how long before the onset of PD the higher rates of major depression begin. In one study this finding was entirely attributable to depression beginning within the 5 years prior to the onset of PD motor symptoms[70], and in another, depression was significantly more common only in the 10 years preceding motor symptom onset (however, the odds ratio for depression in PD was essentially identical both before and after the 10-years-prior mark)[74]. Alone, these two studies might suggest that motor symptoms or impairment evident just prior to diagnosis could explain the association. However, in the sample of Leentjens *et al.*[73], 77% of the time depression prior to the diagnosis of PD started more than a year before diagnosis. In another study the median age of onset of depression was 10 years before parkinsonism developed[70], and in the case–control study of Taylor *et al.*, the mean latency between depression onset and parkinsonism onset was 22 years[69]. Similarly, 10–20 years before motor symptoms begin, future PD patients (and perhaps their relatives) are also more likely than controls to have low novelty-seeking personality traits, anxiety, social phobia and shyness[32,70,75]. In summary, depression and anxiety begin on average between 5 and 20 years prior to PD; i.e. long before patients are aware of physical symptoms or diagnosis.

Age of onset of depression in Parkinson's disease

The typical age of onset of depression in PD patients is difficult to define precisely, given practical research constraints. Depression is known to predate PD motor signs in some patients, as noted above. Therefore, defining the true average age of onset of depression requires examining for depression not only people who already have developed PD but also people who will later develop PD – or else relying on accurate historical detail in people already diagnosed with PD. Neither is simple. However, there is some pertinent information, primarily from studies comparing depression in people with young-onset versus late-onset PD (see section on Epidemiology, below). Cummings[23] cites 14 studies and concludes that there was no clear relationship between depression and the patients' current ages.

Duration of major depressive episodes in Parkinson's disease

Longitudinal data on depression in PD are sparse, but the best evidence is that the great majority of MDEs last for at least 1–2 years[18,76,77].

EPIDEMIOLOGY

Prevalence

Cummings[23] and Slaughter *et al.*[25] reviewed numerous publications on prevalence of depression in people with PD. Taken together, the evidence best supports a cross-sectional prevalence of major depression of about 25%; when combined with dysthymia or minor depression the prevalence is about 40%. Since the review of Slaughter *et al.*, additional studies have given the following estimates for prevalence of major depression in PD: 33%[78], 18%[79] and 21%[22]. Nilsson *et al.*[80] provide additional valuable epidemiological data from a large inpatient population but not a point prevalence estimate.

Different studies have produced remarkably different prevalence estimates, from 4 to 70%, but much of this wide variation arises from studies whose methodology would not now be considered standard. Many newer studies have used accepted definitions and more representative case finding. However, major depression may be less common in community settings or in certain countries[26]. For comparison, in the general population major depression declines somewhat with advancing age, while other depressive disorders become more common[81].

Incidence

The increased risk of major depression continues after diagnosis: the annual incidence of new depressive episodes in PD patients is roughly 10 times higher than in the general older population[82]. Leentjens *et al.*[73] provide data on incidence of depression prior to parkinsonian motor symptoms in people who later will develop PD. Their clinically relevant conclusion is that a non-depressed patient with PD has an 8–9% risk of developing major depression over a 5-year period. Nilsson *et al.*[80] show risk of incident major depression in PD patients of various ages.

Age, sex

A number of studies have compared the prevalence or severity of depression in PD patients whose symptoms started earlier versus later in life[24]. Some of these are summarized in Table 9.3.

Cummings[23] summarized a dozen studies as showing at best questionable evidence for an excess of depression in women versus men with PD. Nilsson and colleagues[80] showed more 'depression' in women with PD, but this was a retrospective study without standardized

Table 9.3 Differences in depression in patients whose parkinsonism began early versus late in life

Reference	*Key findings*
212	Young-onset patients (PD before 50) had more loss of employment, disruption of family life, perceived stigmatization and depression
36	More depression in early-onset patients (PD before 55) even though they had significantly less severe tremor, akinesia and rigidity. The difference in depression frequency persisted after matching for duration of illness. In the younger group depression correlated with cognitive impairment and duration of the disease, whereas in the older group, depression correlated with impairments in activities of daily living
17	Among PD patients, those with major depression had later age of onset (mean 69.6 vs. 63.9 years) but similar disease duration (9.3 vs. 9.0 years)
78	More major depression in early-onset group (PD before 50), but could be explained by longer duration of illness
82	No difference in the age at onset of PD in depressed vs. non-depressed PD patients. Incidence of depression in various age groups was: ≥70 (3/62) = 5%; 60–69 (4/37) = 11%; 50–59 (2/9) = 22%; 40–49 (0/3) = 0%
130	Men with early-onset PD (before 55) had more mood and anxiety disorders

PD, Parkinson's disease

diagnosis and included any depressive diagnosis. The apparent low sex ratio for major depression in PD differs substantially from that in non-PD major depression, which consistently affects women about twice as commonly as men.

Epidemiology of other mood syndromes

There are few good data on the prevalence of dose-related mood fluctuations such as off-period anxiety or depression. Nissenbaum *et al.*[48] found off-period depression in 69% of their patients with motor fluctuations. Our group reported mood fluctuations noted by physicians in 70 of 1063 consecutively evaluated PD patients[37]. Unfortunately, this study was retrospective. Richard *et al.*[54] used a simple and very reliable method of case definition that may allow accurate determination of prevalence in large samples. In this initial report, 44–50% of patients with motor fluctuations (defined as a range within the day of 10 mm or more on a 100 mm visual analog scale) also had mood fluctuations (defined in the same way). A further caveat is that these studies were from identified patient populations rather than an epidemiological sample of PD patients.

Apathy was found in 42% and 45% of PD patients[83,84]. In the study by Starkstein *et al.*[83], the 42% represented 12% in which apathy predominated and 30% in which it was present with depression.

There are few data on the prevalence of pathological crying in PD patients. This may be more common in patients with progressive supranuclear palsy than in Lewy body PD[85].

PATHOPHYSIOLOGY

Biochemistry

Serotonin

There is substantial evidence linking abnormalities in serotonin neurotransmission with depression in PD. At autopsy, PD patients with depression during life had greater loss of serotoninergic neurons in the dorsal raphe nucleus[86]. By contrast, serotonin concentrations in the frontal and temporal poles were similarly decreased in depressed and non-depressed PD patients[87].

The serotonin transporter, which clears serotonin from the synaptic cleft, comes in more than one allele, and in PD the shorter allele is associated with depression[88,89]. The shorter allele is less active biologically, including in human brain tissue *ex vivo*, so higher synaptic serotonin levels would be expected with this allele.

Mayeux *et al.* conducted several remarkable investigations into serotonin function in PD, especially via lumbar cerebrospinal fluid (CSF) monoamine metabolite concentrations[77,90,91]. Depression is significantly and linearly associated with lower concentrations of the serotonin metabolite 5-hydroxyindoleacetic acid (5-HIAA)[91,92]. This was even more pronounced in patients with both depression and dementia[43].

Kostic *et al.*[93] measured brain sensitivity to serotonin using a fenfluramine challenge method. The serum prolactin response to fenfluramine was substantially blunted in people with PD, but was even more abnormal in depressed than in non-depressed patients with PD.

The radiotracer (^{125})I-2β-carbomethoxy-3β-(4-iodophenyl)tropane ([^{125}I]β-CIT) binds to both the dopamine transporter and the serotonin transporter. As the latter predominates in the dorsal midbrain, Murai *et al.*[94] interpreted changes in tracer binding in that region as reflecting primarily serotonin transporter protein concentration. Among 16 PD patients, those with higher depression ratings (items 3 and 4 on the 'mentation, behavior and mood' subscale of the Unified PD Rating Scale (UPDRS) had lower tracer binding in the dorsal midbrain ($r = -0.53$, $p < 0.04$)[94]. However, another study with a similar design showed no correlation[95]. A more specific positron emission tomography (PET) tracer for the serotonin 1A receptor, ^{11}C-WAY 100635, was applied in PD patients, but the analysis focused on motor signs[96].

On a self-rated personality instrument, depressed patients with PD (like patients with non-PD major depression) on average had elevated harm-avoidance scores[97–100]. This definition of this personality construct was originally based on preclinical data linking certain traits to serotonin[101]. However, a PET study showed a significant positive correlation of harm avoidance with right caudate uptake of [^{18}F]fluorodopa, a tracer that reasonably reflects regional dopamine synthesis[102].

Dopamine

Dopaminergic dysfunction can cause depressive symptoms, as illustrated by off-period depression in PD[37], cocaine-withdrawal sadness and dysphoria from dopamine-depleting drugs or neuroleptics. Furthermore, dopamine-influenced neuronal circuits mediate many of the key brain functions that go awry in depression, such as appreciation of rewards[101,103,104]. There is modest evidence for dopaminergic dysfunction in certain animal models of depression[105,106]. Levodopa dose-related mood fluctuations, or the occasional response of depression in PD to dopamimetic therapy, suggest that some depression in PD may be dopaminergic in origin. PD patients during an MDE do not experience the euphoric reaction to methylphenidate that is seen in non-depressed PD patients, non-PD depressed patients, or normal controls[107]; this euphoric effect of stimulants is probably produced by ventral striatal dopamine release[108]. Mood does not correlate with CSF concentrations of homovanillic acid (HVA)[90], but lumbar CSF HVA levels do not correlate tightly with brain dopamine metabolism[109].

In humans, striatal uptake of a dopamine transporter ligand is decreased by about 14% in non-PD major depression[110], with the most significant decrease located in the right caudate. Non-PD patients whose depression improved with sleep deprivation, but not non-responders, showed decreases in striatal dopamine D_2 receptor ligand binding with treatment, consistent with increased striatal dopamine release[111].

At autopsy, PD patients with depression during life had greater loss of ventral tegmental area (VTA) dopaminergic neurons[112,113].

Supportive evidence that dopamine influences mood in PD is found in studies showing that the dopamine D_3 receptor-preferring agonist pramipexole can improve mood[114], whereas dopamine depletion can produce depression (reviewed in reference 26). However, levodopa and mixed dopamine agonists are not potent antidepressants, suggesting that decreased dopamine concentrations do not directly mediate all depression in PD. Additionally, PD patients occasionally have dysphoric reactions to levodopa.

Therefore, unlike in patients with off-period depression, the association of PD plus MDE with progressive cell loss in VTA may not be mediated by momentary dopaminergic tone. There are several alternative explanations. First, in some patients with mood fluctuations, adequate motor benefit is observed at doses lower than those that improve depression or cause mania. In PD patients with major depression there may be more cell loss in VTA than in nigroputaminal dopaminergic neurons (the inverse of the usual situation). Second, more likely, it may be that depression is more sensitive to limbic or ventral striatal dopamine depletion than motor function is to posterior putamen dopamine depletion (there is some evidence to support this in rodents[115]). In either case, dopamine replacement therapy titrated to motor signs (the usual practice) may not reveal a more direct relationship between dopaminergic tone and mood. Third, postsynaptic adaptive changes with illness and/or chronic treatment may differ in different brain regions innervated by dopamine. Fourth, dopaminergic cell loss in VTA may not be causative of depression but rather a marker for other changes, e.g. loss of serotonergic cells in the raphe, which more directly cause depression.

Other

An antidepressant relatively selective for norepinephrine (i.e. desipramine) apparently can treat MDEs in PD[116], although more methodologically adequate studies are needed[117].

The dexamethasone suppression test has consistently revealed more frequent non-suppression of cortisol release in depressed versus non-depressed PD patients[23], demonstrating that the hypothalamic–pituitary–adrenal axis functions abnormally in depression in PD.

McDonald *et al.*[26] have drawn attention to the association of depression in PD with alterations in testosterone, thyroid status and homocysteine (which can be elevated, owing to exogenous levodopa with a peripheral inhibitor of dopa decarboxylase). These factors are not only relevant to clinical management but also may be measured in future depression studies in PD to minimize uncontrolled variability in the research sample.

Neuroanatomy

Structural neuroanatomy

At autopsy, PD patients with depression during life have greater loss of VTA dopaminergic neurons[113,118], and greater loss of serotoninergic neurons in the dorsal raphe nucleus[86]. These changes may relate to the finding during life that depressed PD patients have higher echogenicity in the midbrain by cerebral ultrasound[119,120].

PD patients have smaller than normal caudate nuclei, putamen and thalamus, and patients with non-PD major depression also have reduced striatal volumes[121]; however, depressed PD patients do not have significantly smaller subcortical nuclei than do non-depressed PD patients (reference 122, p. 18).

Functional neuroanatomy

Several studies have addressed the functional neuroanatomy of depression in PD. Ring *et al.*[123] compared resting regional cerebral blood flow (rCBF) in PD patients with and without major depression, using PET and steady-state inhalation of ^{15}O-labeled CO_2. The two groups were matched for age, gender and severity of parkinsonian symptoms. The investigators found bilateral decreases in rCBF in the orbitofrontal and medial prefrontal cortex (approximately in Brodmann areas 9 and 32) in the depressed PD patients. Mayberg *et al.*[124] used [^{18}F]fluorodeoxyglucose (FDG) and PET to study regional glucose metabolism in PD patients with and without depression and in control subjects. PD controls were matched with depressed PD patients on neuropsychological measures of both global and 'frontal lobe' deficits. In this study, there was bilateral hypometabolism of the caudate nuclei and orbitofrontal cortex in PD patients with major depression compared to the two control groups, and the magnitude of the decrease in metabolism correlated with the severity of depression as rated by a standard depression rating scale. Two-dimensional CT measurements of CSF spaces suggested that global atrophy was unlikely to explain these decreases. Similar changes are seen in depressed patients with other basal ganglia diseases, including caudate stroke and Huntington's disease[125]. Jagust *et al.*[126] investigated 20 non-demented PD patients using single photon emission computed tomography (SPECT) and a qualitative measure of rCBF. Although this study was not designed to compare PD patients with and without major depression, they did find that depressive symptoms, measured by a standard depression inventory, were associated with relative hypoperfusion of the dorsolateral prefrontal cortex and predicted greater impairment on neuropsychological measures sensitive to frontal lobe injuries. In a recent report on PD patients without major depression, sadness was associated with decreased glucose metabolism in the lateral and medial prefrontal cortex[127]. Without further data it is possible that some of these findings may reflect a partial volume effect due to preferential focal atrophy in the depressed groups rather than an actual decline in rCBF per gram of cortex. However, results in these studies are relatively consistent, showing decreased metabolism and blood flow in the prefrontal cortex and caudate nuclei.

Recently, Stefurak *et al.* presented preliminary results of an FDG PET study of depression in PD, before and after antidepressant treatment[128]. Compared to controls, patients had decreased glucose metabolism at baseline in the anterior and posterior cingulate, insula and area 9, and increased metabolism in the globus pallidus and putamen. Response of depression to treatment was accompanied by increased regional glucose metabolism in Brodmann areas 9, 10 and 32, and decreases including in the posterior cingulate (Brodmann areas 25 and 31). Those who later would respond to treatment had higher baseline metabolism in areas 24a and 24b (anterior cingulate).

Tessitore *et al.*[129] studied not major depression but perception of affect in PD. Patients and matched healthy controls performed a task that involved perception of fearful stimuli during functional magnetic resonance imaging (fMRI) scanning. Controls had a bilateral amygdala response that was not seen in PD patients when off dopamimetic medication, but an amygdala response was partially present when patients were on medication.

Clinical clues to neuroanatomy

Cummings[23] summarizes studies reporting motor or cognitive features associated with depression in PD patients (Table 9.4). All findings are relative rather than non-overlapping, and are of little use clinically except perhaps for prognosis. However, such comparisons may give clues as to the functional anatomy of depressive symptoms in PD. To summarize, depressed PD patients have more postural instability and gait disorder but less tremor than in euthymic PD. They also have worse cognition in many domains. The association with left-brain hemiparkinsonism[130] is fascinating, given other known associations of mood with lateralized neurological disease (e.g. ictal crying or laughing in partial seizures; major depression after stroke)[131]; however, the difference in prevalence is modest, and there are potential ascertainment biases in clinical samples.

Table 9.4 Differences in motor signs and cognition in Parkinson's disease (PD) with depression (versus PD without depression)

Cardinal signs of PD
more bradykinesia
more rigidity
more postural instability
more freezing
less tremor
More neuropsychological impairment
worse on frontal lobe tasks: executive function, fluency, working memory
slowing of central processing
worse abstract reasoning
worse verbal auditory attention
More right body (left brain) hemiparkinsonism
Earlier onset

Lesions or electrical stimulation of specific subcortical nuclei can reduce parkinsonian or hyperkinetic abnormal movements in PD, but can also affect mood[132–135]. In some cases the effects are dissociated so that mood is affected without altered motor function or vice versa[136].

Neuroanatomical findings – summary

In summary, there is evidence that depression in PD is related to lesions or altered functioning of specific brain regions, most notably the caudate nucleus, globus pallidus, subthalamic nucleus, amygdala and orbital frontal or medial frontal (anterior cingulate) cortex. These nuclei are heavily interconnected[137]. They are largely distinct anatomically from similar circuits in more motor areas of the brain, yet they are similar in cell types and course spatially very near each other in the pallidum, thalamus and subthalamic nucleus, so that the disease processes may be expected often to affect both motor- and limbic-related regions[138].

Possible neuroanatomical substrates

Studies outside the PD population have shed some light on what brain systems may be most likely to mediate expression of major depression in PD. Excellent reviews have summarized this knowledge recently[139–141]. A few observations from these may be especially pertinent to PD, given the points made earlier in this section.

(1) A variety of research approaches have linked non-PD major depression to alterations in serotoninergic function.

(2) Neuronal circuits influenced by dopamine release in the nucleus accumbens are required for positive behavioral reinforcement with rewards and for noticing deviations from expected consequences of behavior.

(3) The basal ganglia influence not only the primary motor cortex and closely related areas but also the orbital, lateral and medial prefrontal cortex and certain nuclei of the amygdala. These areas in turn innervate or receive projections from other mood-associated regions such as the hypothalamus and posterior cingulate cortex. They are thus well poised to affect mood, motivation, reward learning and visceral concomitants of mood.

Pathophysiology is not etiology

It is worth pointing out in closing this section that most of these pathophysiological mechanisms do not clarify etiology: toxin exposures that kill selected cell populations may cause altered neurotransmitter profiles leading to depression and parkinsonism, but conversely, sad thoughts or increased cortisol release may produce regional alterations in brain metabolism or cause selective loss of certain brain cells. Etiology is best addressed by other study designs, discussed next.

ETIOLOGY

Etiological possibilities

Major depression and PD are both common in late life, and some people would be expected to have both, even if there were no etiological link. However, as noted above, they occur together much more often than can be explained by pure coincidence. The cause of the excess of major depression in PD has been discussed with substantial interest and vigor, and occasionally with pertinent data. There are only a few possible models of causality.

(1) Parkinsonism causes depression:

  (a) Directly, perhaps via an intervening variable such as disability or awareness of illness; or

  (b) By interaction with independent individual traits that confer vulnerability to major depression (e.g. fear of dependence, or genes increasing depression risk).

(2) Major depression causes PD:

   (a) Directly (i.e. the experience of untreated depression injures midbrain neurons); or
   (b) By interaction with individual vulnerability to PD, e.g. by increasing behavioral exposure to certain toxins.

(3) A common etiological factor causes both parkinsonism and depression:
   (a) Directly, e.g. a toxin destroys brain cells, producing parkinsonism or major depression depending stochastically on what regions of the brain are more affected in a given individual; or
   (b) Through interaction with syndrome-specific risk factors (i.e. the common etiological factor produces PD or depression or both, depending on individual vulnerabilities peculiar to each syndrome).

Many studies of depression in PD can distinguish these possibilities only weakly. However, other observations directly bear on this question and are discussed in this section. For instance, studying the temporal precedence of PD versus depression can address the first two possibilities, and positive genetic studies can help to address all three.

Depression precedes parkinsonism

As noted in the section on natural history, the excess of major depression in PD begins many years *prior* to the onset of motor signs or diagnosis. This is not explained by skewed recall of depression in people with current PD, because several of these studies diagnose depression solely on the basis of well-kept contemporaneous records from years before the onset of PD. This observation is also not explained by subjects' being aware of, or disabled by, mild motor symptoms that precede a formal diagnosis of PD, because in some studies the excess of depression in those destined to develop PD was evident as much as 20 years prior to the diagnosis of PD. Furthermore, by comparison with an unrelated antecedent illness (osteoarthritis), Nilsson *et al.*[71] showed that the excess co-occurrence of depression and PD was not simply a marker for chronic increased care-seeking, that takes people to a doctor more often for depressive symptoms earlier in life and for parkinsonian symptoms later in life. In these patients, who have neither motor disability nor awareness of a chronic illness, the excess prevalence of major depression cannot result from the non-specific stress of a progressive illness or its interaction with individual vulnerability. The only explanations that fit the data are that either major depression causes PD, or that major depression can be an early manifestation of whatever also leads to the loss of midbrain dopaminergic neurons.

Chronic, especially untreated, major depression (in non-PD populations) has been associated with loss of certain brain cells[139], including glia[142] and dopamine-influenced nuclei in the amygdala[143]. However, the great majority of people with major depression never develop clinical parkinsonism, and autopsy studies of major depression do not show significant nigral cell loss at autopsy. For instance, in one large-sample study, the risk of new PD diagnosis for those with depression was 0.14% per year, higher than the 0.04% per year onset of PD in non-depressed people but still making PD a rare complication of MDE[72]. Furthermore, the excess incidence of new (first-ever) depression continued after development of PD. Thus, the experience of major depression itself is very unlikely to cause PD.

Family and genetic studies

Two studies found no higher rate of depression (or other psychiatric illnesses) reported in the family members of depressed versus non-depressed PD patients[144,145]. Again, this differs from idiopathic (non-PD) major depression, which clearly is familial, and which in fact runs in families specifically because of genetic influences[146]. This observation suggests two likely conclusions: (1) PD and depression are not both caused by the same genes; and (2) depression in PD is not caused by the interaction of experiencing PD (e.g. chronic illness, disability or handicap) with a genetic diathesis to ordinary (non-PD) major depression.

The study of Leentjens *et al.*[147] appears to differ: in a sample of PD patients a family history of depression was an independent predictor of depression in the probands. Twenty-six PD patients had a positive family history of depression (in first- or second-degree relatives) and 40 PD patients had a current MDE. If the first group is a subset of the second, this indicates that 65% of depressed PD patients have a close relative with major depression. However, even with this assumption, this percentage is difficult to interpret vis-à-vis etiology without a control population studied contemporaneously. Such a study – a direct comparison of family history of major depression in depressed people with versus without PD (matched for age and other pertinent variables) – would help shed light on the etiology.

Two studies have demonstrated an association between depressive and anxious symptoms in PD patients with a specific gene (the short allele of the serotonin transporter promoter)[88,89]. This exciting finding confirms that at least some of the risk for depression in PD occurs prior to any PD motor symptoms, but as this allele also may be more common in non-PD depression, it does not discriminate between the possibilities numbered (1) (b) and (3) (b) in the introduction to this section (p. 212–213).

Environmental effects on depression

The etiology of depression in PD would be clarified by its association, compared with non-depressed PD patients, with a specific environmental influence that could not be caused by the illness. Examples would include parental educational attainment, maternal influenza during gestation, or occupational exposures. More information is available on the association of PD depression with environmental factors that might themselves be caused by the illness, e.g. disability, unemployment, or divorce. Because of this possible bidirectional causation, such studies argue less persuasively for any one model of etiology.

Other etiologically relevant data

Several other kinds of evidence have been frequently adduced to argue for or against a particular etiological model for depression in PD. Although relevant, these data are less compelling, as they generally address a particular theory of how depression arises.

Many authors have focused on the relationship of depression in PD to illness severity or disability, in part due to an assumption that if having PD causes major depression, it is probably mediated by such a variable. (Note that this itself assumes that life stressors cause major depression, whereas the causal relationship between life stressors and major depression is bidirectional and complex[148].) As it happens, there is no simple relationship in PD patients between illness severity or disability and depression; for instance, most studies have found that patients who are more disabled are not on average more depressed than minimally disabled patients[17,23,24,30,97,130,149]. Menza and Mark[97] did find significance in a correlation of PD severity and depression, but this relationship explained only 9% of the variance in depression. Tandberg *et al.*[17] found that depression rating scale scores correlated strongly with a number of measures of PD motor severity and duration (including levodopa dose and residence in a nursing home).

Additionally, depression is more common or severe in PD than in patients equally disabled by non-CNS chronic illness[23,80,97,150], although not all studies have found this[30]. On balance, these observations support the conclusion that the experience of chronic disabling illness, with or without additional individual vulnerability, cannot account for all depression in PD. These data also argue against hypothesis (3) (a) above (p. 213), since anything that directly caused both PD and depression should equally raise the risk for both.

Similarly, if the experience of chronic, disabling illness caused major depression, chronicity would be expected to relate to increasing prevalence; studies have not consistently found such a relation[14,82].

However, as pointed out by Brown and Jahanshahi[24], these correlative observations do not themselves rule out the possibility that individual experience of illness may contribute to the etiology of major depression via a more sophisticated model. They also highlight the apparent association of depression with speed of progression of motor signs[23], but this observation does not distinguish between various theories of etiology. In this context it is interesting to note that PD patients with and without depression have similar levels of social support[35].

Another similar line of evidence also addresses the theory that the experience of illness causes depression in vulnerable patients with PD in the same way that other psychosocial stressors may cause depression in the non-PD population. Specifically, if this model is true, major depression in PD would be likely to have similar characteristics to major depression in the general population. Equivalently, if MDEs in PD differ substantially from major depression in other people, it is probably not 'ordinary' major depression precipitated by a specific stressor (living with PD). In fact, as noted above, PD does differ from ordinary major depression in terms of family history, sex ratio, symptom profile and probably chronicity.

One would also like to know whether clinical features separate MDEs that are etiologically related to PD from those that may be coincidentally co-morbid. In the absence of firm knowledge of etiological markers, one could estimate this based on clinical features. For instance, is the symptom profile or sex ratio more similar to non-PD major depression in PD patients whose first MDE begins late in life or who have no relative with major depression? The authors are not aware of such data. However, in the general elderly population, both major depression and dysthymia that begin in late life differ importantly from the same disorders when they begin earlier in life[151], suggesting that this line of research may be fruitful but may require an age-matched control population.

Finally, another model of depression etiology in PD posits that depression follows from personality features, some of which are themselves strongly related to future development of PD. In the study of Menza and Mark, the personality trait called harm avoidance correlated with depressive symptoms and could account for 31% of the variance in depression[97]. However, harm avoidance is state dependent, decreasing with remission of depression[152], so this correlation has unclear significance for etiology.

Etiology of major depression in Parkinson's disease: summary

Table 9.5 summarizes the foregoing discussion in terms of a 'scorecard' for the various etiological possibilities with which this section started, sorted by type of evidence.

Although one might reasonably argue about some of the table entries, the scorecard suggests that the most likely cause for the depression–PD connection is that both syndromes arise from similar causes, with either appearing first in a given individual. However, additional independent causative factors probably augment or mitigate the risk that an individual will develop major depression, parkinsonism, neither, or both. These conclusions do not identify the shared etiological factor(s) nor pathophysiological final common pathways. However, one may reasonably search for such

Table 9.5 Etiology 'scorecard' for major depression in Parkinson's disease (PD)

	Having PD causes major depression		*Having major depression causes PD*		*A common etiological factor causes both PD and major depression*	
Evidence	*Direct*	*Interaction*	*Direct*	*Interaction*	*Direct*	*Interaction*
Prevalence	–	+	–		–	
Temporal precedence	×	×	–	+	+	+
Family history		– ★			– ★	
Serotonin transporter gene allele association		+				+
Primary environmental risk factors ★★						
No simple association of depression with disability	–	–			–	
Depression symptom profile in PD versus non-PD		–	–	–		

+, The data argue for the cause identified at the head of this column; –, the data argue against this cause; ×, the data appear substantially to exclude this cause; ★, family history evidence is pertinent primarily to a genetic diathesis or cause; ★★, the authors are not aware of pertinent data on environmental risk factors for depression in PD that are not themselves potentially produced by parkinsonism or depression shared causative factors among the known risk factors for either PD or major depression, such as genes (probably plural), aging, chemical toxins, or psychologically stressful life events.

Etiology of other depressive syndromes

Less information is available about the etiology of depressive symptoms outside of MDEs in PD. Separate consideration of the etiology for such symptoms appears to be appropriate, based on the data of Starkstein and colleagues[14], who demonstrated differences in cognition, ADL and motor symptoms between PD patients with major depression versus minor depression. Many of the studies cited above do provide data based on symptom severity rather than syndromal major depression and thus probably include many non-MDE patients in the depressed group. One could theorize that, in many PD patients, non-MDE depression may include appropriate or normative reactions to physical illness, and be more affected by external factors such as social support or physical therapy.

Finally, the etiology is likely to be different for PD-specific mood syndromes such as off-period depression. Circumstantial evidence strongly suggests that, in patients with dose-related mood fluctuations, levodopa affects mood via direct effects on mood-related neuronal circuits rather than indirectly via relief of motor side-effects (reviewed in reference 37).

DIAGNOSIS

Diagnosing depression in PD is acknowledged to be a problematic area and has generated some controversy. In December 2003, the National Institute of Neurological Disorders and Stroke (NINDS) convened a meeting of two dozen experts on PD and psychiatric diagnosis to discuss the issue. Some of the issues discussed are noted below.

Gathering information needed for clinical diagnosis and quantifying severity

Patients and caretakers do not necessarily agree on the patient's depressive symptoms, and in clinical practice it is important to attempt to gather information from both.

Self-report questionnaires such as the Beck Depression Inventory (BDI)[153] have been reasonably criticized on the grounds that they include questions on vegetative symptoms or activities that are affected directly by parkinsonian motor symptoms. Additionally, they do not distinguish between momentary and steady sadness and cannot establish diagnosis. However, in practice, scores on these questionnaires correspond reasonably well to other measures of depression severity[154,155] and are excellent choices to screen for clinically relevant depressive symptoms. Some experts have recommended screening every PD patient for depression with a low-cost instrument such as this. For screening, a score of 9 or higher has been recommended as suggesting clinical attention[156]. The Neuropsychiatric Inventory of Cummings *et al.*[157] has been used increasingly in clinical studies; it is essentially a caretaker self-report measure recorded by an interviewer and has excellent reliability.

The Hamilton Depression Rating Scale (HDRS)[158] ensures that information is recorded about all major depressive symptoms, and allows for clinical judgment in assessing symptoms and judging their severity. It includes many questions on physical symptoms, so that it may be somewhat influenced by motor signs even with an expert rater. The HDRS has been studied in comparison to clinical diagnosis in

PD, and best discriminated depressed from non-depressed patients at a cut-off of 14 or higher (on the 17-item version of the rating scale)[159]. Standardized interviews such as the Structured Clinical Interview for DSM-IV (SCID)[160] provide for gathering data comprehensively while allowing the patient to clarify certain points during the course of the evaluation.

Diagnostic criteria

Given that depressive symptoms outside major depression are common and contribute substantially to perceived quality of life in PD, and that diagnosis can be difficult in PD, some have advocated abandoning a categorical diagnosis of depression in favor of simply rating depressive symptom severity in PD. However, the use of operationalized diagnostic criteria is strongly preferred by many experts for different reasons, and is generally accepted as at least of practical importance, given concerns of third parties such as insurers (for payment of professional claims) or the US Food and Drug Administration (for approval of a drug, biological product or device).

Two significant questions arise in respect of diagnostic rules. One is what set of symptoms, signs, course and other features define the diagnosis, and the other is how to decide whether or not a given criterion is met in a particular patient. Changing either of these could substantially change which patients are or are not diagnosed with an illness.

As to the first question of what symptoms or signs are important in defining depression, the DSM-IV criteria are widely disseminated and generally reasonable, and no compelling reason yet exists for abandoning them in PD. The DSM-IV syndrome-level etiological judgment rule applies at this stage but can be discarded for research purposes. The main data-based objections to using the DSM-IV definition are that apathy (i.e. abnormal lack of motivation) and fatigue both exist independently of depressed mood in PD patients[84,155,161–164]. Reasonable conclusions are that pure apathy should not be 'counted' towards the anhedonia criterion for depression in PD, and that the low energy criterion is suspect. In these cases the argument is that judgments of etiology are irrelevant because, even if present without other cause, either PD or major depression could cause them. One could argue either for dropping one required symptom or for requiring five symptoms other than fatigue. (However, fatigue is more common with depressed mood even in PD patients, so perhaps no change in criteria is needed[163].) Other DSM-IV criterion A symptoms for a major depressive episode, by contrast, appear to be highly correlated with depressed mood in PD[155], and are probably reasonable.

However, there is substantial controversy as to whether the DSM-IV criteria should be changed in PD in its symptom-level etiological judgment rule. Cole *et al.*[130] demonstrated how much difference this attribution could make. Current or lifetime depression was diagnosed either excluding or including all symptoms that were present but that might possibly be attributed to PD. Depending on this choice of diagnostic criteria, the rate of current major depression varied from 10 to 13%, and the rate of lifetime major depression varied from 19 to 32%. Table 9.6 describes Cole's findings for the 11 patients diagnosed with major depression.

The following is an example of the symptom-level etiological judgment. In a patient whose sleep is troubled because of obstructive sleep apnea or REM sleep behavior disorder, DSM-IV dictates that this does not

Table 9.6 The effect on diagnosis of the 'exclusive' approach to the symptom-level etiological rule of DSM-IV. Thirty-one outpatients with PD were diagnosed with or without the exclusion of fatigue and psychomotor retardation from consideration[130]. Five symptoms are required for the diagnosis of major depression; note that patients 1, 6, 7, 8 and 9 changed from 'cases' to non-cases with the exclusion of psychomotor retardation and fatigue

	Patient no.										
	Current depression				*Lifetime depression*						
Symptoms	*1*	*2*	*3*	*4*	*5*	*6*	*7*	*8*	*9*	*10*	*11*
1. Depressed mood	×	×	×	×	×	×	×	×		×	×
2. Anhedonia	×	×	×	×	×	×	×	×	×	×	×
3. Psychomotor retardation			×	×		×	×	×	×	×	×
4. Fatigue	×	×	×	×	×	×		×	×	×	×
5. Sleep disorder			×	×	×	×		×	×		×
6. Appetite disturbance			×		×			×	×	×	×
7. Low self-esteem or guilt	×	×	×	×	×				×	×	
8. Trouble concentrating		×	×	×			×			×	×
9. Hopelessness/suicidality	×	×	×				×				
Total symptoms	5	6	9	7	6	5	5	6	6	7	7

'count' as insomnia toward a diagnosis of depression. On the other hand, unquestioned self-report is unreliable: many patients attribute insomnia to physical symptoms even when, on further inquiry, they are awakening at 04.00 for no clear reason, worrying about the future. In our opinion, expert judgment of each criterion is required for just this reason, to differentiate superficially similar but distinguishable symptoms such as worry versus akathisia, sedation versus fatigue, or loss of enjoyment versus disability. Expert judgment also helps identify signs of depression on examination: facial hypomimia can be distinguished easily from frowning, crying and absence of humor; and we believe that an examiner very familiar with both euthymic PD patients and severe non-PD major depression can usually distinguish retardation from typical bradykinesia and akinesia[165].

Nevertheless, this is all opinion, some experts disagree and certainly this is an impractical strategy for most epidemiological studies. Fortunately, increasingly data are accruing to help determine whether the optimal strategy is to include specific depressive symptoms never, sometimes, or always. For instance, studies can soon hope to reveal whether disagreements between strategies are common or rare, whether one or another strategy best predicts treatment response, and so on.

Differential diagnosis

After careful history taking and examination, and wrestling with the difficult issues above, the distinction between major depression (or bipolar disorder) and dysthymia or minor depression becomes simple. Again, note that all three of these diagnoses require symptoms 'most of the day, more days than not'. This oft-overlooked criterion helps distinguish major depression from other depressive phenomena seen in PD.

Off-period depression was introduced on p. 204 above. Once one is aware of the syndrome, the primary difficulty is in verifying the patient's report. The repeated visual analog scale approach of Richard *et al.*[54] may be useful for research and in doubtful situations. Patients will probably complete a couple of days of such ratings reliably before ennui and frustration set in. Alternative approaches include corroboration of the history by a reliable third party, or repeated observations of the patient, e.g. by including a mood rating with ratings of motor function in about six physical therapy assessments during the course of a single day.

Two other syndromes that may be confused with major depression are pathological crying and apathy. Pathological crying, also known as organic emotionalism, pathological laughing and weeping, or pseudobulbar affect, arises from bilateral functional lesions of supranuclear cranial nerve motor pathways[166,167]. It can be seen in uncomplicated PD as well as during electrical stimulation of motor circuits[168]. The primary clinical feature to distinguish this from sadness is that pathological crying is largely a motor phenomenon (e.g. it can be seen in motor neuron disease) and, although an affective display is involved, the patient may not actually feel sad at the time, or at least the display is considerably out of proportion to any true emotional stimulus.

The 'most of the day, more days than not' criterion of DSM-IV also excludes a diagnosis of major depression in many patients with subsyndromal depression (i.e. bothersome or impairing depressive symptoms that do not meet criteria for minor depression or dysthymia). These patients have lower perceived quality of life than those without depressive symptoms, and would probably benefit from professional help. Included in this group, however, are patients who feel reasonably well most of the time but who become frustrated at moments of confronting disability or become sad when musing on their future. This seems an appropriate enough response to suggest that treatment may include faith, social support, activity therapy and physical or occupational therapy. Whether this is true and whether somatic treatments are equally effective remains to be tested.

Apathy is common in PD and is also in the differential diagnosis[161,169]. It occurs independently of major depression in PD[84,161,162], and can respond differentially to treatment[170,171]. Apathy is even more strongly associated with dementia than is depression[161,162]. These observations argue for requiring a syndrome with many depressive symptoms, rather than severity of one symptom, for diagnosis of major depression. In practice these are patients brought in by their families; the families are bothered by the patient's recent giving up of previously pleasurable activities, or the patient's sitting in a chair all day doing nothing in particular, while the patient is hardly bothered at all and does not understand all the fuss. The patient denies sadness, does not look sad, and does not have other cognitive or affective symptoms typical of depression. This contrasts with the usual situation in major depression, in which the patient feels miserable. As William James observed, typical depression is not only a 'sense of incapacity for joyous feeling', it 'is positive and active anguish, a sort of psychical neuralgia wholly unknown to healthy life' (reference 172, Lectures VI and VII, Section II).

Laboratory tests for diagnosing major depression in Parkinson's disease

In most patients, clinical judgment suffices for prognostic and treatment decisions for depression in PD. As with most proposed diagnostic tests, association with illness is inadequate and a useful test must have favorable test statistics, realistically influence clinical decision making, and have a favorable ratio of potential benefit to cost, risk, or inconvenience[173]. However, treatment for depression can improve quality of life substantially. Some patients have difficulty providing an accurate history, and the clinical diagnostic issues noted above beg for help. If a laboratory test for depression had appropriate test statistics and minimal cost or risk, it might occasionally be useful clinically and used more often in research.

As noted above (p. 209), the dexamethasone suppression test (DST) is more often abnormal in depressed PD patients. However, sensitivity and specificity were variable across studies and modest at best (75% sensitivity and 72% specificity in each of two studies of the six reviewed by Cummings[23].) Accepting these favorable numbers and assuming a prevalence for major depression of 20%, the positive predictive value (PPV) of the DST would be only 40%, meaning that non-suppression would be more likely to be a false positive than a true indicator of major depression. By contrast, with these same assumptions, the negative predictive value (NPV) of the DST would be 92%, so that normal suppression on the DST could be taken as reasonably excluding major depression in a PD patient, and the test might have some clinical utility. These conclusions depend on the test statistics and prevalence in the patients actually tested; using for instance a prevalence of 20% and the less favorable test statistics reported by Kostic *et al.*[174], the DST would have a PPV of only 11% and an NPV of only 64%.

Depressed PD patients share the shortened latency to REM sleep of non-PD depressed patients (mean 41 min versus 129 min[175]). In this study sensitivity was 75% and specificity was 78%, so that, assuming a population prevalence of major depression of 20%, the PPV is only 46% but the NPV is 93%. If one assumes instead that a test would be ordered only in patients with depressive symptoms, and that half of these have major depression, the PPV and NPV would both be about 76%.

Thyroid functional challenge test abnormalities were not specific to depression in PD[90]. Most other laboratory tests studied in PD depression have not been reported in such a way as to quantify their expected utility for diagnosis, and most of these would be impractical in any event (e.g. functional imaging data, 5-HIAA abnormalities in CSF).

Areas of consensus agreement regarding diagnosis

The above review and opinion hides the fact that there is substantial agreement regarding some aspects of assessment in PD, including the following. (1) Ideally, information should be obtained from multiple sources prior to diagnosis, such as the patient, the family or caretaker, and clinical observation. (2) The BDI and other well-validated self-report measures are appropriate for screening, for judging treatment response, and for some research goals. (3)

Pathological crying and off-period depression occur in PD and should not be mistaken for major depression. (4) Apathy and fatigue can arise in PD whether or not any other symptoms of depression occur. (5) The syndrome-level etiological judgment required by DSM-IV cannot be definitively made in individual patients with PD, on the basis of current knowledge. (6) For categorical diagnosis of major depression, no alternative definition has proven benefits compared to the DSM-IV definition. (7) At present, laboratory tests for depression have no important role in PD.

TREATMENT OF DEPRESSION IN PARKINSON'S DISEASE

Given the data above, suggesting that major depression in PD is most likely to be a different disorder from that seen in non-PD populations, and given the known loss of monoaminergic nuclei in PD, treatments effective for idiopathic major depression may or may not be effective in PD. Additionally, the presence of PD may increase the risk/benefit ratio for antidepressant treatments. Therefore, gathering data about antidepressant treatment in the PD population itself assumes great importance.

Major depression

As with any treatment, benefits and risks can be summarized on three dimensions: efficacy, safety and other factors (cost, time to onset of action, etc.). Ideally this section would contain data on each of these points for each treatment. Unfortunately, available data are severely limited. A meta-analysis[117] of treatment data for major depression in 1995 identified no high-quality studies and concluded, 'There are virtually no empirical data on the treatment of depression in PD. Further studies are urgently needed. The situation has changed little since then; the Movement Disorder Society task force on treatment identified no new randomized controlled trial (RCT) for treatment of major depression[176]. The authors of a 2003 review similarly concluded that large RCTs are urgently needed[177,178].

Mixed serotonin–norepinephrine antidepressants

An RCT with nortriptyline showed evidence for efficacy for depression and some positive effect on parkinsonism, but was poorly tolerated[179]. This study had methodological imperfections, safer treatment alternatives are now available for most patients. Venlafaxine is in common use in PD but there are few data on efficacy or tolerability in PD.

Serotoninergic antidepressants

Selective serotonin reuptake inhibitors (SSRIs) are the most commonly prescribed antidepressant treatment in current practice[180]. Open prospective studies have indicated antidepressant benefit for the SSRIs paroxetine, sertraline, fluvoxamine and citalopram (summarized in reference 26). However, a parallel-group RCT of citalopram showed no benefit for the active drug[181]. Selegiline (deprenyl), an inhibitor of monoamine oxidase type B, is often prescribed in PD, raising concerns that concomitant use of SSRIs may lead to the potentially very serious serotonin syndrome. However, according to a large survey of American neurologists, serotonin syndrome is rarely diagnosed[182]. This may reflect the safety of this combination, or it may reflect the retrospective nature of the data and a lack of careful surveillance for serotonin syndrome in clinical practice. SSRIs may on average slightly exacerbate parkinsonism, but in clinical practice any such effect is usually negligible in comparison to the overall disease severity and the effect of antiparkinsonian treatment.

Noradrenergic antidepressants

Older RCTs studied imipramine and desipramine as antidepressants in PD. Desipramine improved depression and rigidity in 25–50% of PD patients in a double-blind RCT[116], and similar results were seen with desipramine's less selective parent compound imipramine[183]. The studies suffered from the methodology available at the time for assessing depression and parkinsonism, but suggest that noradrenergic antidepressants may be effective in PD. A more recent open trial suggests benefit for the newer noradrenergic antidepressant reboxetine[184]. Atomoxetine may also prove effective, although it is not marketed as an antidepressant, and very limited clinical experience suggests that it may not be as well tolerated in PD as are some other antidepressants. Uncontrolled reports on mirtazapine suggest that it may not only help major depression in PD but also improve rest tremor[185]. However, an unpublished report and clinical experience suggest that (like most antidepressants) it may worsen REM sleep behavior disorder in PD patients. Nevertheless, in clinical practice most PD patients tolerate this drug well and enjoy the sedation that bothers many younger patients.

Other antidepressant treatments

Bupropion has indirect positive effects on dopaminergic transmission and thus may seem to be a good choice for treatment of depression in PD. However, only some patients respond, and, at our center, some patients with PD do not tolerate bupropion, owing to exacerbation of 'internal tremor' or anxiety.

The D_3-receptor-preferring dopamine agonist pramipexole is commonly used in treating PD. At doses of 1–5 mg/day it proved more effective than placebo in idiopathic (non-PD) major depression, and at a higher dose it was similar in efficacy to fluoxetine[114]. Open trials in the non-PD depressed population have reported some benefit with the D_2-like dopamine receptor agonist ropinirole. These results suggest that D_2-like agonists may have specific benefit on mood in PD independent of any indirect effects on mood via reduction in burden of illness. This latter effect is seen in most trials of effective antiparkinsonian treatments even when patients with major depression are excluded.

Selegiline may have some antidepressant effect[186,187], but in PD patients this appears to be modest in degree.

Electroconvulsive therapy improves both mood and motor function in PD, appears to be well tolerated and is the treatment of choice for some PD patients with severe depression or depression with psychosis[176,188–191]. We recommend unilateral lead placement and twice- rather than thrice-weekly treatment to reduce the risk of confusion with electroconvulsive therapy (ECT) in PD patients[192].

The group at Emory University have followed up on the facts that many older men have reduced testosterone synthesis, that low testosterone level is associated with low mood, and that PD patients are older[26]. They have shown that male PD patients with low testosterone levels have improved mood with exogenous testosterone supplementation[193,194]. Hypothyroidism is also prevalent in PD and replacement may be expected to improve mood in such patients[26].

An interesting line of research has shown mild elevations of plasma homocysteine concentrations in (non-PD) dementia and depression, and that PD patients in general also have higher than usual homocysteine levels[195]. In part this may be exacerbated by treatment with levodopa and peripheral dopa decarboxylase inhibitors[196], especially in patients who are pharmacogenetically predisposed[26].

An open-label trial with *S*-adenosylmethionine appeared very positive for treatment-resistant depression in PD[197]; it has been used as treatment for depression in the general population as well.

Transcranial magnetic stimulation (TMS) has been proposed as a possible antidepressant treatment in PD[198]. A small open trial has suggested that it may improve both mood and motor symptoms[199], although the benefit on movement is not consistently demonstrated[200].

As noted above, mood benefits have been reported for certain neurosurgical treatments in PD. However, these studies often exclude patients with significant psychopathology and were not designed to test antidepressant efficacy.

In idiopathic major depression, especially in mild to moderate depression, specific psychotherapies have proven efficacy and are acceptable first-line treatments. These include cognitive behavioral therapy (CBT) and interpersonal therapy. Many patients with PD are able to participate actively in such treatments, and, at our center, some appear to have benefited from CBT. Additionally, behavioral treatments have proved effective for specific parkinsonian motor signs[201,202], suggesting that behavioral therapies for depression probably will be proved no less efficacious in PD. Unfortunately, as yet no quality data have assessed the efficacy or safety of any psychotherapy for depression in PD.

The NINDS is now sponsoring a multicenter randomized placebo-controlled trial of paroxetine or venlafaxine for treatment of depression in PD (the Study of Antidepressants in PD; SAD-PD study), and this important research may be expected to settle some of the questions raised above.

Other depressive syndromes

Optimal treatment for off-period depression is not clear. The most parsimonious approach of course is to attempt to reduce motor fluctuations by means such as increased dose of levodopa, use of slow-release formulations, addition of inhibitors of monoamine oxidase or catechol *O*-methyl transferase, or use of long half-life agonists. Unfortunately, two problems appear[37]. First, the patients with mood fluctuations often have motor fluctuations despite intensive treatment management. Smoothing out levodopa delivery and dopamine synthesis are already goals of neurological treatment, and these patients usually have advanced disease. Second, these patients have a high risk of psychosis (and apparently also of other psychiatric treatment complications such as dose-related gambling, hypersexuality, or repetitive stereotyped movements), and the pharmacological interventions mentioned further increase these risks. Some patients have had benefit after deep brain stimulators were placed and provided a smoother motor course during the day, but others have had motor benefit without improvement in mood. No one knows whether typical antidepressants work better than a placebo for this problem, but depending on severity they may be appropriate.

Apathy sometimes responds to treatment with stimulants[203] or dopaminomimetics[204].

Pathological crying often improves with initiation of antiparkinsonian medications. In stroke patients it can improve with sub-antidepressant doses of an antidepressant[205].

Fatigue has substantial negative impact on quality of life[206], and effective treatment would be important. Some basic elements of physical fatigue respond to dopamine replacement therapy, but not 'mental fatigue', which is more closely related to depression[207,208].

Most PD patients with depressive symptoms do not have major depression. Treatments for major depression probably will also help in many patients with dysthymia, as this obtains in the non-PD population. Treatments are poorly defined for minor depression, subsyndromal depression or what might be called normal sadness, and there are few hints as to whether any specific benefit is likely. In the non-PD population, even major depression, if of only mild severity, responds no better to antidepressants than to a placebo[209], and many PD patients are already on polydrug treatment regimens. In the absence of data, at our center treatment of depressive symptoms that are not major depression often includes attempts to reduce the burden of illness via occupational and physical therapy, referral to the American Parkinson Disease Association local chapter, reassurance, caretaker education and respite, encouraging activity and reconnection to social support, and for some patients supportive or cognitive–behavioral psychotherapy. Many of these patients are already treated with antidepressant medications and the decision on whether to continue or stop them depends on a clinical judgment of individual likelihood of benefit and risks.

CONCLUSIONS

Depressed mood is the strongest predictor of quality of life in PD. Major depressive episodes occur very often in PD and the high rate of depression and anxiety precedes by many years the onset of motor symptoms. Additionally, depression in PD differs from ordinary major depression in symptom profile and demographics. These facts suggest that MDEs in PD will prove to be a different nosological entity than major depression in the non-PD population. Clinical features peculiar to PD complicate the clinical diagnosis of major depression, yet no compelling substitute exists at present for clinical diagnosis. The relative lack of data to guide treatment is remarkable and is only now being addressed in a multicenter RCT. Future etiological studies may profit from a change of focus. Examples might include the following: family studies of depression with diagnosis or symptom rating done blind to whether or not the proband has PD; genetic linkage or association studies; epidemiological or identical twin studies of environmental risks for depression in PD; and studies in which known contributors to depression in PD, such as testosterone concentration or serotonin transporter allele, are accounted for prior to group comparisons.

ACKNOWLEDGMENTS

Preparation of this chapter was supported in part by the NIH (R01 NS044598; R01 NS046487; R01 NS40808), the Dana Foundation, the Washington University Center for Aging, and the American Parkinson Disease Association (APDA) Advanced Research Center at Washington University.

REFERENCES

1. Parkinson J. An essay on the shaking palsy, 1817. J Neuropsychiatry Clin Neurosci 2002; 14: 223–236.

2. Goetz CG, Chmura TA, Lanska DJ. The history of Parkinson's disease: part 2 of the MDS-sponsored History of Movement Disorders Exhibit, Barcelona, June, 2000. Mov Disord 2001; 16: 156–161.

3. Cummings JL. Understanding Parkinson disease. JAMA 1999; 281: 376–378.

4. Hornykiewicz O. Die Topische Lokalisation und das Verhalten von Noradrenalin und Dopamin in der Substantia Nigra des normalen und Parkinsonkranken Menschen. Wien Klin Wochenschr 1963; 75: 309–312.

5. Roe DL. From DOPA to Parkinson's disease: the early history of dopamine research. J Hist Neurosci 1997; 6: 291–301.

6. Rajput AH, Rozdilsky B, Rajput A. Accuracy of clinical diagnosis in parkinsonism – a prospective study. Can J Neurol Sci 1991; 18: 275–278.

7. Samii A, Nutt JG, Ransom BR. Parkinson's disease. Lancet 2004; 363: 1783–1793.

8. Twelves D, Perkins KS, Counsell C. Systematic review of incidence studies of Parkinson's disease. Mov Disord 2003; 18: 19–31.

9. Shulman LM, Taback RL, Bean J, et al. Comorbidity of the nonmotor symptoms of Parkinson's disease. Mov Disord 2001; 16: 507–510.

10. Marsh L, Berk A. Neuropsychiatric aspects of Parkinson's disease: recent advances. Curr Psychiatry Rep 2003; 5: 68–76.

11. Factor SA, Feustel PJ, Friedman JH, et al. Longitudinal outcome of Parkinson's disease patients with psychosis. Neurology 2003; 60: 1756–1761.

12. Goetz CG, Stebbins GT. Mortality and hallucinations in nursing home patients with advanced Parkinson's disease. Neurology 1995; 45: 669–671.

13. Goetz CG, Stebbins GT. Risk factors for nursing home placement in advanced Parkinson's disease. Neurology 1993; 43: 2227–2229.

14. Starkstein SE, Mayberg HS, Leiguarda R, et al. A prospective longitudinal study of depression, cognitive decline, and physical impairments in patients with Parkinson's disease. J Neurol Neurosurg Psychiatry 1992; 55: 377–382.

15. The Global Parkinson's Disease Survey Steering Committee. Factors impacting on quality of life in Parkinson's disease: results from an international survey. Mov Disord 2002; 17: 60–67.

16. Weintraub D, Moberg PJ, Duda JE, et al. Effect of psychiatric and other nonmotor symptoms on disability in Parkinson's disease. J Am Geriatr Soc 2004; 52: 784–788.

17. Tandberg E, Aarsland D, Laake K, et al. Risk factors for depression in Parkinson's disease. Arch Neurol 1997; 54: 625–630.

18. Starkstein SE, Bolduc PL, Mayberg HS, et al. Cognitive impairments and depression in Parkinson's disease: a follow up study. J Neurol Neurosurg Psychiatry 1990; 53: 597–602.

19. Mayberg HS, Solomon DH. Depression in Parkinson's disease: a biochemical and organic viewpoint. Adv Neurol 1995; 65: 49–60.

20. Norman S, Troster AI, Fields JA, et al. Effects of depression and Parkinson's disease on cognitive functioning. J Neuropsychiatry Clin Neurosci 2002; 14: 31–36.

21. Shulman LM, Taback RL, Rabinstein AA, et al. Non-recognition of depression and other non-motor symptoms in Parkinson's disease. Parkinsonism Relat Disord 2002; 8: 193–197.

22. Weintraub D, Moberg PJ, Duda JE, et al. Recognition and treatment of depression in Parkinson's disease. J Geriatr Psychiatry Neurol 2003; 16: 178–183.

23. Cummings JL. Depression and Parkinson's disease: a review. Am J Psychiatry 1992; 149: 443–454.

24. Brown R, Jahanshahi M. Depression in Parkinson's disease: a psychosocial viewpoint. Adv Neurol 1995; 65: 61–84.

25. Slaughter JR, Slaughter KA, Nichols D, et al. Prevalence, clinical manifestations, etiology, and treatment of depression in Parkinson's disease. J Neuropsychiatry Clin Neurosci 2001; 13: 187–196.

26. McDonald WM, Richard IH, DeLong MR. Prevalence, etiology, and treatment of depression in Parkinson's disease. Biol Psychiatry 2003; 54: 363–375.

27. Gallo JJ, Lebowitz BD. The epidemiology of common late-life mental disorders in the community: themes for the new century. Psychiatr Serv 1999; 50: 1158–1166.

28. American Psychiatric Association. Diagnostic and Statistical Manual of Mental Disorders, 4th edn. Text Revision. Washington, DC: American Psychiatric Association, 2000.

29. Yamamoto M. Depression in Parkinson's disease: its prevalence, diagnosis, and neurochemical background. J Neurol 2001; 248 (Suppl 3): III-5–11.

30. Gotham A-M, Brown RG, Marsden CD. Depression in parkinson's disease: a quantitative and qualitative analysis. J Neurol Neurosurg Psychiatry 1986; 49: 381–389.

31. Stenager EN, Wermuth L, Stenager E, et al. Suicide in patients with Parkinson's disease. An epidemiological study. Acta Psychiatr Scand 1994; 90: 70–72.

32. Meah Y, Miyawaki E, Kormos TC, et al. Depression in Parkinson's disease: a meta-analysis regarding severity. Mov Disord 1995; 11: 159.

33. Henderson R, Kurlan R, Kersun JM, et al. Preliminary examination of the comorbidity of anxiety and depression in Parkinson's disease. J Neuropsychiatry Clin Neurosci 1992; 4: 257–264.

34. Menza MA, Robertson-Hoffman DE, Bonapace AS. Parkinson's disease and anxiety: comorbidity with depression. Biol Psychiatry 1993; 34: 465–470.

35. Ehmann TS, Beninger RJ, Gawel MJ, et al. Coping, social support, and depressive symptoms in Parkinson's disease. J Geriatr Psychiatry Neurol 1990; 3: 85–90.

36. Starkstein SE, Berthier ML, Bolduc PL, et al. Depression in patients with early versus late onset of Parkinson's disease. Neurology 1989; 39: 1441–1445.

37. Racette BA, Hartlein JM, Hershey T, et al. Clinical features and comorbidity of mood fluctuations in Parkinson's disease. J Neuropsychiatry Clin Neurosci 2002; 14: 438–442.

38. Uekermann J, Daum I, Peters S, et al. Depressed mood and executive dysfunction in early Parkinson's disease. Acta Neurol Scand 2003; 107: 341–348.

39. Taylor AE, Saint-Cyr JA, Lang AE, et al. Parkinson's disease and depression. A critical re-evaluation. Brain 1986; 109: 279–292.

40. Stern Y, Marder K, Tang MX, et al. Antecedent clinical features associated with dementia in Parkinson's disease. Neurology 1993; 43: 1690–1692.

41. Hughes TA, Ross HF, Musa S, et al. A 10-year study of the incidence of and factors

predicting dementia in Parkinson's disease. Neurology 2000; 54: 1596–1602.

42. Anderson KE. Dementia in Parkinson's disease. Curr Treat Options Neurol 2004; 6: 201–207.

43. Sano M, Stern Y, Williams J, et al. Coexisting dementia and depression in Parkinson's disease. Arch Neurol 1989; 46: 1284–1286.

44. Sadek N, Bona J. Subsyndromal symptomatic depression: a new concept. Depress Anxiety 2000; 12: 30–39.

45. Lebowitz BD, Pearson JL, Schneider LS, et al. Diagnosis and treatment of depression in late life. Consensus statement update. JAMA 1997; 278: 1186–1190.

46. Judd LL, Rapaport MH, Paulus MP, et al. Subsyndromal symptomatic depression: a new mood disorder? J Clin Psychiatry 1994; 55 (Suppl): 18–28.

47. Lyness JM, King DA, Cox C, et al. The importance of subsyndromal depression in older primary care patients: prevalence and associated functional disability. J Am Geriatr Soc 1999; 47: 647–652.

48. Nissenbaum H, Quinn NP, Brown RG, et al. Mood swings associated with the 'on–off' phenomenon in Parkinson's disease. Psychol Med 1987; 17: 899–904.

49. Hardie RJ, Lees AJ, Stern GM. On–off fluctuations in Parkinson's disease: a clinical and neuropharmacological study. Brain 1984; 107: 487–506.

50. Menza MA, Sage J, Marshall E, et al. Mood changes and 'on–off' phenomena in Parkinson's disease. Mov Disord 1990; 5: 148–151.

51. Maricle RA, Nutt JG, Carter JH. Mood and anxiety fluctuation in Parkinson's disease associated with levodopa infusion: preliminary findings. Mov Disord 1995; 10: 329–332.

52. Maricle RA, Nutt JG, Valentine RJ, et al. Dose–response relationship of levodopa with mood and anxiety in fluctuating Parkinson's disease: a double-blind, placebo-controlled study. Neurology 1995; 45: 1757–1760.

53. Maricle RA, Valentine RJ, Carter J, et al. Mood response to levodopa infusion in early Parkinson's disease. Neurology 1998; 50: 1890–1892.

54. Richard IH, Justus AW, Kurlan R. Relationship between mood and motor fluctuations in Parkinson's disease. J Neuropsychiatry Clin Neurosci 2001; 13: 35–41.

55. Damásio AR, Lobo-Antunes J, Macedo C. Psychiatric aspects in Parkinsonism treated with L-dopa. J Neurol Neurosurg Psychiatry 1971; 34: 502–507.

56. Keshavan MS, David AS, Narayanen HS, et al. 'On–off' phenomena and manic-depressive mood shifts: case report. J Clin Psychiatry 1986; 47: 93–94.

57. Lees AJ. The on–off phenomenon. J Neurol Neurosurg Psychiatry 1989; (Suppl): 29–37.

58. Friedenberg DL, Cummings JL. Parkinson's disease, depression, and the on–off phenomenon. Psychosomatics 1989; 30: 94–99.

59. Goodwin FK. Biochemical and pharmacological studies. In: Goodwin FK, Jamison KR, eds. Manic-Depressive Illness. New York: Oxford University Press, 1990: 416–502.

60. Riley DE, Lang AE. The spectrum of levodopa-related fluctuations in Parkinson's disease. Neurology 1993; 43: 1459–1464.

61. Vázquez A, Jiménez-Jiménez FJ, Garcia-Ruiz P, et al. 'Panic attacks' in Parkinson's

disease. A long-term complication of levodopa therapy. Acta Neurol Scand 1993; 87: 14–18.

62. Siemers ER, Shekhar A, Quaid K, et al. Anxiety and motor performance in Parkinson's disease. Mov Disord 1993; 8: 1–6.

63. Cantello R, Gilli M, Riccio A, et al. Mood changes associated with 'end-of-dose deterioration' in Parkinson's disease: a controlled study. J Neurol Neurosurg Psychiatry 1986; 49: 1182–1190.

64. Witjas T, Kaphan E, Azulay JP, et al. Nonmotor fluctuations in Parkinson's disease: frequent and disabling. Neurology 2002; 59: 408–413.

65. Carter JH, Lyons KS, Stewart BJ, et al. Do the clinical features of Parkinson's disease explain spouse caregiver strain? Neurology 2002; 58: A468 (abstract).

66. Koller WC, Montgomery EB. Issues in the early diagnosis of Parkinson's disease. Neurology 1997; 49: S10–S25.

67. Santamaria J, Tolosa E, Valles A. Parkinson's disease with depression: a possible subgroup of idiopathic parkinsonism. Neurology 1986; 36: 1130–1133.

68. Gonera EG, van't Hof M, Berger HJ, et al. Symptoms and duration of the prodromal phase in Parkinson's disease. Mov Disord 1997; 12: 871–876.

69. Taylor CA, Saint-Hilaire MH, Cupples LA, et al. Environmental, medical, and family history risk factors for Parkinson's disease: a New England-based case control study. Am J Med Genet 1999; 88: 742–749.

70. Shiba M, Bower JH, Maraganore DM, et al. Anxiety disorders and depressive disorders preceeding Parkinson's disease: a case–control study. Mov Disord 2000; 15: 669–677.

71. Nilsson FM, Kessing LV, Bolwig TG. Increased risk of developing Parkinson's disease for patients with major affective disorder: a register study. Acta Psychiatr Scand 2001; 104: 380–386.

72. Schuurman AG, Van Den Akker M., Ensinck KT, et al. Increased risk of Parkinson's disease after depression: a retrospective cohort study. Neurology 2002; 58: 1501–1504.

73. Leentjens AF, Van Den AM, Metsemakers JF, et al. Higher incidence of depression preceding the onset of Parkinson's disease: a register study. Mov Disord 2003; 18: 414–418.

74. Behari M, Srivastava AK, Das RR, et al. Risk factors of Parkinson's disease in Indian patients. J Neurol Sci 2001; 190: 49–55.

75. Bell IR, Amend D, Kaszniak AW, et al. Trait shyness in the elderly: evidence for an association with Parkinson's disease in family members and biochemical correlates. J Geriatr Psychiatry Neurol 1995; 8: 16–22.

76. Gotham AM, Brown RG, Marsden CD. 'Frontal' cognitive function in patients with Parkinson's disease 'on' and 'off' levodopa. Brain 1988; 111: 299–321.

77. Mayeux R, Stern Y, Sano M, et al. The relationship of serotonin to depression in Parkinson's disease. Mov Disord 1988; 3: 237–244.

78. Kostic VS, Filipovic SR, Lecic D, et al. Effect of age at onset on frequency of depression in Parkinson's disease. J Neurol Neurosurg Psychiatry 1994; 57: 1265–1267.

79. Liu CY, Wang SJ, Fuh JL, et al. The correlation of depression with functional activity in Parkinson's disease. J Neurol 1997; 244: 493–498.

80. Nilsson FM, Kessing LV, Sorensen TM, et al. Major depressive disorder in Parkinson's

disease: a register-based study. Acta Psychiatr Scand 2002; 106: 202–211.

81. Romanoski AJ, Folstein MF, Nestadt G, et al. The epidemiology of psychiatrist-ascertained depression and DSM-III depressive disorders. Results from the Eastern Baltimore Mental Health Survey Clinical Reappraisal. Psychol Med 1992; 22: 629–655.

82. Dooneief G, Mirabello E, Bell K, et al. An estimate of the incidence of depression in idiopathic parkinson's disease. Arch Neurol 1992; 49: 305–307.

83. Starkstein SE, Mayberg HS, Preziosi TJ, et al. Reliability, validity, and clinical correlates of apathy in Parkinson's disease. J Neuropsychiatry Clin Neurosci 1992; 4: 134–139.

84. Isella V, Melzi P, Grimaldi M, et al. Clinical, neuropsychological, and morphometric correlates of apathy in Parkinson's disease. Mov Disord 2002; 17: 366–371.

85. Menza MA, Cocchiola J, Golbe LI. Psychiatric symptoms in progressive supranuclear palsy. Psychosomatics 1995; 36: 550–554.

86. Paulus W, Jellinger K. The neuropathologic basis of different clinical subgroups of Parkinson's disease. J Neuropathol Exp Neurol 1991; 50: 743–755.

87. D'Amato RJ, Zweig RM, Whitehouse PJ, et al. Aminergic systems in Alzheimer's disease and Parkinson's disease. Ann Neurol 1987; 22: 229–236.

88. Menza MA, Palermo B, DiPaola R, et al. Depression and anxiety in Parkinson's disease: possible effect of genetic variation in the serotonin transporter. J Geriatr Psychiatry Neurol 1999; 12: 49–52.

89. Mossner R, Henneberg A, Schmitt A, et al. Allelic variation of serotonin transporter expression is associated with depression in Parkinson's disease. Mol Psychiatry 2001; 6: 350–352.

90. Mayeux R, Stern Y, Williams JB, et al. Clinical and biochemical features of depression in Parkinson's disease. Am J Psychiatry 1986; 143: 756–759.

91. Mayeux R, Stern Y, Cote L, et al. Altered serotonin metabolism in depressed patients with parkinson's disease. Neurology 1984; 34: 642–646.

92. Mayeux R. The 'serotonin hypothesis' for depression in Parkinson's disease. Adv Neurol 1990; 53: 163–166.

93. Kostic VS, Lecic D, Doder M, et al. Prolactin and cortisol responses to fenfluramine in Parkinson's disease. Biol Psychiatry 1996; 40: 769–775.

94. Murai T, Muller U, Werheid K, et al. In vivo evidence for differential association of striatal dopamine and midbrain serotonin systems with neuropsychiatric symptoms in Parkinson's disease. J Neuropsychiatry Clin Neurosci 2001; 13: 222–228.

95. Kim SE, Lee WY, Choe YS, et al. SPECT measurement of iodine-123-beta-CIT binding to dopamine and serotonin transporters in Parkinson's disease: correlation with symptom severity. Neurol Res 1999; 21: 255–261.

96. Doder M, Rabiner EA, Turjanski N, et al. Tremor in Parkinson's disease and serotonergic dysfunction: an 11C-WAY 100635 PET study. Neurology 2003; 60: 601–605.

97. Menza MA, Mark MH. Parkinson's disease and depression: the relationship to disability and personality. J Neuropsychiatry Clin Neurosci 1994; 6: 165–169.

98. Menza M. The personality associated with Parkinson's disease. Curr Psychiatry Rep 2000; 2: 421–426.

99. Menza MA, Golbe LI, Cody RA, et al. Dopamine-related personality traits in Parkinson's disease. Neurology 1993; 43: 505–508.

100. Tomer R, Aharon-Peretz J. Novelty seeking and harm avoidance in Parkinson's disease: effects of asymmetric dopamine deficiency. J Neurol Neurosurg Psychiatry 2004; 75: 972–975.

101. Cloninger CR. A systematic method for clinical description and classification of personality variants: a proposal. Arch Gen Psychiatry 1987; 44: 573–588.

102. Kaasinen V, Nurmi E, Bergman J, et al. Personality traits and brain dopaminergic function in Parkinson's disease. Proc Natl Acad Sci USA 2001; 98: 13272–13277.

103. Schultz W, Dayan P, Montague PR. A neural substrate of prediction and reward. Science 1997; 275: 1593–1599.

104. Schultz W. Dopamine neurons and their role in reward mechanisms. Curr Opin Neurobiol 1997; 7: 191–197.

105. Gambarana C, Ghiglieri O, Tagliamonte A, et al. Crucial role of D1 dopamine receptors in mediating the antidepressant effect of imipramine. Pharmacol Biochem Behav 1995; 50: 147–151.

106. Willner P. Sensitization of dopamine D2- or D3-type receptors as a final common pathway in antidepressant drug action. Clin Neuropharmacol 1995; 18: S49–S56.

107. Cantello R, Aguggia M, Gilli M, et al. Major depression in Parkinson's disease and the mood response to intravenous methylphenidate: possible role of the 'hedonic' dopamine synapse. J Neurol Neurosurg Psychiatry 1989; 52: 724–731.

108. Drevets WC, Gautier C, Price JC, et al. Amphetamine-induced dopamine release in human ventral striatum correlates with euphoria. Biol Psychiatry 2001; 49: 81–96.

109. Hildebrand J, Bourgeois F, Buyse M, et al. Reproducibility of monoamine metabolite measurements in human cerebrospinal fluid. Acta Neurol Scand 1990; 81: 427–430.

110. Meyer JH, Kruger S, Wilson AA, et al. Lower dopamine transporter binding potential in striatum during depression. Neuroreport 2001; 12: 4121–4125.

111. Ebert D, Feistel H, Kaschka W, et al. Single photon emission computerized tomography assessment of cerebral dopamine D2 receptor blockade in depression before and after sleep deprivation – preliminary results. Biol Psychiatry 1994; 35: 880–885.

112. Torack RM, Morris JC. The association of ventral tegmental area histopathology with adult dementia. Arch Neurol 1988; 45: 497–501.

113. Brown AS, Gershon S. Dopamine and depression. J Neural Transm Gen Sect 1993; 91: 75–109.

114. Corrigan MH, Denahan AQ, Wright CE, et al. Comparison of pramipexole, fluoxetine, and placebo in patients with major depression. Depress Anxiety 2000; 11: 58–65.

115. Kelly PH, Seviour PW, Iversen SD. Amphetamine and apomorphine responses in the rat following 6-OHDA lesions of the nucleus accumbens septi and corpus striatum. Brain Res 1975; 94: 507–522.

116. Laitinen L. Desipramine in treatment of Parkinson's disease. A placebo-controlled study. Acta Neurol Scand 1969; 45: 109–113.

117. Klaassen T, Verhey FR, Sneijders GH, et al. Treatment of depression in Parkinson's disease: a meta-analysis. J Neuropsychiatry Clin Neurosci 1995; 7: 281–286.

118. Torack RM, Morris JC. The association of ventral tegmental area histopathology with adult dementia. Arch Neurol 1988; 45: 497–501.

119. Becker T, Becker G, Seufert J, et al. Parkinson's disease and depression: evidence for an alteration of the basal limbic system detected by transcranial sonography. J Neurol Neurosurg Psychiatry 1997; 63: 590–596.

120. Berg D, Supprian T, Hofmann E, et al. Depression in Parkinson's disease: brainstem midline alteration on transcranial sonography and magnetic resonance imaging. J Neurol 1999; 246: 1186–1193.

121. McDonald WM, Krishnan KR. Magnetic resonance in patients with affective illness. Eur Arch Psychiatry Clin Neurosci 1992; 241: 283–290.

122. Lisanby SH, McDonald WM, Massey EW, et al. Diminished subcortical nuclei volumes in Parkinson's disease by MR imaging. J Neural Transm Suppl 1993; 40: 13–21.

123. Ring HA, Bench CJ, Trimble MR, et al. Depression in Parkinson's disease: a positron emission study. Br J Psychiatry 1994; 165: 333–339.

124. Mayberg HS, Starkstein SE, Sadzot B, et al. Selective hypometabolism in the inferior frontal lobe in depressed patients with Parkinson's disease. Ann Neurol 1990; 28: 57–65.

125. Mayberg HS. Frontal lobe dysfunction in secondary depression. J Neuropsychiatry Clin Neurosci 1994; 6: 428–442.

126. Jagust WJ, Reed BR, Martin EM, et al. Cognitive function and regional cerebral blood flow in Parkinson's disease. Brain 1992; 115: 521–537.

127. Mentis MJ, McIntosh AR, Perrine K, et al. Relationships among the metabolic patterns that correlate with mnemonic, visuospatial, and mood symptoms in Parkinson's disease. Am J Psychiatry 2002; 159: 746–754.

128. Stefurak TL, New P, Mahurin RK, et al. Response specific regional metabolic changes with fluoxetine treatment in depressed Parkinson's patients. Presented at Mental and Behavioral Dysfunction in Movement Disorders, Montreal, Canada, 13 October 2001.

129. Tessitore A, Hariri AR, Fera F, et al. Dopamine modulates the response of the human amygdala: a study in Parkinson's disease. J Neurosci 2002; 22: 9099–9103.

130. Cole SA, Woodard JL, Juncos JL, et al. Depression and disability in Parkinson's disease. J Neuropsychiatry Clin Neurosci 1996; 8: 20–25.

131. Sackeim HA, Greenberg MS, Weiman AL, et al. Hemispheric asymmetry in the expression of positive and negative emotion: Neurologic evidence. Arch Neurol 1982; 39: 210–218.

132. Miyawaki E, Perlmutter JS, Troster AI, et al. The behavioral complications of pallidal stimulation: a case report. Brain Cogn 2000; 42: 417–434.

133. Bejjani BP, Damier P, Arnulf I, et al. Transient acute depression induced by high-frequency deep-brain stimulation. N Engl J Med 1999; 340: 1476–1480.

134. Berney A, Vingerhoets F, Perrin A, et al. Effect on mood of subthalamic DBS for Parkinson's disease: a consecutive series of

24 patients. Neurology 2002; 59: 1427–1429.

135. Ollivier P, Witjas T, Azulay JP. Does the dopaminergic status influence the non-motor performances of Parkinsonian patients in pre-surgical evaluation? Mov Disord 2002; 17: S94 (abstract).

136. Green J, McDonald WM, Vitek JL, et al. Neuropsychological and psychiatric sequelae of pallidotomy for PD: clinical trial findings. Neurology 2002; 58: 858–865.

137. Öngür D, Price JL. The organization of networks within the orbital and medial prefrontal cortex of rats, monkeys and humans. Cereb Cortex 2000; 10: 206–219.

138. Mega MS, Cummings JL. Frontal–subcortical circuits and neuropsychiatric disorders. J Neuropsychiatry Clin Neurosci 1994; 6: 358–370.

139. Sheline YI. Neuroimaging studies of mood disorder effects on the brain. Biol Psychiatry 2003; 54: 338–352.

140. Drevets WC. Neuroimaging and neuropathological studies of depression: implications for the cognitive-emotional features of mood disorders. Curr Opin Neurobiol 2001; 11: 240–249.

141. Mayberg HS. Modulating dysfunctional limbic–cortical circuits in depression: towards development of brain-based algorithms for diagnosis and optimised treatment. Br Med Bull 2003; 65: 193–207.

142. Öngür D, Drevets WC, Price JL. Glial reduction in the subgenual prefrontal cortex in mood disorders. Proc Natl Acad Sci USA 1998; 95: 13290–13295.

143. Klimek V, Schenck J, Han H, et al. Dopaminergic abnormalities in amygdaloid nuclei in major depression: a postmortem study. Biol Psychiatry 2002; 52: 740–748.

144. Mayeux R, Stern Y, Rosen J, et al. Depression, intellectual impairment, and Parkinson disease. Neurology 1981; 31: 645–650.

145. Starkstein SE, Preziosi TJ, Bolduc PL, et al. Depression in Parkinson's disease. J Nerv Ment Dis 1990; 178: 27–31.

146. Sullivan PF, Neale MC, Kendler KS. Genetic epidemiology of major depression: review and meta-analysis. Am J Psychiatry 2000; 157: 1552–1562.

147. Leentjens AFG, Lousberg R, Verhey FRJ. Markers for depression in Parkinson's disease. Acta Psychiatr Scand 2002; 106: 196–201.

148. Kendler KS, Karkowski LM, Prescott CA. Causal relationship between stressful life events and the onset of major depression. Am J Psychiatry 1999; 156: 837–841.

149. Ringman JM, Diaz-Olavarrieta C, Rodriguez Y, et al. The prevalence and correlates of neuropsychiatric symptoms in a population with Parkinson's disease in Mexico. Neuropsychiatry Neuropsychol Behav Neurol 2002; 15: 99–105.

150. Ehmann TS, Beninger RJ, Gawel MJ, et al. Depressive symptoms in Parkinson's disease: a comparison with disabled control subjects. J Geriatr Psychiatry Neurol 1990; 3: 3–9.

151. Devanand DP, Adorno E, Cheng J, et al. Late onset dysthymic disorder and major depression differ from early onset dysthymic disorder and major depression in elderly outpatients. J Affect Disord 2004; 78: 259–267.

152. Black KJ, Sheline YI. Personality disorder scores improve with effective pharmacotherapy of depression. J Affect Disord 1997; 43: 11–18.

153. Beck AT, Ward CH, Mendelson M, et al. An inventory for measuing depression. Arch Gen Psychiatry 1961; 4: 561–571.

154. Levin BE, Llabre MM, Weiner WJ. Parkinson's disease and depression: psychometric properties of the Beck Depression Inventory. J Neurol Neurosurg Psychiatry 1988; 51: 1401–1404.

155. Starkstein SE, Preziosi TJ, Forrester AW, et al. Specificity of affective and autonomic symptoms of depression in Parkinson's disease. J Neurol Neurosurg Psychiatry 1990; 53: 869–873.

156. Leentjens AF, Verhey FR, Luijckx GJ, et al. The validity of the Beck Depression Inventory as a screening and diagnostic instrument for depression in patients with Parkinson's disease. Mov Disord 2000; 15: 1221–1224.

157. Cummings JL, Mega M, Gray K, et al. The Neuropsychiatric Inventory: comprehensive assessment of psychopathology in dementia. Neurology 1994; 44: 2308–2314.

158. Hamilton M. A rating scale for depression. J Neurol Neurosurg Psychiatry 1960; 23: 56–62.

159. Leentjens AF, Verhey FR, Lousberg R, et al. The validity of the Hamilton and Montgomery–Asberg depression rating scales as screening and diagnostic tools for depression in Parkinson's disease. Int J Geriatr Psychiatry 2000; 15: 644–649.

160. Spitzer R, Gibbon M, Williams J. Structured Clinical Interview for Axis I DSM-IV Disorders. New York: Biometrics Research Department, New York State Psychiatric Institute, 1994.

161. Starkstein SE, Mayberg HS, Preziosi TJ, et al. Reliability, validity, and clinical correlates of apathy in Parkinson's disease. J Neuropsychiatry Clin Neurosci 1992; 4: 134–139.

162. Levy ML, Cummings JL, Fairbanks LA, et al. Apathy is not depression. J Neuropsychiatry Clin Neurosci 1998; 10: 314–319.

163. Friedman J, Friedman H. Fatigue in Parkinson's disease. Neurology 1993; 43: 2016–2018.

164. Karlsen K, Larsen JP, Tandberg E, et al. Fatigue in patients with Parkinson's disease. Mov Disord 1999; 14: 237–241.

165. Farber NB, Black KJ. Psychiatric disorders associated with general medical conditions. In: Rubin EH, Zorumski CF, eds. Adult Psychiatry. Bristol, UK: Blackwell, 2004.

166. Black DW. Pathological laughter. A review of the literature. J Nerv Ment Dis 1982; 170: 67–71.

167. Dark FL, McGrath JJ, Ron MA. Pathological laughing and crying. Aust NZ J Psychiatry 1996; 30: 472–479.

168. Okun MS, Raju DV, Walter BL, et al. Pseudobulbar crying induced by stimulation in the region of the subthalamic nucleus. J Neurol Neurosurg Psychiatry 2004; 75: 921–923.

169. Pluck GC, Brown RG. Apathy in Parkinson's disease. J Neurol Neurosurg Psychiatry 2002; 73: 636–642.

170. Aarsland D, Larsen JP, Lim NG, et al. Range of neuropsychiatric disturbances in patients with Parkinson's disease. J Neurol Neurosurg Psychiatry 1999; 67: 492–496.

171. Funkiewiez A, Ardouin C, Caputo E, et al. Long term effects of bilateral subthalamic nucleus stimulation on cognitive function, mood, and behaviour in Parkinson's disease. J Neurol Neurosurg Psychiatry 2004; 75: 834–839.

172. James W. The Varieties of Religious Experience. Cambridge, MA: Harvard University Press, 1985.

173. Nierenberg AA, Feinstein AR. How to evaluate a diagnostic marker test: lesions from the rise and fall of the dexamethasone suppression test. JAMA 1988; 259: 1699–1702.

174. Kostic VS, Covickovic-Sternic N, Beslac-Bumbasirevic L, et al. Dexamethasone suppression test in patients with Parkinson's disease. Mov Disord 1990; 5: 23–26.

175. Kostic VS, Susic V, Przedborski S, et al. Sleep EEG in depressed and nondepressed patients with Parkinson's disease. J Neuropsychiatry Clin Neurosci 1991; 3: 176–179.

176. MDS Task Force. Treatment of depression in idiopathic Parkinson's disease. Mov Disord 2002; 17: S112–S119.

177. Chung TH, Deane KH, Ghazi-Noori S, et al. Systematic review of antidepressant therapies in Parkinson's disease. Parkinsonism Relat Disord 2003; 10: 59–65.

178. Ghazi-Noori S, Chung TH, Deane KH, et al. Therapies for depression in Parkinson's disease. Cochrane Database Syst Rev 2003; CD003465.

179. Andersen J, Aabro E, Gulmann N, et al. Anti-depressive treatment in Parkinson's disease. A controlled trial of the effect of nortriptyline in patients with Parkinson's disease treated with L-DOPA. Acta Neurol Scand 1980; 62: 210–219.

180. Richard IH, Kurlan R, Parkinson Study Group. A survey of antidepressant drug use in Parkinson's disease. Neurology 1997; 49: 1168–1170.

181. Wermuth L, Sorensen PS, Timm B, et al. Depression in idiopathic Parkinson's disease treated with citalopram: a placebo-controlled trial. Nord J Psychiatry 1998; 52: 163–69.

182. Richard IH, Kurlan R, Tanner C, et al. Serotonin syndrome and the combined use of deprenyl and an antidepressant in Parkinson's disease. Neurology 1997; 48: 1070–1077.

183. Strang RR. Imipramine in treatment of parkinsonism: a double-blind placebo study. Br Med J 1965; 5452: 33–34.

184. Lemke MR. Effect of reboxetine on depression in Parkinson's disease patients. J Clin Psychiatry 2002; 63: 300–304.

185. Gordon PH, Pullman SL, Louis ED, et al. Mirtazapine in Parkinsonian tremor. Parkinsonism Relat Disord 2002; 9: 125–126.

186. Mendlewicz J, Youdim MBH. L-Deprenil, a selective monoamine oxidase type b inhibitor, in the treatment of depression: a double blind evaluation. Br J Psychiatry 1983; 142: 508–511.

187. Bodkin JA, Kwon AE. Selegiline and other atypical monoamine oxidase inhibitors in depression. Psychiatric Annals 2001; 31: 385–391.

188. Andersen K, Balldin J, Gottfries CG, et al. A double-blind evaluation of electroconvulsive therapy in Parkinson's disease with 'on–off' phenomena. Acta Neurol Scand 1987; 76: 191–199.

189. Fall PA, Ekman R, Granerus AK, et al. ECT in Parkinson's disease. Changes in motor symptoms, monoamine metabolites and neuropeptides. J Neural Transm Park Dis Dement Sect 1995; 10: 129–140.

190. Wengel SP, Burke WJ, Pfeiffer RF, et al. Maintenance electroconvulsive therapy for intractable Parkinson's disease. Am J Geriatr Psychiatry 1998; 6: 263–269.

191. Moellentine C, Rummans T, Ahlskog JE, et al. Effectiveness of ECT in patients with parkinsonism. J Neuropsychiatry Clin Neurosci 1998; 10: 187–193.

192. Figiel GS, Hassen MA, Zorumski CF, et al. ECT-induced delirium in depressed patients with Parkinson's disease. J Neuropsychiatry Clin Neurosci 1991; 3: 405–411.

193. Okun MS, McDonald WM, DeLong MR. Refractory nonmotor symptoms in male patients with Parkinson disease due to testosterone deficiency: a common unrecognized comorbidity. Arch Neurol 2002; 59: 807–811.

194. Okun MS, Walter BL, McDonald WM, et al. Beneficial effects of testosterone replacement for the nonmotor symptoms of Parkinson disease. Arch Neurol 2002; 59: 1750–1753.

195. Reutens S, Sachdev P. Homocysteine in neuropsychiatric disorders of the elderly. Int J Geriatr Psychiatry 2002; 17: 859–864.

196. Muller T, Werne B, Fowler B, et al. Nigral endothelial dysfunction, homocysteine, and Parkinson's disease. Lancet 1999; 354: 126–127.

197. Di Rocco A, Rogers JD, Brown R, et al. *S*-Adenosyl-methionine improves depression in patients with Parkinson's disease in an open-label clinical trial. Mov Disord 2000; 15: 1225–1229.

198. Wassermann EM, Lisanby SH. Therapeutic application of repetitive transcranial magnetic stimulation: a review. Clin Neurophysiol 2001; 112: 1367–1377.

199. Dragasevic N, Potrebic A, Damjanovic A, et al. Therapeutic efficacy of bilateral prefrontal slow repetitive transcranial magnetic stimulation in depressed patients with Parkinson's disease: an open study. Mov Disord 2002; 17: 528–532.

200. Ghabra MB, Hallett M, Wassermann EM. Simultaneous repetitive transcranial magnetic stimulation does not speed fine movement in PD. Neurology 1999; 52: 768–770.

201. Muller V, Mohr B, Rosin R, et al. Short-term effects of behavioral treatment on movement initiation and postural control in Parkinson's disease: a controlled clinical study. Mov Disord 1997; 12: 306–314.

202. Ramig LO, Fox C, Sapir S. Parkinson's disease: speech and voice disorders and their treatment with the Lee Silverman Voice Treatment. Semin Speech Lang 2004; 25: 169–180.

203. Chatterjee A, Fahn S. Methylphenidate treats apathy in Parkinson's disease. J Neuropsychiatry Clin Neurosci 2002; 14: 461–462.

204. Czernecki V, Pillon B, Houeto JL, et al. Motivation, reward, and Parkinson's disease: influence of dopatherapy. Neuropsychologia 2002; 40: 2257–2267.

205. Black KJ. Diagnosing depression after stroke. South Med J 1995; 88: 699–708.

206. Herlofson K, Larsen JP. The influence of fatigue on health-related quality of life in patients with Parkinson's disease. Acta Neurol Scand 2003; 107: 1–6.

207. Lou JS, Kearns G, Oken B, et al. Exacerbated physical fatigue and mental fatigue in Parkinson's disease. Mov Disord 2001; 16: 190–196.

208. Lou JS, Kearns G, Benice T, et al. Levodopa improves physical fatigue in Parkinson's disease: a double-blind, placebo-controlled, crossover study. Mov Disord 2003; 18: 1108–1114.

209. Elkin I, Shea MT, Watkins JT, et al. National Institute of Mental Health Treatment of Depression Collaborative Research Program. General effectiveness of treatments. Arch Gen Psychiatry 1989; 46: 971–982.

210. Caap-Ahlgren M, Dehlin O. Insomnia and depressive symptoms in patients with Parkinson's disease. Relationship to health-related quality of life. An interview study of patients living at home. Arch Gerontol Geriatr 2001; 32: 23–33.

211. Starkstein SE, Preziosi TJ, Robinson RG. Sleep disorders, pain, and depression in Parkinson's disease. Eur Neurol 1991; 31: 352–355.

212. Schrag A, Hovris A, Morley D, et al. Young- versus older-onset Parkinson's disease: impact of disease and psychosocial consequences. Mov Disord 2003; 18: 1250–1256.

10

Depression in Alzheimer's disease

Paul B Rosenberg and Constantine G Lyketsos

ALZHEIMER'S DISEASE

Epidemiology

Alzheimer's disease (AD) is a major cause of disability and a burden to society and caregivers. There are currently an estimated 4.5 million patients with AD in the USA, a number predicted to increase approximately 3-fold by 2050[1]. In late stages of the disease, AD patients become fully dependent for their activities of daily living (ADL) and many require nursing home admission. The combined cost of medical and family care for AD is estimated at $100 billion/year in the USA alone[2,3], with the greatest financial cost incurred by the time spent on caregiving by families[4]. It can be fairly argued that the greatest costs are emotional, with the increasing burden of caregiving leading to depression, guilt, anger and worsening health for the caregiver[5].

Clinical course

AD is largely a disease of old age. Estimates of its prevalence increase from about 1% at ages 60–64 to 43–68% at ages >95[6,7]. The onset of AD is usually insidious, with the initial presentation typically involving problems with recall or changes in personality. Patients are often described at these early stages as forgetful, worried, inattentive, or depressed, and are frequently diagnosed with depression before cognitive symptoms are apparent. Patients with early AD typically have neuropsychological deficits in verbal episodic memory and executive function[8,9] with deficits in verbal fluency, comprehension, gnosis and praxis developing as the disease progresses[10]. AD is also associated with increased mortality, with the mean duration of survival after onset estimated at 3–3.5 years after diagnosis[11]. AD is often accompanied by motor symptoms including parkinsonism.

Neurobiological findings

The brain of AD patients is characterized by cortical atrophy, deep cortical sulci, enlarged ventricles and loss of total brain mass[12]. The atrophy tends to be symmetrical without hemispheric differences or focal lobar tissue loss[13]. Neuronal dysfunction in widespread regions of the neocortex is accompanied by extracellular deposition of β-amyloid fibrils and intraneuronal accumulations of neurofibrillary tangles[14]. β-Amyloid fibrils are insoluble aggregates of amyloid-β, a 40–42 amino acid peptide product of the metabolism of amyloid precursor protein (APP) by the β- rather than the α-secretase cleavage pathway, followed by further cleavage by γ-secretase. Amyloid-β 1-42 forms these insoluble aggregates much more readily than amyloid-β 1-40, and in rare familial autosomal dominant variants of AD APP has a single base-pair mutation that leads to altered metabolic processing and a relative increase in the proportion of amyloid-β 1-42 compared to 1-40[14]. β-Amyloid plaques form throughout the cortex but are especially concentrated in temporal and parietal association cortices. β-Amyloid aggregates, amyloid-β monomers or oligomers are suspected to be neurotoxic moieties and the cause of neuronal loss in AD[14]. Neurofibrillary tangles consist of intraneuronal deposition of phosphorylated tau protein initially found in the hippocampus, entorhinal cortex and limbic areas, but spreading to many cortical areas in advanced disease[15,16].

Neurotoxicity appears to be mediated by the activation of microglia by amyloid-β, followed by activation of the complement cascade and release of pro-inflammatory cytokines such as interleukin (IL)-1, IL-6 and tumor necrosis factor (TNF)-α[17]. The presence of these inflammatory mediators suggests that neuronal death may in part be due to autoimmune mechanisms. Neuronal dysfunction and cognitive deficits in AD are not closely correlated with the number or regional distribution of plaques or tangles, however, but may instead be associated with activated microglia releasing cytokines and activating complement[15–18]. The most prominent change in neurotransmitter synthesis and release is decrease in cholinergic neurons originating in the nucleus basalis of Meynert providing excitatory innervation to widespread areas of the cortex[19]. The latter finding provides the rationale for some current treatments of AD with acetylcholinesterase inhibitors. In addition, there are changes in serotonergic, dopaminergic and *N*-methyl-D-aspartate (NMDA) neurotransmission, which may lead to vulnerability to neuropsychiatric complications of AD.

The hippocampi and entorhinal cortices of AD patients are decreased in volume compared to those of controls on structural magnetic resonance imaging (MRI)[20]. Further studies have also suggested a decreased volume of the anterior and posterior cingulate cortex. Functional imaging (single photon emission computed tomography (SPECT) and positron emission tomography (PET) scans) show a pattern of temporoparietal hypometabolism. These symmetrical and posterior deficits distinguish AD from frontotemporal dementias which demonstrate hypometabolism of the frontal cortex[20].

EPIDEMIOLOGY OF DEPRESSION IN ALZHEIMER'S DISEASE

Neuropsychiatric complications are common in AD. They include apathy, depression, hallucinations, delusions, anxiety, agitation, insomnia, catastrophic reactions and mania[21–24]. Mood disorders are the most common neuropsychiatric complications of dementia according to the population-based Cache County survey, where 21% of participants with dementia exhibited apathy and 18% exhibited depression[25]. The annual incidence of depressive symptoms in AD was estimated to be 18% in this cohort. Similar estimates of incidence and prevalence have been reported from population-based studies in Great Britain[21–24] and the Cardiovascular Health Study[26]. The latter study reported prevalence rates of depressive symptoms in dementia to be 4- to 5-fold higher than in cognitively intact individuals; this increased prevalence was similar for other dementias. Risk factors for developing depression in AD include a history of depressive episodes[27,28], family history of depression[29,30], presenile onset of dementia[31] and possibly low educational level[32].

ASSESSMENT OF DEPRESSION IN ALZHEIMER'S DISEASE

Diagnostic dilemmas

Clinicians have long felt that applying strict DSM criteria for major depression in AD leads to inclusion of only a narrow and particularly acute group of patients, and excludes many patients with clinically significant depressive symptoms. Patients with AD often have fluctuating behavioral symptoms which do not fit the DSM criteria of 'most of the day, nearly every day', yet clearly are detrimental to functioning. The strict time and acuity constraints of DSM may be too narrow for this population. Similarly, the requirement that patients have five of nine criteria may be too strict, particularly because depressed AD patients are less likely than depressed patients without dementia to complain of depressed mood, or have suicidal ideation, or prominent sleep and appetite changes (see below). Formulating concepts of mood, hopelessness and suicidality involves higher-order cognitive functions and the ability to integrate emotions and behavioral reactions over time. This requires the ability to recall recent events and to think abstractly, which are typically deficient in AD. AD patients will often answer questions about their mood with clichéd, 'empty' responses, possibly due to decreased verbal fluency[33]. These cognitive deficits probably introduce bias into diagnostic schemes that rely heavily on patient self-report. This may account for frequent discrepancies between patient and caregiver report of mood symptoms in AD (see below).

Another consideration that applies in general to the diagnosis of depression in elderly patients is the overlap of medical and psychiatric symptoms in a population with a high prevalence of chronic disease. DSM criteria and rating scales such as the Hamilton Depression Rating Scale (HAM-D) are heavily weighted with neurovegetative signs of mood disorder such as insomnia, anorexia, weight loss, fatigue and psychomotor retardation. If the examiner judges that a neurovegetative symptom is more likely to be due to medical illness it is not counted towards a depressive diagnosis, although this distinction is not as clear on rating scales. In practice, this attributional judgment is often difficult and, given the high prevalence of chronic illness, fatigue and insomnia in the elderly population in general[34,35], it is not clear that these neurovegetative symptoms have adequate specificity for diagnosis of depression in the elderly.

The differing spectrum of symptoms

Depression in Alzheimer's disease has a different constellation of symptoms than major depressive disorder as defined in DSM-IV. Several studies have compared symptom clusters in depressed patients with and without dementia. Janzing *et al.*[36] found that demented patients had more 'motivation' symptoms and fewer 'mood' symptoms than a non-demented group, despite similar severity of depressive symptoms. Their cohort had the advantage of including subsyndromal depressive symptoms and may be particularly representative of AD patients in the community. Li *et al.*[37] similarly noted that motivational symptoms predominate among depressive symptoms of AD. Zubenko *et al.*[38] compared depressed AD and non-demented patients with the Clinical Assessment of Depression in Dementia (CADD) incorporating questions from the HAM-D, Neuropsychiatric Inventory (NPI) and Structured Clinical Interview for Diagnosis. They found depressed AD patients to be more likely to complain of difficulties with concentration and to have delusions/hallucinations, and to be less likely to feel excessively guilty, as well as a trend toward less suicidal ideation. These findings concur with results[39,40] showing that psychosis and depression are associated in AD patients. It is noteworthy that, in these studies, insomnia and weight loss are equally common in AD and non-demented patients, in sharp distinction to the results reported by Purandare *et al.*[41], in which mood symptoms and neurovegetative signs were among the best questions to discriminate between the two groups. However, Purandare *et al.* were examining the baseline prevalence of depressive symptoms in AD patients in general, and their findings may not generalize to patients in AD with depressive syndromes. Rubin *et al.*[42] similarly reported that indecision discriminates between depressed and non-depressed AD, along with depressed mood and fatigue.

Rating scales

There are three major approaches to rating depressive symptoms in AD. The first approach is to include depression in a general inventory of neuropsychiatric difficulties such as the Behavioral Pathology in Alzheimer's Disease Rating Scale (BEHAVE-AD)[43], the NPI[44] and the CERAD Behavior Rating Scale for AD[45]. These scales have the advantage of anchoring depressive symptoms as one of a group of problem behaviors or affects that AD patients exhibit and that are troubling to caregivers. However, these scales merely screen for depression with a small number of questions and are not sensitive or specific enough to detect change in a clinical trial.

A second approach is to adapt rating instruments used for younger and cognitively healthy patients such as the HAM-D[46–48], Geriatric Depression Scale (GDS)[49], Zung Scale[50], Montgomery–Asberg Depression Rating Scale (MADRS)[51,52] and Beck Depression Inventory (BDI)[53]. Each of these scales has been used with a reasonable degree of validity and reliability in mild-to-moderate AD. However, they have disadvantages including: (1) being heavily weighted toward patient self-report, which notoriously underrates depressive symptoms in AD (see below)[54,55]; (2) the HAM-D being substantially influenced by somatic symptoms, which often overlap with medical illness in the elderly; and (3) each of these scales involving a significant degree of verbal comprehension (particularly the BDI), which is often impaired in AD. Naarding *et al.*[56] reported that to get the maximum specificity and sensitivity from the HAM-D requires different cut-offs for AD, stroke and Parkinson's disease. The MADRS has been developed to be sensitive to change and has shown treatment effects in medication trials[51,52].

A third approach is to use a 'dementia-specific' depression scale. The most widely used scale is the Cornell Scale for Depression in Dementia (CSDD), which combines observation, patient input and caregiver input into a consensus determined by the interviewer[57]. The CSDD has 19 items, seven of which involve somatic/neurovegetative signs; this relatively light weighting towards somatic symptoms minimizes overlap with chronic medical illness. It is easily administered to patients with a wide range of cognitive abilities, is sensitive to treatment effects[46,51,58] and is practical for screening for depression in nursing home populations[59]. A cut-off score of 13 or more is suggested for a major depressive episode (MDE), but a score of 6 or more is suggested for depression in AD. The Dementia Mood Assessment Scale[60] is a similar scale with good reliability and face validity in demented patients, although it has not been widely used. The CADD[38] is a promising new instrument that assesses both diagnosis and severity, incorporating items from the HAM-D and NPI, with robust face validity and interrater reliability.

Caregiver bias in depression assessment

Clearly caregiver input is needed for the objective assessment of depressive symptoms in dementia. Patients may lack a clear sense of time passing, of their functional level, and may not recall significant and recurrent symptoms. The CSDD and NPI are rating scales that encourage or require caregiver input in the assessment. However, it is possible that caregivers may give a biased assessment, and in general clinicians often find patients underreporting and caregivers overreporting depressive symptoms (and neuropsychiatric symptoms in general in AD). There are some data that support these opinions. Teri and Truax[54] found a moderate correlation between caregiver depression and patient depression; in addition, caregivers meeting criteria for major depression in their study were more likely to report patient symptoms severe enough to diagnose major depression in the patient. Non-depressed caregivers appeared more neutral and better able to distinguish severity of mood symptoms in AD. Rubin *et al.*[42] reported a sharper distinction between patient and caregiver reports of depression in AD, with factor analysis deriving different factors for each. Their results concur with those of Burke *et al.*[55], reporting that caregivers report more depressive symptoms than AD patients. Neundorfer *et al.*[61] found a closer relationship between caregiver and patient mood in depressed AD patients; the rate of change in caregiver mood was also associated with the rate of change in patient mood and decline in instrumental ADL.

These studies point out the close relationship between caregiver mood, caregiver assessment of patient mood and AD patients' self-assessment of mood. In the absence of a 'gold standard' validation for depression of AD they do not clarify whether patients underreport or caregivers overreport mood symptoms in AD. Thus, the clinician is advised to utilize *both* caregiver and patient input in assessing mood symptoms in AD, but not to take an average of the two. Rather, the clinician should take into account apparent caregiver distress, assess whether this may be biasing the caregiver report, and integrate these findings with observation and patient report into a coherent whole. This process may be significantly more complex than a typical assessment of mood disorder in the young adult, and 'depression of Alzheimer's disease' diagnostic criteria (below) have been proposed to clarify this process.

Apathy versus depression

Apathy is an evolving concept in neuropsychiatry, generally defined as lack of motivation or 'will'. It is a notable source of stress for family caregivers, who may feel emotionally drained by feeling that they have to supply motivation to the patient for everyday activities. The distinction between apathy and depression is less clear; in practice apathy is often treated as a symptom of mood disorder rather than a separate syndrome. In the Cache County study it was the most frequent neuropsychiatric symptom of dementia[25] and was often present without prominent mood disturbance. Starkstein *et al.*[62] reported that 37% of a large clinical sample of AD patients were apathetic. Many (24%) had apathy coexisting with major depression. Apathy is associated with poorer cognitive functioning and worse ADL impairment[63], and may be a symptom of preclinical AD[64]. Since apathy is closely associated with measures of executive dysfunction[65] rather than mood measures, it has not been included as a diagnostic criterion for 'depression of Alzheimer's disease' (below).

Proposed criteria for 'depression of Alzheimer's disease'

In light of the above considerations regarding the diagnosis of depression in AD, the National Institutes of Mental Health (NIMH) convened a consensus conference of experts in the field to develop coherent diagnostic criteria[66,67]. Table 10.1 presents the proposed NIMH criteria for 'depression of Alzheimer's disease' with comments to clarify the diagnostic process for clinical use. These criteria define an atypical syndrome of depression common in AD, incorporating many of the unique phenomenological findings outlined above and avoiding criteria that have notable overlap with cognitive deficits and/or chronic medical illness.

CAUSES OF DEPRESSION IN ALZHEIMER'S DISEASE

Patients with depressive syndromes usually have multiple risk factors rather than a single clear

Table 10.1 Operationalized criteria for depression of Alzheimer's disease (dAD)

General comments: The process of rating the dAD criteria

Rating of the dAD criteria is best carried out using all available clinical information, as appropriate to each clinical or research situation. Sources to include consist of patient report, caregiver report, review of medical records and examiner/rater observations. The process must include an interview with the patient and separately with his/her caregiver or another knowledgeable informant. After completing both interviews the interviewer must make clinical judgments reconciling any differences between the sources of information, and then rate the items below.

The rater must often make a judgment as to whether the patient's self-report is valid or influenced by impaired cognition. The rater should make this judgment based on the severity of cognitive impairment, corroboratory information from caregiver(s) and the medical record, and observations made during clinical interview.

Similarly, the rater must make a judgment as to whether the caregiver's report of symptoms is valid or appears influenced by mood factors (e.g. caregiver depression), cultural factors (e.g. caregiver's belief in how the participant should act or how elderly people should behave in general), denial (e.g. caregiver's desire to minimize symptoms) and/or caregiver education (e.g. caregiver's degree of knowledge and sophistication regarding dementia and its complications).

Elements of dAD (in italics are the criteria from reference 66 and comments from reference 68)

A. Three (or more) of the following symptoms have been present during the same 2-week period and represent a change from previous functioning: at least one of the symptoms is either (1) depressed mood or (2) decreased interest and pleasure. To be rated as present, each pertinent symptom must be present most days, for hours per day, during the previous two weeks.

continued

Table 10.1 *continued*

Comment: The rater should not rate as present items that in his/her judgment are clearly due to a medical condition other than Alzheimer disease. For example, there are many medical conditions that can cause fatigue and/or psychomotor retardation, including malignancy, congestive heart failure, chronic obstructive pulmonary disease, neuromuscular conditions, overuse of sedative medications, metabolic abnormalities (hypoglycemia, hyperglycemia, hypernatremia), and conditions associated with neurological deficits (particularly sensorimotor deficits, apraxia or agnosia). The 'mask facies' and diminished facial affective responses seen in Parkinson's disease may mimic depressed mood or anhedonia. Anorexia or weight loss may be due to malignancy or gastrointestinal pathology. Frequent awakenings may be caused by pain, by respiratory distress due to sleep apnea, or need to toilet or to be cleaned (listed under item 5). Endocrine abnormalities such as hypothyroidism, apathetic hyperthyroidism, Cushing's disease, Addison's disease and paracrine disturbances of malignancy may mimic many of the depressive symptoms listed.

Criterion	*Comment*
1. Clinically significant depressed mood (e.g. depressed, sad, hopeless, discouraged, tearful)	*The rater may distinguish between 'not caring' (apathy) from more acute/distressed sad mood (depression), and may need to distinguish the patient's statements from behavior observed by others. The transient affective lability of 'emotional incontinence' probably should not lead to a positive rating if it is not associated with changes in mood outside the periods of mood 'incontinence'.*
2. Decreased positive affect or pleasure in response to social contacts and usual activities	*To illustrate this item, the examiner might wish to elicit examples from the patient and caregiver of activities that the patient has usually found pleasurable in the past or that most people find pleasurable (e.g. a good meal, visiting with grandchildren, attending a family event, etc.). Reports of behavior tend to be most important in rating this item, often contradicting patient's recall of events. Useful questions include whether the participant is still 'having as much fun as he/she used to', 'still enjoying life', or 'has a zest for life'.*
3. Social isolation or withdrawal	*The examiner assesses both the patient's desire to seek out people and their sociability when with other people. Social isolation/withdrawal does not lead to a positive rating if they are lifelong traits or appear to be motivated by an effort to avoid tasks that challenge cognition.*
4. Disruption in appetite	*Because a patient's cognitive deficits may especially interfere with recall and interpretation of appetite or weight change, it is essential to corroborate information on this item with caregiver reports, and if possible, documentation of weight changes.*
5. Disruption in sleep	*The examiner rates change from usual sleep habits and/or a sense of poor quality of sleep. It may be difficult to assess sleep accurately if the patient sleeps alone. Awakenings solely for purposes of toileting or hygiene if the patient is able to fall back to sleep rapidly (within 30 minutes) are best not used to make a positive rating.*

continued

Table 10.1 *continued*

Criterion	*Comment*
6. Psychomotor changes (e.g. agitation or retardation)	*Agitation: This may include inability to sit still, pacing and/or hand-wringing. This does not imply disruptive or oppositional behavior or combativeness/violent behavior. Agitation and irritability (criterion 7) are often seen together. Efforts are best made to distinguish agitation from anxiety, the latter being common but not specifically part of the dAD criteria. Note that agitation can coexist with fatigue/loss of energy.* *Retardation: The rater will need to assess the contribution of medical conditions to this item.*
7. Irritability	*Irritability is rated when it represents a change from the patient's usual personality traits. Typical descriptions include having a short temper, 'flying off the handle', or 'going off over little things'. This item particularly requires sensitivity to cultural differences and diverse use of language. It excludes simple frustration over cognitive deficits or difficulties performing daily tasks. Delirium can mimic this symptom.*
8. Fatigue or loss of energy	*This symptom may be most apparent after taking a medical review of systems, and is often mimicked by medical illnesses. The examiner has to judge whether the degree of fatigue seems out of proportion to the burden of medical illness.*
9. Feelings of worthlessness, hopelessness, or excessive or inappropriate guilt	*Feelings of pessimism, negativity, lowered self-esteem, a sense of foreshortened future, self-blame, decreased confidence and delusional guilt might be included in this rating. These feelings may be elicited when patients encounter their cognitive limitations during mental status examination. It is best to confirm that they are present at other times, so as to avoid making a positive rating entirely based on frustration with failing cognitive testing.*
10. Recurrent thoughts of death, suicidal intention, plan, or attempt	*This rating includes thoughts of wanting to die whether or not the patient appears to have 'intent'. It is important to distinguish desire to die from a decision to forego aggressive medical care in accordance with spiritual, moral and ethical beliefs.*

B. All criteria are met for diagnosis of dementia of the Alzheimer type (DSM-IV-TR).

C. The symptoms cause clinically significant distress or disruption in functioning.

It is important for this item to take note of changes, not lifelong predispositions. The examiner might consider typical areas of functioning such as social, occupational and psychological, as well as the social context of the patient to decide what areas of functioning are relevant. For example, if the patient is a nursing-home resident with advanced dementia, the rater might consider the ability to assist with activities of daily living, participation in recreational activities, whether he/she acknowledges and

continued

Table 10.1 *continued*

participates in visits by family and friends, etc. For a patient with early or questionable dementia residing alone in the community the examiner might consider, for example, their ability to participate in paid or volunteer work, travel independently, socialize with friends or family outside the home, pay bills and maintain independence in other instrumental activities of daily living. The appropriateness of areas of functioning will also depend on cultural norms (e.g. an elderly man may never have cooked regularly for himself, and so lack of ability to cook is not a significant finding) or personal history (e.g. occupational and educational history may affect the range of current interests, activities and hobbies). The examiner is encouraged to make rating this item as specific to the individual as possible.

D. The symptoms do not occur exclusively during the course of a delirium.

Delirium can mimic virtually all the symptoms of dAD and is probably the most important condition to rule out. The attentional disturbance of delirium is particularly likely to mimic anhedonia, social withdrawal, appetite disturbance, fatigue and psychomotor agitation or retardation. Delirious patients are notoriously impulsive and are thus prone to emotional over-reaction to minor stressors, irritability, feeling of worthlessness and suicidal thoughts/acts. In addition, sleep disturbances are very common in delirium.

E. The symptoms are not due to the direct physiological effects of a substance (e.g. a drug of abuse or a medication).

When considering the effects of abused substances (such as alcohol, stimulants, opioids, benzodiazepines, barbiturates), the examiner is reminded that patients can appear depressed both while intoxicated and while withdrawing from these substances, and that the mood disorder of withdrawal can be protracted (i.e. lasting weeks or months) particularly in fragile elderly persons. When considering prescription medications, the commonest to mimic depression are sedating medications (i.e. opioid analgesics, benzodiazepines, anesthetics, barbiturates, anticonvulsants, antidepressants, antipsychotics). The examiner must also consider the possibility of idiosyncratic reactions to any medication. The presence of 'therapeutic' levels does not rule out medication toxicity, because many medications with an established range of therapeutic serum levels can cause mimicking of depression when at supratherapeutic levels or in the high-therapeutic part of the range (e.g. digoxin, lithium, phenytoin, carbamazepine, valproic acid, amitriptyline and nortriptyline).

F. The symptoms are not better accounted for by other conditions such as major depressive disorder, bipolar disorder, bereavement, schizophrenia, schizoaffective disorder, psychosis of Alzheimer disease, anxiety disorder, or substance-related disorder.

Specify if:

Co-occurring onset: if onset antedates or co-occurs with the AD symptoms

Post-AD onset: if onset occurs after AD symptoms

Specify:

With psychosis of Alzheimer's disease

With other significant behavioral signs and symptoms

With history of mood disorder

cause. AD patients are no exception, and on careful scrutiny the clinician can usually identify psychosocial and neurobiological factors making patients vulnerable to depression. Understanding possible causes of depression may help the clinician optimize treatment and particularly help with psychoeducation of the patient and caregivers.

Psychosocial factors

Many people believe that depression is a natural and inevitable reaction to the increasing disabilities of AD (e.g. 'Who wouldn't be depressed?') However, given that the majority of AD patients do not have clinically significant depressed mood, depression is not an inevitable reaction. For example, it is often assumed that awareness of deficits is a major risk for depression, and that lack of awareness in advanced disease is protective against depression. AD patients tend to have insight into their cognitive deficits early in the course of disease[69]. There are several studies showing no association between insight into AD and depression (for example, see reference 70). However, some studies have supported the idea that insight is associated with depression[71] and particularly with hopelessness[72].

Other psychosocial stressors that might lead to depression include loss of occupational or social roles and decreased ability to socialize. This role change is often subtle, insidious and drawn out in time. Families may make strenuous efforts to help the patient believe that they still maintain significant role functioning. For example, many AD patients sign checks that their spouses have made out, maintaining the illusion that they are still in charge of the finances. Another stressor is the AD patient's awareness that someone else has assumed a significant role in their lives and their care. Even at relatively advanced stages of AD patients may have a vague feeling that their spouse is taking care of them against the 'natural order of things' and that they should be 'doing more'. This may lead to a sense of grieving, to a mourning of losses and possibly to depression.

In addition to these general issues of role change and self-esteem, the specific cognitive deficits of AD can be detrimental to a patient's self-image. One of the most poignant examples is expressive aphasia, which affects a substantial proportion of AD patients[73]. When patients cannot speak fluently they often feel frustrated and embarrassed in social situations, and may shun social contacts to avoid these conflicts. In later stages of dementia many patients suffer from agnosia[74]. This cognitive deficit may make a patient prone to wandering, requiring physical containment or restraint, at times through a move to a locked-door environment. Patient awareness of the locked nature of the environment may be an additional demoralizing factor. Disorientation itself can lead to profound mood changes; it is not uncommon for an AD patient to be tearful, sad and beseeching, because they cannot fathom where they live, or why they live there, or what the purpose of living there is, particularly if they are in a dementia unit or skilled nursing setting. Thus, there is a constant interaction between the patient's encounters with his/her functional limitations and vulnerability to depressed mood.

Biological factors

Neurochemistry

Several pathological studies have suggested that neuronal degeneration in aminergic brain-stem nuclei is associated with depression in AD. Zubenko and Moossy[74] studied 37 brains from patients with pathological diagnoses of AD and found that depressive symptoms in these patients were associated with decreased numbers of pigmented cell bodies in the locus coeruleus (LC) and substantia nigra (SN). These pigmented cell bodies are thought to supply the majority of the catecholaminergic innervation from the brain stem to the rest of the brain. Indeed, these changes in cell number were associated with a marked decrease in norepinephrine in the neocortex and hippocampus, along with a non-significant trend toward decreased serotonin in the middle frontal cortex[75].

The decrease in cell counts in the LC, but not in the SN, in depressed AD patients has been replicated in two comparable studies[76,77]. These were considered supportive of a hypothesis that depression in AD is associated with decreased monoaminergic innervation throughout the brain, similar to the hypotheses that drove the development of new antidepressant medications in the 1980s and 1990s. However, a recent report has not replicated these findings[78] and highlighted the methodological problems of earlier studies: since cell loss in the LC and SN is associated with progression of AD itself, it is crucial to control for the clinical stage of dementia, which is challenging in advanced disease or when the patient is near death. Thus, the monoaminergic theory of depression in AD is uncertain.

Genetics

The cognitive symptoms of AD appear to have different genetic vulnerabilities. The e4 allele of the apolipoprotein E gene has been well established as a risk factor for the development of AD and for earlier age of onset of AD[79]. While apoE4 has been reported to mediate the effect of antidepressant treatment in depressed elderly individuals without dementia[80], it does not appear to be a risk factor for depression in AD and may add separately to the risk of AD from depression[81]. In fact, the apoE2 allele, which is protective against the development of AD, may be associated with depression in AD[82].

Neuroimaging

Given the findings in many neurological diseases that frontal lobe involvement is associated with mood symptoms, it is not surprising that initial results with AD show similar results. O'Brien *et al.*[83] reported an association between frontal lobe white matter hypointensities and depression in AD[84]. Hirono *et al.* found frontal lobe hypometabolism and depressive symptoms to be associated in a cohort of 53 AD patients. In addition, anterior cingulate hypometabolism was also associated with depressive symptoms, which corroborates the report of Migneco *et al.*[85] associating apathy with depression in AD. Further imaging studies have suggested that depressive mood symptoms are associated with frontal lobe hypometabolism[86], while a decrease in motivated behaviors is associated with anterior cingulate hypometabolism[87].

Inflammatory markers

The potential connection between inflammatory markers and depression in AD is intriguing though still entirely speculative. Given the growing evidence that neurotoxicity in AD is mediated by inflammatory mechanisms in the brain, including the release of pro-inflammatory cytokines by activated microglia[17], and given the association of elevated serum levels of these cytokines with depressed mood in a large elderly community sample[88], it is conceivable that these cytokines mediate both cognitive impairment and mood disorder in AD. Central nervous system (CNS) or peripheral cytokine synthesis and release may be a future target for treatment of AD and depression in AD.

CLINICAL APPROACH TO 'DEPRESSION OF ALZHEIMER'S DISEASE'

Diagnosis

Table 10.1 presents the operationalized diagnostic criteria for depression of AD, which offer a broader and more inclusive approach to diagnosis than DSM criteria. To highlight the process of diagnosis and interview:

- It is essential to obtain information from both the patient and a knowledgeable caregiver who spends significant time with the patient. It may be necessary to interview the caregiver on the telephone after the visit. A depression evaluation is best not considered complete without caregiver input.
- The clinician should interview the patient and caregiver separately whenever possible. This allows both to express their feelings and concerns candidly without worrying about causing greater conflict or friction, particularly in a family dyad.
- Medical record review, as well as a brief physical and neurological examination, are important parts of the evaluation of depression in AD. It is essential to rule out common causes of acute mental status changes in demented patients including urinary tract infection, pneumonia, metabolic disturbances and effects of medications. Clinicians should be particularly alert to the cognitive toxicity of anticholinergic medications[89,90] and sedation from benzodiazepines or opioid analgesics.
- The CSDD is recommended as a guiding scale if the interviewer chooses to use a rating scale. Proper administration of the CSDD involves interviewing patient and caregiver, then making an informed judgment as to the final scores and not simply averaging patient and caregiver scores. The interviewer should take into account the patient's cognitive level and ability to recall mood and behavior; for example, for a severely demented patient, caregiver input may be more heavily weighted than in early disease.

TREATMENT

Medications

The results of controlled trials of antidepressant medications for depression in AD are summarized in Table 10.2. Note that this is a scattered literature with virtually no replication studies, and only suggests guidelines for treatment. The efficacy of antidepressants in these studies is mixed, and there are several negative studies. This may be the result of using inclusion criteria more suitable for younger populations, possibly leading to heterogeneous patient samples. Two trials of tricyclic antidepressants (TCAs) reported a negative effect on cognition and they should be used with care in AD. Selective serotonin reuptake inhibitors (SSRIs) appear better tolerated but not consistently effective. Two head-to-head comparisons of TCA versus SSRI reported no difference in efficacy but there was a suggestion that SSRIs were better tolerated. Newer agents such as escitalopram, venlafaxine, mirtazapine and bupropion have

Table 10.2 Controlled studies of treatments for depression in dementia

Reference	*Design*	*Total number*	*Depression diagnosis*	*Treatments*	*Effect of treatment on mood outcomes*	*Effect of treatment on non-mood outcomes*	*Safety findings*
Reifler *et al.* (1989)[100]	8-weeks parallel, masked RCT	61	DSM-III major depression, HAM-D > 14	Imipramine (mean 82–83 mg/day) Placebo	0	– (Cognitive function declined with drug)	Drowsiness and dizziness equally reported for both groups
Nyth and Gottfries (1990)[52]	4-week parallel, masked RCT	98	None. Baseline PDD	Citalopram (30 mg/day max) Placebo	0	+ (Irritability improved with drug)	Mild, expected side-effects more common on citalopram
Nyth *et al.* (1992)[101]	6-week parallel, masked RCT	149 (but only 29 with depression and AD)	HAM-D > 13 74% met criteria for DSM-III major depression (total sample)	Citalopram (30 mg/day max) Placebo	+	+ (Cognitive function improved with drug)	Side-effects reported > 10% for entire sample: tiredness, sedation, tension
Petracca *et al.* (1996)[48]	12-week crossover RCT	21	DSM-III-R major depression or dysthymia, HAM-D > 10	Clomipramine (100 mg/day max) Placebo	+	– (Cognitive function declined with drug)	Dry mouth, dizziness, sleep problems, tremor more frequent in clomipramine group
Roth *et al.* (1996)[47]	6-week parallel, masked RCT	726	DSM-III major depressive episode, HAM-D > 13	Moclobemide (400 mg/day max) Placebo	+	+ (Cognitive function improved with drug)	No significant differences in side-effects, ECG, vital signs
Magai *et al.* (2000)[102]	8-week parallel, masked RCT	31	DSM-IV major or minor depression, CSDD, > 2, Gestalt scale > 0	Sertraline (100 mg/day max) Placebo	0	+ ('Knit-brow face' improved with drug (trend))	Not reported

continued

Table 10.2 *continued*

Reference	*Design*	*Total number*	*Depression diagnosis*	*Treatments*	*Effect of treatment on mood outcomes*	*Effect of treatment on non-mood outcomes*	*Safety findings*
Petracca *et al.* (2001)[48]	6-week parallel, masked RCT	41	DSM-IV major or minor depression, HAMD > 13	Fluoxetine (40 mg/day max) Placebo	0	0	Mild tremor more common on fluoxetine
Lyketsos *et al.* (2003)[46]	12-week parallel, masked RCT	44	DSM-IV major depressive episode	Sertraline (150 mg/day max) Placebo	+	+ (Placebo group declined more than drug group in ADL(trend))	No difference between sertraline and placebo
Taragano *et al.* (1997)[103]	45-day parallel, masked RCT	37	DSM-III major depressive episode	Fluoxetine (25 mg/day max) Amitriptyline (10 mg/day max)	+ (No difference between drugs)	+ (Cognitive function improved in both groups)	42% completion for amitriptyline; 78% for fluoxetine
Katona *et al.* (1998)[51]	8-week parallel, masked RCT	198	RDC major or minor depression MADRS > 19	Paroxetine (40 mg/day max) Imipramine (100 mg/day max)	+ (No difference between drugs)	N/A	Paroxetine better tolerated, but marginally so
Teri *et al.* (1997)[98]	Parallel RCT	72	DSM-III-R major depressive disorder or minor depressive disorder	(1) Patient intervention: 'pleasant events' (2) Caregiver intervention: 'problem solving' (3) Wait-list control (4) Typical care	+ (Behavioral treatments both better than wait-list or typical care)	+ (Caregiver depression symptoms improved with both behavioral treatments)	0

continued

Table 10.2 *continued*

Reference	*Design*	*Total number*	*Depression diagnosis*	*Treatments*	*Effect of treatment on mood outcomes*	*Effect of treatment on non-mood outcomes*	*Safety findings*
Teri *et al.* (2003)[58]	Parallel RCT	153 total AD (unknown no. in depressed subgroup)	CSDD > 5	Behavioral intervention with caregivers plus exercise program vs. usual care	+	N/A	0

RCT; randomized controlled trial; ADL, activities of daily living; PDD, primary degenerative dementia; AD, Alzheimer's disease; ECG, electrocardiogram; N/A, not available

not been studied, but are also well tolerated and are being widely used by clinicians.

Based on these results, the following approach to antidepressant medication treatment of depression in AD is recommended[91,92](summarized in Table 10.3):

- Start with an SSRI, usually sertraline or citalopram (escitalopram is a reasonable alternative), because these have the best efficacy and safety data in AD. Clinicians are advised to start at a low dose but should be sure to reach an adequate dose if tolerated (e.g. 100 mg of sertraline or 30–40 mg of citalopram) (dose recommendations summarized in Table 10.3). Other SSRIs might be used, but they have less clear data of efficacy for depression in AD.

- Start at the low doses recommended in Table 10.3 and increase dose every 1–2 weeks until maximum recommended dose is reached or side-effects limit dose titration. After reaching maximum dose, wait at least 4 weeks to assess response. If no response is seen at 4 weeks of maximum dose, it is reasonable to switch medications or add an antipsychotic or anticonvulsant (see below). If partial response is seen at 4 weeks of maximum dose, continue to a total of 9–12 weeks at full dose, because this duration is often needed to observe full response.

- Since evaluation of depression in AD requires observation, patient interview and caregiver interviewer, it is advisable to use a depression severity rating scale such as CSDD for the reasons outlined above. If a scale is not used, it is important to pick target symptoms (which may differ between patients). Particularly emphasize symptoms causing maximum distress for

Table 10.3 Antidepressant medications for depression in Alzheimer's disease

Drug	*Initial dose*	*Maximum dose*	*Side-effects*	*Comment*
Fluoxetine	10 mg	20–40 mg	GI distress, anxiety, insomnia	Long-acting with consequent risk of side-effects
Sertraline	25 mg	150 mg	GI distress, anxiety, insomnia	
Paroxetine	10 mg	20–40 mg	GI distress, anxiety, insomnia, confusion, dry mouth	May be calming and helpful for sleep
Citalopram	10 mg	20–40 mg	GI distress, anxiety, insomnia	
Escitalopram	5 mg	20 mg	GI distress, anxiety, insomnia	Enantiomer of citalopram
Venlafaxine (long-acting)	37.5 mg	225 mg	Hypertension	More stimulating than SSRIs
Bupropion (long-acting)	75 mg	450 mg	Seizures at supratherapeutic doses only	Dopaminergic effect may also be more stimulating than SSRIs
Mirtazapine	7.5 mg	30 mg	Weight gain, sedation	Use at bedtime Widely used as hypnotic
Nortriptyline	10 mg	100 mg	Constipation, dry mouth	Best choice among tricyclic antidepressants due to favorable side-effect profile
Methylphenidate	5 mg	20 mg divided between breakfast and midday	Insomnia, dyskinetic movements	May be helpful for apathy and fatigue Limited research base

GI, gastrointestinal; SSRI, selective serotonin reuptake inhibitors patient and caregiver or which seem to be adversely affecting the patient's functional level.

- If the patient's mood does not significantly improve in 12 weeks, taper off medication and switch to another agent. The second agent may be another SSRI or agent from another drug class (there is no clear evidence for which to choose).
- If mood improves but agitation is still a prominent symptom, consider adding an anticonvulsant. Depakote has the best clinical evidence base, and initial doses of 250 mg bid or 500 mg q hs are recommended, titrating dose to reach a level of 50–100 μg/ml or treatment response which may be achieved at the lower end of this range[93].
- If delusions and hallucinations are present but not causing severe impairment, consider treating with an antidepressant alone. Since delusions so frequently accompany depression of AD, consider the possibility that antidepressant treatment may relieve the symptom without the need for antipsychotic treatment as well. For example, a randomized clinical trial of citalopram versus perphenazine demonstrated comparable efficacy in treating agitation and psychosis in demented nursing-home patients[94].
- In severe and/or treatment-refractory depression of AD, consider electroconvulsive therapy (ECT). Despite concerns about worsening cognitive impairment or extended post-ictal delirium, ECT was well tolerated in a cohort of 31 patients with depression and dementia, and mood improved to a clinically significant extent[95].

Behavioral and cognitive therapies

It is apparent from Table 10.2 that medication therapies are not uniformly effective or tolerated in depression of AD, and that psychosocial interventions are needed. In addition, medication therapies alone are often not acceptable to patients. For example, in a medication-alone study of fluoxetine, no patients completed the trial[96]. Clearly, traditional psychotherapeutic techniques must be modified to suit the needs of cognitively impaired patients and their caregivers. The most common treatments have been a combination of behavioral interventions for the patient and problem-solving strategies for the caregiver[97]. Table 10.2 presents the two controlled trials of behavioral treatments. Teri *et al.*[98] studied two different behavioral treatments, one based on problem-solving strategies and the other using the Pleasant Events Schedule[99]. They reported improved patient (and caregiver) mood in both behavioral treatment groups versus standard care or waiting-list control subjects. Their group also studied an intervention that combined an individualized exercise program and problem-solving strategies for the caregiver, and found mood improvement both in a general AD population and in a depressed subgroup[58]. While these findings are not definitive, they suggest:

(1) When practical, develop a therapeutic activities program based on the individualized needs and resources of the patient and caregiver. This may be implemented with attendance at adult day care or a senior citizens' center. Consider activities such as exercise, reminiscence therapy, focusing on pleasant events such as with the Pleasant Events Schedule[99], music, etc. Try to target these activities to patient's cognitive strengths and weaknesses (for

example, patients with substantial language deficits may enjoy music activities while being frustrated by verbal activities).

(2) Encourage modest or moderate exercise within the patient's mobility limitations.

(3) Reduce caregiver stress by:

   (a) Problem-solving interventions addressing problem behaviors that are most distressing to the caregiver;

   (b) Validating their accomplishments and pointing out how they are doing a good job (particularly important for family caregivers);

   (c) Encouraging realistic expectations;

   (d) Addressing safety issues (leaving the stove on, wandering into hazardous environments, driving);

   (e) Allowing them therapeutic time and emotional space to express feeling of frustration, anger, disappointment and grief;

   (f) Addressing the possibility of placement in assisted living or nursing home.

To summarize, good care of depression in AD starts with good dementia care.

CONCLUSION

Depression is a major neuropsychiatric complication of AD and a source of distress to patients and caregivers alike. The depressive syndromes of AD are not identical to those of younger cognitively intact persons and have a unique constellation of symptoms which make them often difficult to diagnose in clinical practice. We present proposed criteria for the atypical depressive syndrome termed 'depression of Alzheimer's disease'. Caregiver input is essential in the assessment process, but caregiver depression and burden must be taken into account as possible biases in reporting. The neurobiological bases of depression in AD are still being explored, but are not identical to the bases of the cognitive symptoms. There is growing evidence that antidepressant medications may be effective in treating depression in AD, but it is essential to include behavioral interventions for the patient and caregiver as well.

REFERENCES

1. Hebert LE, Scherr PA, Bienias JL, et al. Alzheimer disease in the US population: prevalence estimates using the 2000 census. Arch Neurol 2003; 60: 1119–1122.

2. Schumock GT. Economic considerations in the treatment and management of Alzheimer's disease. Am J Health Syst Pharm 1998; 55 (Suppl 2): 17–21.

3. Fillit HM. The pharmacoeconomics of Alzheimer's disease. Am J Manag Care 2000; 6: S1139–S1144.

4. Souetre E, Thwaites RM, Yeardley HL. Economic impact of Alzheimer's disease in the United Kingdom. Cost of care and disease severity for non-institutionalised patients with Alzheimer's disease. Br J Psychiatry 1999; 174: 51–55.

5. Rabins PV, Lyketsos CG, Steele CD. Practical Dementia Care. New York: Oxford University Press, 1999.

6. Kukull WA, Bowen JD. Dementia epidemiology. Med Clin North Am 2002; 86: 573–590.
7. Franceschi M, Colombo B, Rossi P, Canal N. Headache in a population-based elderly cohort. An ancillary study to the Italian Longitudinal Study of Aging (ILSA). Headache 1997; 37: 79–82.
8. Collie A, Maruff P. The neuropsychology of preclinical Alzheimer's disease and mild cognitive impairment. Neurosci Biobehav Rev 2000; 24: 365–374.
9. Binetti G, Magni E, Padovani A, et al. Executive dysfunction in early Alzheimer's disease. J Neurol Neurosurg Psychiatry 1996; 60: 91–93.
10. Petersen RC, Smith GE, Waring SC, et al. Mild cognitive impairment: clinical characterization and outcome. Arch Neurol 1999; 56: 303–308.
11. Wolfson C, Wolfson DB, Asgharian M, et al. Clinical Progression of Dementia Study Group. A reevaluation of the duration of survival after the onset of dementia. N Engl J Med 2001; 344: 1111–1116.
12. Petrella JR, Coleman RE, Doraiswamy PM. Neuroimaging and early diagnosis of Alzheimer disease: a look to the future. Radiology 2003; 226: 315–336.
13. Rombouts SA, Barkhof F, Witter MP, Scheltens P. Unbiased whole-brain analysis of gray matter loss in Alzheimer's disease. Neurosci Lett 2000; 285: 231–233.
14. Selkoe DJ. Deciphering the genesis and fate of amyloid β-protein yields novel therapies for Alzheimer disease. J Clin Invest 2002; 110: 1375–1381.
15. Guillozet AL, Weintraub S, Mash DC, Mesulam MM. Neurofibrillary tangles, amyloid, and memory in aging and mild cognitive impairment. Arch Neurol 2003; 60: 729–736.
16. Haroutunian V, Purohit DP, Perl DP, et al. Neurofibrillary tangles in nondemented elderly subjects and mild Alzheimer disease. Arch Neurol 1999; 56: 713–718.
17. McGeer PL, McGeer EG. Inflammation, autoxicity, and Alzheimer disease. Neurobiol Aging 2001; 22: 799–809.
18. Gomez-Isla T, Hollister R, West H, et al. Neuronal loss correlates with but exceeds neurofibrillary tangles in Alzheimer's disease. Ann Neurol 1997; 41: 17–24.
19. Terry AV Jr, Buccafusco JJ. The cholinergic hypothesis of age and Alzheimer's disease-related cognitive deficits: recent challenges and their implications for novel drug development. J Pharmacol Exp Ther 2003; 306: 821–827.
20. Good DC. Dementia and ageing. Br Med Bull 2003; 65: 159–168.
21. Burns A, Jacoby R, Levy R. Psychiatric phenomena in Alzheimer's disease I: disorders of thought content. Br J Psychiatry 1990; 157: 72–76.
22. Burns A, Jacoby R, Levy R. Psychiatric phenomena in Alzheimer's disease II: disorders of perception. Br J Psychiatry 1990; 157: 76–81.
23. Burns A, Jacoby R, Levy R. Psychiatric phenomena in Alzheimer's disease III: disorders of mood. Br J Psychiatry 1990; 157: 81–86.
24. Burns A, Jacoby R, Levy R. Psychiatric phenomena in Alzheimer's disease IV: disorders of behavior. Br J Psychiatry 1990; 157: 86–94.
25. Steinberg MS, Sheppard JM, Tschanz JT, et al. The incidence of mental and behavioral disturbances in dementia: the Cache

County Study. J Neuropsychiatry Clin Neurosci 2003; 15: 340–345.

26. Lyketsos CG, Lopez O, Jones B, et al. Prevalence of neuropsychiatric symptoms in dementia and mild cognitive impairment: results from the cardiovascular health study. JAMA 2002; 288: 1475–1483.

27. Migliorelli R, Teson A, Sabe L, et al. Prevalence and correlates of dysthymia and major depression among patients with Alzheimer's disease. Am J Psychiatry 1995; 152: 37–44.

28. Garre-Olmo J, Lopez-Pousa S, Vilalta-Franch J, et al. Evolution of depressive symptoms in Alzheimer disease: one-year follow-up. Alzheimer Dis Assoc Disord 2003; 17: 77–85.

29. Pearlson GD, Ross CA, Lohr WD, et al. Association between family history of affective disorder and the depressive syndrome of Alzheimer's disease. Am J Psychiatry 1990; 147: 452–456.

30. Lyketsos CG, Tune LE, Pearlson G, Steele C. Major depression in Alzheimer's disease. An interaction between gender and family history. Psychosomatics 1996; 37: 380–384.

31. Loreck DJ, Folstein MF. Depression in Alzheimer's disease. In Starkstein SC, Robinson RG, eds. Depression in Neurologic Disease, 1st edn. Baltimore: Johns Hopkins University Press, 1993.

32. Hargrave R, Reed B, Mungas D. Depressive syndromes and functional disability in dementia. J Geriatr Psychiatry Neurol 2000; 13: 72–77.

33. Cerhan JH, Ivnik RJ, Smith GE, et al. Diagnostic utility of letter fluency, category fluency, and fluency difference scores in Alzheimer's disease. Clin Neuropsychol 2002; 16: 35–42.

34 Buysse DJ. Insomnia, depression and aging. Assessing sleep and mood interactions in older adults. Geriatrics 2004; 59: 47–51.

35 Grimby A, Svanborg A. Morbidity and health-related quality of life among ambulant elderly citizens. Aging (Milano) 1997; 9: 356–364.

36. Janzing JG, Hooijer C, van 't Hof MA, Zitman FG. Depression in subjects with and without dementia: a comparison using GMS-AGECAT. Int J Geriatr Psychiatry 2002; 17: 1–5.

37. Li YS, Meyer JS, Thornby J. Longitudinal follow-up of depressive symptoms among normal versus cognitively impaired elderly. Int J Geriatr Psychiatry 2001; 16: 718–727.

38. Zubenko GS, Zubenko WN, McPherson S, et al. A collaborative study of the emergence and clinical features of the major depressive syndrome of Alzheimer's disease. Am J Psychiatry 2003; 160: 857–866.

39. Tractenberg RE, Weiner MF, Patterson MB, et al. Comorbidity of psychopathological domains in community-dwelling persons with Alzheimer's disease. J Geriatr Psychiatry Neurol 2003; 16: 94–99.

40. Bassiony MM, Warren A, Rosenblatt A, et al. The relationship between delusions and depression in Alzheimer's disease. Int J Geriatr Psychiatry 2002; 17: 549–556.

41. Purandare N, Burns A, Craig S, et al. Depressive symptoms in patients with Alzheimer's disease. Int J Geriatr Psychiatry 2001; 16: 960–964.

42. Rubin EH, Veiel LL, Kinscherf DA, et al. Clinically significant depressive symptoms and very mild to mild dementia of the Alzheimer type. Int J Geriatr Psychiatry 2001; 16: 694–701.

43. Reisberg B, Auer SR, Monteiro M. Behavioral pathology in Alzheimer's disease (BEHAVE-AD) rating scale. Int Psychogeriatr 1996; 3: 301–308.

44. Cummings JL, Mega M, Gray K, et al. The Neuropsychiatric Inventory: comprehensive assessment of psychopathology in dementia. Neurology 1994; 44: 2308–2314.

45. Weiner MF, Doody RS, Sairam R, et al. Prevalence and incidence of major depressive disorder in Alzheimer's disease: findings from two databases. Dement Geriatr Cogn Disord 2002; 13: 8–12.

46. Lyketsos CG, DelCampo L, Steinberg M, et al. Treating depression in Alzheimer's disease. Efficacy and safety of sertraline therapy and the benefits of depression reduction: the DIADS. Arch Gen Psychiatry 2003; 60: 737–746.

47. Roth M, Mountjoy CQ, Amrein R, and the International Collaborative Study Group. Moclobemide in elderly patients with cognitive decline and depression. Br J Psychiatry 1996; 168: 149–157.

48. Petracca GM, Chemerinski E, Starkstein SE. A double-blind, placebo-controlled study of fluoxetine in depressed patients with Alzheimer's disease. Int Psychogeriatr 2001; 13: 233–240.

49. Parmelle, PA, Lawton MP, Katz IR. Psychometric properties of the Geriatric Depression Scale among the institutionalized aged. Psychol Asses 1989; 1: 331–338.

50. Gottlieb Gl, Gur RE, Gur RC. Reliability of psychiatric scales in patients with dementia of the Alzheimer type. Am J Psychiatry 1988; 145: 857–860.

51. Katona CL, Hunter BN, Bray J. A double-blind comparison of the efficacy and safety of paroxetine and imipramine in the treatment of depression with dementia. Int J Geriatr Psychiatry. 1998; 13: 100–108.

52. Nyth AL, Gottfries CG. The clinical efficacy of citalopram in treatment of emotional disturbances in dementia disorders. A Nordic multicentre study. Br J Psychiatry 1990; 157: 894–901.

53. Wagle AC, Ho LW, Wagle SA, Berrios GE. Psychometric behaviour of BDI in Alzheimer's disease patients with depression. Int J Geriatr Psychiatry 2000; 15: 63–69.

54. Teri L, Truax P. Assessment of depression in dementia patients: association of caregiver mood with depression ratings. Gerontologist 1994; 34: 231–234.

55. Burke WJ, Roccaforte WH, Wengel SP, et al. Disagreement in the reporting of depressive symptoms between patients with dementia of the Alzheimer type and their collateral sources. Am J Geriatr Psychiatry 1998; 6: 308–319.

56. Naarding P, Leentjens AFG, van Kooten F, Verhey FRJ. Disease-specific properties of the Hamilton Rating Scale for Depression in patients with stroke, Alzheimer's dementia, and Parkinson's disease. J Neuropsychiatry Clin Neurosci 2002; 14: 329–334.

57. Alexopoulos GS, Abrams RC, Young RC, Shamoian CA. Cornell Scale for Depression in Dementia. Biol Psychiatry 1988; 23: 271–284.

58. Teri L, Gibbons LE, McCurry SM, et al. Exercise plus behavioral management in patients with Alzheimer disease: a randomized controlled trial. JAMA 2003; 290: 2015–2022.

59. Kurlowicz LH, Evans LK, Strumpf NE, Maislin G. A psychometric evaluation of the Cornell Scale for Depression in Dementia

in a frail, nursing home population. Am J Geriatr Psychiatry 2002; 10: 600–608.

60. Sunderland T, Alterman IS, Yount D, et al. A new scale for the assessment of depressed mood in demented patients. Am J Psychiatry 1988; 145: 955–959.

61. Neundorfer MM, McClendon MJ, Smyth KA, et al. A longitudinal study of the relationship between levels of depression among persons with Alzheimer's disease and levels of depression among their family caregivers. J Gerontol B Psychol Sci Soc Sci 2001; 56: 301–313.

62. Starkstein SE, Petracca G, Chemerinski E, Kremer J. Syndromic validity of apathy in Alzheimer's disease. Am J Psychiatry 2001; 158: 872–877.

63. Landes AM, Sperry SD, Strauss ME, Geldmacher DS. Apathy in Alzheimer's disease. J Am Geriatr Soc 2001; 49: 1700–1707.

64. Berger AK, Fratiglioni L, Forsell Y, et al. The occurrence of depressive symptoms in the preclinical phase of AD: a population-based study. Neurology 1999; 53: 1998–2002.

65. Stout JC, Wyman MF, Johnson SA, et al. Frontal behavioral syndromes and functional status in probable Alzheimer disease. Am J Geriatr Psychiatry 2003; 11: 683–686.

66. Olin JT, Katz IR, Meyers BS, et al. Provisional diagnostic criteria for depression of Alzheimer disease: rationale and background. Am J Geriatr Psychiatry 2002; 10: 129–141.

67. Olin JT, Schneider LS, Katz IR, et al. Provisional diagnostic criteria for depression of Alzheimer disease. Am J Geriatr Psychiatry 2002; 10: 125–128.

68. Rosenberg PB, Onyike CU, Katz IR, et al. Depression of Alzheimer's Disease Study. Clinical application of operationalized criteria for 'Depression of Alzheimer's Disease'. Int J Geriatr Psychiatry 2005; 20: 119–27

69. McDaniel KD, Edland SD, Heyman A. Relationship between level of insight and severity of dementia in Alzheimer disease. CERAD Clinical Investigators. Consortium to Establish a Registry for Alzheimer's Disease. Alzheimer Dis Assoc Disord 1995; 9: 101–104.

70. Verhey FR, Ponds RW, Rozendaal N, Jolles J. Depression, insight, and personality changes in Alzheimer's disease and vascular dementia. J Geriatr Psychiatry Neurol 1995; 8: 23–27.

71. Harwood DG, Sultzer DL, Wheatley MV. Impaired insight in Alzheimer disease: association with cognitive deficits, psychiatric symptoms, and behavioral disturbances. Neuropsychiatry Neuropsychol Behav Neurol 2000; 13: 83–88.

72. Harwood DG, Sultzer DL. 'Life is not worth living': hopelessness in Alzheimer's disease. J Geriatr Psychiatry Neurol 2002; 15: 38–43.

73. Storey E, Slavin MJ, Kinsella GJ. Patterns of cognitive impairment in Alzheimer's disease: assessment and differential diagnosis. Front Biosci 2002; 7: 155–184.

74. Zubenko GS, Moossy J. Major depression in primary dementia: clinical and neuropathological correlates. Arch Neurol 1988; 45: 1182–1186.

75. Zubenko GS, Moossy J, Kopp U. Neurochemical correlates of major depression in primary dementia. Arch Neurol 1990; 47: 209–214.

76. Forstl H, Burns A, Luthert P, et al. Clinical and neuropathological correlates of depression in Alzheimer's disease. Psychol Med 1992; 22: 877–884.

77. Zweig RM, Ross CA, Hedreen JC, et al. The neuropathology of aminergic nuclei in Alzheimer's disease. Ann Neurol 1988; 24: 233–242.

78. Hoogendijk WJ, Sommer IE, Pool CW, et al. Lack of association between depression and loss of neurons in the locus coeruleus in Alzheimer disease. Arch Gen Psychiatry 1999; 56: 45–51.

79. Tanzi RE, Bertram L. New frontiers in Alzheimer's disease genetics. Neuron 2001; 32: 181–184.

80. Murphy GM, Kremer C, Rodrigues H, Schatzberg AF; Mitrazapine versus Paroxetine Study Group. The apolipoprotein E epsilon4 allele and antidepressant efficacy in cognitively intact elderly depressed patients. Biol Psychiatry 2003; 54: 665–673.

81. Butters MA, Sweet RA, Mulsant BH, et al. APOE is associated with age-of-onset, but not cognitive functioning, in late-life depression. Int J Geriatr Psychiatry 2003; 18: 1075–1081.

82. Holmes C, Russ C, Kirov G, et al. Apolipoprotein E: depressive illness, depressive symptoms, and Alzheimer's disease. Biol Psychiatry 1998; 43: 159–164.

83. O'Brien J, Perry R, Barber R, et al. The association between white matter lesions on magnetic resonance imaging and noncognitive symptoms. Ann NY Acad Sci 2000; 903: 482–489.

84. Hirono N, Mori E, Ishii K, et al. Frontal lobe hypometabolism and depression in Alzheimer's disease. Neurology 1998; 50: 380–383.

85. Migneco O, Benoit M, Koulibaly PM, et al. Perfusion brain SPECT and statistical parametric mapping analysis indicate that apathy is a cingulate syndrome: a study in Alzheimer's disease and nondemented patients. Neuroimage 2001; 13: 896–902.

86. Lai T, Payne ME, Byrum CE, et al., Reduction of orbital frontal cortex volume in geriatric depression. Biol Psychiatry 2000; 48: 971–975.

87. Sewards TV, Sewards MA. Representations of motivational drives in mesial cortex, medial thalamus, hypothalamus and midbrain. Brain Res Bull 2003; 61: 25–49.

88. Penninx BW, Kritchevsky SB, Yaffe K, et al. Inflammatory markers and depressed mood in older persons: results from the Health, Aging and Body Composition study. Biol Psychiatry 2003; 54: 566–572.

89. Mulsant BH, Pollock BG, Kirshner M, et al. Serum anticholinergic activity in a community-based sample of older adults: relationship with cognitive performance. Arch Gen Psychiatry 2003; 60: 198–203.

90. Sunderland T, Tariot PN, Cohen RM, et al. Anticholinergic sensitivity in patients with dementia of the Alzheimer type and age-matched controls. A dose–response study. Arch Gen Psychiatry 1987; 44: 418–426.

91. Alexopoulos GS, Katz IR, Reynolds CF, et al. Pharmacotherapy of depressive disorders in older patients. Postgrad Med October 2001 (Special Report): 1–86.

92. Lyketsos CG, Lee HB. Diagnosis and treatment of depression in Alzheimer's disease. Dement Geriatr Cogn Disord 2004; 17: 55–64.

93. Porsteinsson AP, Tariot PN, Jakimovich LJ, et al. Valproate therapy for agitation in dementia: open-label extension of a double-blind trial. Am J Geriatr Psychiatry 2003; 11: 434–440.

94. Pollock BG, Mulsant BH, Rosen J, et al. Comparison of citalopram, perphenazine,

and placebo for the acute treatment of psychosis and behavioral disturbances in hospitalized, demented patients. Am J Psychiatry 2002; 159: 460–465.

95. Rao V, Lyketsos CG. The benefits and risks of ECT for patients with primary dementia who also suffer from depression. Int J Geriatr Psychiatry 2000; 15: 729–735.

96. Stevens T, Katona C, Manela M, et al. Drug treatment of older people with affective disorders in the community: lessons from an attempted clinical trial. Int J Geriatr Psychiatry 1999; 14: 467–472.

97. Teri L, Logsdon RG, McCurry SM. Nonpharmacologic treatment of behavioral disturbance in dementia. Med Clin North Am 2002; 86: 641–656.

98. Teri L, Logsdon RG, Uomoto J, McCurry SM. Behavioral treatment of depression in dementia patients: a controlled clinical trial. J Gerontol B Psychol Sci Soc Sci 1997; 52: 159–166.

99. Logsdon RG, Teri L. The Pleasant Events Schedule-AD: psychometric properties and relationship to depression and cognition in Alzheimer's disease patients. Gerontologist 1997; 37: 40–45.

100. Reifler BV, Teri L, Raskind M, et al. Double-blind trial of imipramine in Alzheimer's disease patients with and without depression. Am J Psychiatry 1989; 146; 45–49.

101. Nyth AL, Gottfries CG, Lyby K, et al. A controlled multicenter clinical study of citalopram and placebo in elderly depressed patients with and without comcomitant dementia. Acta Psychiatr Scand 1992; 86: 138–145.

102. Magai C, Kennedy G, Cohen CI, Gomberg D. A controlled clinical trial of sertraline in the treatment of depression in nursing home patients with late-stage Alzheimer's disease. Am J Geriatr Psychiatry 2000; 8: 66–74.

103. Taragano FE, Lyketsos CG, Mangone CA, et al. A double-blind, randomized, fixed-dose trial of fluoxetine vs. amitriptyline in the treatment of major depression complicating Alzheimer's disease. Psychosomatics 1997; 38: 246–252.

11

Depression and multiple sclerosis

Andria L Ford and Robert T Naismith

INTRODUCTION

The presence of depression in the setting of multiple sclerosis (MS) is well established and is more commonly seen than in the general population. Neuropsychiatric disturbances documented in MS include major depression, dysthymia, bipolar disorder, panic and generalized anxiety disorders, euphoria, pathological laughing and weeping (pseudobulbar affect), and other frontal lobe syndromes[1]. Recognizing depression in MS is vital, because it is potentially treatable, and clearly has a negative impact on quality of life[2–4]. It may also reduce adherence to treatment[5] and is a risk factor for suicide[6]. Investigating depression in MS remains challenging, because its etiology is probably related to organic brain disease as well as a reaction to receiving this diagnosis and coping with accumulating disability. Confounding depressive symptoms in MS are common, and include fatigue, pain, insomnia, cognitive difficulties and substance abuse.

This chapter focuses on the literature available on major depressive disorder in MS, with respect to the prevalence, pathogenesis, quality of life, correlation with imaging and treatment options.

PREVALENCE

Traditionally, studies have estimated the prevalence of depression in MS to be between 37 and 54%[7–10]. Depression was diagnosed using scales such as the Beck Depression Inventory (BDI)[11], the Schedule for Affective Disorders and Schizophrenia[12], the structured interview for the DSM-III-R[13] and the Composite International Diagnostic Interview Short Form for Major Depression (CIDI-SFMD)[14]. Many of these studies were performed in academic referral centers and may be susceptible to certain biases. A population-based study by Patten *et al.*[15] in an MS clinic sample found a lifetime prevalence of 22.8% compared with 16% in the general population[16]. Risk factors in this study included being female, being aged less than 35, having a family history of major depression, high stress and heavy caffeine use.

In an attempt to alleviate the health-care utilization bias which confounded the prior prevalence studies, Patten *et al.*[17] subsequently took advantage of the Canadian general health survey conducted between 2000 and 2001. This study allowed a population-based estimation of major depression in a community-based sample which included individuals with MS as well as other chronic medical conditions. The sample included 115 071 survey members, of which 322 individuals had MS and 9019 individuals had major depression diagnosed by the CIDI-SFMD. The annual prevalence of major depression in those with MS was 15.7%, compared to 7.4% in the non-MS population. Being younger and female were each associated with a higher rate of depression. In those with MS aged 18–45, prevalence was 25.7% compared to 8.4% for those over 45. Women with MS had a prevalence of 16.7% compared to the prevalence in men which was 13.1%. When confounding symptoms of fatigue and cognitive symptoms were excluded, overall prevalence rates were lower in both the MS and non-MS populations, but the MS group continued to show a greater prevalence at 6.8% versus 3.2% in those without MS.

The prevalence of depression has also been found to be more common in MS than in other chronic medical illnesses and other central nervous system illnesses. In Patten's population-based study, the prevalence of depression in the MS population versus the population with other chronic medical conditions showed that MS had a greater prevalence at 15.7% versus 9.1%[17]. Schiffer and Babigian[18] compared depression in MS, temporal lobe epilepsy and amyotrophic lateral sclerosis, finding that the prevalence of depressive affective disorders was higher in MS than in the other two diseases.

SUICIDE

Major depression is a known risk factor for suicide, and this risk is shared by MS patients. Multiple studies have found an increased number of attempts and completions of suicide relative to the general population[6,19–21]. Sadovnick *et al.*[20] found completed suicide in MS patients to be 7.5 times that for the age-matched population. Recently, Feinstein[6] studied 140 MS clinic patients and found that the lifetime prevalence of suicidal ideation was 28.6%. Additionally, 6.4% of the population had attempted suicide. He compared patients who endorsed suicidal intent at some point in their life and those who did not, finding no differences in demographics, physical disability, disease duration and course, or cognitive dysfunction. Risk factors included living alone, a family history of mental illness, major depression, anxiety disorder and alcohol abuse. Not surprisingly, these are similar to suicide risk factors in the general population.

A Swedish study by Fredrikson *et al.*[22] examined a large MS hospital population with 12 834 patients, followed over 10 years. Of the 5052 who had died, suicide was determined to be an underlying cause of death in 90 cases (1.8%). Time between the initial admission with an MS diagnosis and the date of death was 5.8 years for the suicide cases, which was significantly shorter than the mean time of 7.9 years for MS cases who died from other causes. Using a standardized mortality ratio, suicide risk was highest in the first year after initial admission with an MS diagnosis and among younger men.

Suicide in other neurological illnesses has been studied. A sample of 144 364 patients with Parkinson's disease who were over 40 years old were assessed for cause of death, and 122 of these were found to have died by suicide. The rate was only 0.08% compared to 0.8% in the general population[23]. Stenager and Stenager[24] reviewed many studies estimating suicide risk associated with MS, epilepsy, Huntington's chorea, spinal cord lesions, cranial trauma, brain tumors, Parkinson's disease, vascular disorders and migraine. There was a significantly increased suicide risk in MS and spinal cord injury patients and in certain groups for those with epilepsy, but the risk in other neurological illnesses was uncertain, owing to methodological problems in study design.

PATHOGENESIS

No unifying theory exists explaining why individuals with MS are more likely to experience depression. Multiple ideas have been postulated and studied with variable results. One of the leading theories implicates demyelinating lesions in the frontotemporal region, suggesting that depression represents a manifestation of organic brain disease. Magnetic resonance imaging (MRI) has been used to investigate potential sites of associated lesion burden; these studies are discussed in more detail below.

Many have sought a genetic link that might tie together etiologies between depression and MS. However, it would seem as though patients with MS and depression are less likely to have a first-degree relative with depression, suggesting that there is a weaker genetic component than that of depression seen in the general population[7,25].

The hypothalamic–pituitary–adrenal (HPA) axis has been implicated in MS and depression[26,27]. As MS is thought to be primarily an immune-mediated illness, alterations in glucocorticoid secretion maintained by the HPA axis may affect the pathogenesis of MS. HPA axis activation has been shown to correlate with neurological disability as defined by the Extended Disability Status Scale (EDSS)[28]. Subsequently it was shown in a placebo-controlled trial that treatment with the antidepressant moclobemide normalized the HPA axis in non-depressed patients with an acute exacerbation in the course of relapsing–remitting MS[29].

Circulating cytokines may also contribute to depressive symptomatology. A few studies have demonstrated an association between tumor necrosis factor-α and depression in acute exacerbations of MS[30–32]. Both decreased cerebrospinal fluid (CSF) somatostatin and increased melatonin have also been implicated[33,34]. Autoantibodies to serotonin were measured by enzyme-linked immunosorbent assay (ELISA) in patients with schizophrenia, schizoaffective psychosis, depression, Alzheimer's disease, chronic alcoholism, rheumatoid arthritis and MS. Decreased antibody reactivity was seen only in the depression (decreased IgM) and MS (decreased IgG) groups[35].

QUALITY OF LIFE, DURATION OF ILLNESS AND PERCEPTION OF DISABILITY

There has been increasing appreciation of quality of life in MS studies. The EDSS has been the traditional disability measure in MS, and places great emphasis on ambulation in its scoring. Other disabling problems, such as mood, cognition and fatigue, on the other hand, are not factored as prominently. Although disease severity as measured by the EDSS correlates modestly with quality of life, depression is consistently a stronger predictor of quality of life than is the EDSS[2–4,36].

Studies on the duration of illness in relation to quality of life have shown significant variability[2–4,37–39]. This is not surprising, as quality of life may change with different stages of MS. A recent diagnosis of a chronic debilitating illness brings uncertainty of the future and may increase the risk of depression. Later in the course of MS, depression may arise when a transition is made from a relapsing course to a progressive course. In a community sample of 739 individuals with MS, shorter disease duration was associated with increased depressive symptoms. In the same community, patients with increased severity of MS and higher EDSS had greater depressive symptoms, while a subtype of MS (i.e. relapsing and progressive) had no association with depression[40].

In the course of MS, depression appears adversely to affect the perception of one's disability[41]. In one study, individuals who were mildly depressed significantly overestimated their memory difficulties, while those who were not depressed underestimated their actual memory and attention deficits. Those who were moderately depressed were found to estimate their memory difficulties accurately[42].

FATIGUE, SLEEP DIFFICULTIES, SUBSTANCE ABUSE AND PAIN AS POTENTIAL CONFOUNDERS IN THE DIAGNOSIS OF DEPRESSION

Fatigue is present in approximately 75–90% of patients with MS[43–46]. Although many have fatigue in the absence of depression, it may be difficult to distinguish the two as related or independent problems. Multiple studies have attempted to delineate the relationship, showing that fatigue and depression are highly correlated, even when the depression scale is corrected for items overlapping with fatigue[46,47]. Mood and disability are consistent predictors of fatigue[46]. When divided into physical and mental fatigue symptoms, depression correlates with mental fatigue and not physical fatigue[48]. Despite their close relationship, both fatigue and depression are independently associated with quality of life[49]. Mohr *et al.*[50] hypothesized that decreased mood causes greater fatigue and that treatment of the depression would show improvement in fatigue. Three treatment arms were used, including cognitive behavioral therapy, supportive group therapy and sertraline. In all treatment arms, global fatigue severity was significantly reduced in association with decreased scores on the BDI. The relationship was found to be entirely secondary to changes in mood and not related to change in cognitive or vegetative symptoms.

Sleep disturbance is another common complaint in MS, and is three times more prevalent in MS than in controls[51]. The etiology, whether related to depression or other pathophysiological processes, is unclear. For example, chronic pain or urinary frequency might result in diminished sleep. Regardless, sleep disturbance and depression frequently coincide. Lesions in the supplementary motor cortex of those with MS and sleep disturbance have been implicated as well[51].

Alcohol and drug use in MS have not been well studied. Substance abuse leads to increased prevalence of depression and may amplify existing cognitive and motor impairments. A community-based MS population survey showed that alcohol abuse and drug misuse were associated with younger age, less severe disability and increased depressive symptoms[52]. The same MS population was surveyed on levels of pain, another potential confounder of depression in MS. Forty-four per cent reported significant pain interfering with daily activities. While pain was associated with illness severity, marital status and self-ratings of overall health, it was not associated with depressive symptoms[53].

CORRELATION WITH STRUCTURAL CENTRAL NERVOUS SYSTEM CHANGES USING MAGNETIC RESONANCE IMAGING

Investigating the relationship between cognitive dysfunction and MS as predicted by MRI changes has become of growing interest. Initial studies showed that, when prominent spinal cord demyelination was compared to prominent cerebral demyelination, depression was significantly more common in the cerebral group[8,54]. An early study showed that lesions in the left arcuate fasciculus were associated with affective symptoms and somatic complaints on the BDI, but did not correlate with performance difficulties and cognitive distortions[55,56]. In another study, T1 hypointensities and atrophy seemed to correlate better than T2 lesion burden or enhancing lesions. The severity of depression was predicted by superior frontal, superior parietal and temporal hypointense T1 lesions, lateral and third ventricular enlargement and frontal atrophy[57].

Both lesion load and volume loss have been the major focus in recent studies. Many have tried to determine whether lesion load or atrophy is more predictive of depressive symptoms, and whether laterality plays a role. The majority of studies demonstrate lesion load to be the most predictive, although this may partly be due to methodological problems. Specifically, right temporal lesion load tends to be most significantly associated with depressive symptoms. In general, lesion load in the right hemisphere is more predictive than on the left[58–64]. A few studies found that lesion load had no significance and that atrophy, specifically left frontal and right temporal volume loss, was the most predictive of depression[62–65]. A recent controlled study compared MS patients with and without depression who were matched with respect to demographics, disease and cognitive measures. The two independent structural predictors of depression were left medial inferior prefrontal cortex T2 lesion volume and left anterior temporal CSF volume. The two variables accounted for 42% of the variance in depression score[65].

The importance of determining such a relationship has already been demonstrated by one study that indicated that lesion location may influence the efficacy of treatment response in MS patients with depression. Mohr *et al.*[66] took moderately depressed MS patients and administered antidepressant therapy for 16 weeks. At treatment cessation, higher BDI scores correlated with discrete lesions in the temporal lobes. At 6 months after treatment cessation, BDI scores were predicted by greater total lesion volume as well as many specific regions, suggesting that treatment gains may have been lost in patients with overall greater lesion load. In the immediate post-treatment analyses, cognitive dysfunction was not found to be a mediator of poor treatment response. However, at 6-month follow-up, neuropsychological functioning was a large predictor of failure to maintain treatment gains.

RISK OF INTERFERON TREATMENT ON DEPRESSION

The interferons used to treat multiple sclerosis have a long list of possible side-effects, depression included[67]. The initial large clinical trials on interferon had contradictory results with respect to depression as a side-effect. While most showed no difference between the two treatment arms[68–72], the North American study on interferon β-1b of 372 patients with relapsing–remitting MS had four suicide attempts and one completed suicide in the interferon arm while the placebo arm had none[73]. This was not statistically significant, but potentially concerning. On the other hand, the European trial on interferon β-1b treatment of secondary progressive MS found depression to be more common in the placebo group[74].

While the initial trials sufficiently demonstrated that depression is not a significant concern when initiating interferon therapy, there were subsequent studies that documented increased depression at 6 months after initiating therapy[5,75]. Since then, we have been reassured by studies that have demonstrated that major depression post-treatment is linked to major depression prior to starting treatment[70,76,77]. Those who were depressed 2 weeks prior to starting interferon showed significant reduction in depression at the initiation of therapy and subsequently returned to their baseline level of depression by 2 months of therapy[77]. The remaining studies have consistently shown that there is no increased incidence of depression in those treated with interferon[78–81]. In some instances, increased rates of fatigue and depression have been associated with discontinuance of therapy[75]. It should also be noted that the majority of the studies have been performed in groups of patients with relapsing–remitting MS and, in at least one of the studies, patients with severe depression were excluded.

TREATMENT OPTIONS AND THEIR EFFICACY

Not only will treatment of depression improve quality of life and reduce the risk of suicide, it has been shown to improve adherence to interferon treatments[5]. Current practice is to treat patients for depression in the same way that depression in the general population is managed. Traditional treatment modalities such as pharmacological therapy including serotonin reuptake inhibitors (SSRIs) and tricyclic antidepressants (TCAs), cognitive behavioral therapy (CBT), support groups and, rarely, electroconvulsive therapy (ECT) are all used and can be efficacious[50,82–91]. When comparing pharmacological therapy, CBT and support groups, it has been shown that medications and CBT were equally efficacious, while support groups may have been less helpful[85,86,92].

The TCA desipramine was studied in a double-blinded, placebo-controlled trial and noted to be moderately efficacious. However, the anticholinergic side-effects, particularly constipation and urinary retention, proved to limit the dose of desipramine in half of the treatment group[87]. The SSRI sertraline, and the monoamine oxidase A inhibitor moclobemide, were studied in open-label, non-controlled trials and found to be efficacious and well-tolerated[88,89]. The SSRIs also have an activating effect which may help with fatigue, although no studies have properly evaluated this[90].

There are a few case reports on the utility of ECT; however, the effects of ECT on MS have not been properly studied[91,93]. Six weeks of outpatient rehabilitation demonstrated improvement of BDI scores and quality of life with respect to emotional and mental health, while the EDSS appeared unchanged[94]. Cannabinoids have not proved to be helpful in irritability or depression[95]. Yoga and aerobic exercise were studied in a randomized controlled trial and were not clearly shown to improve mood[96].

SUMMARY

Undeniably, depression in MS is common and deserves further study. The increased prevalence and perhaps the earlier age of onset of neurological disability seen in the MS population contribute to the increased risk of suicide. The etiology of depression is probably multifactorial, including psychological, biochemical and structural components. In MRI studies, both lesion load and atrophy, specifically in the right temporal lobe, seem to be associated with increased depressive symptoms. Interferon therapy is unlikely to cause or exacerbate depression. Conversely, treating depression may improve compliance with interferon and other medical therapy. Most importantly, recognition and treatment of depression in MS may result in significant improvement in the patient's quality of life.

REFERENCES

1. Feinstein A. The neuropsychiatry of multiple sclerosis. Can J Psychiatry 2004; 49: 157–163.
2. Fruehwald S, Loeffler-Stastka H, Eher R, et al. Depression and quality of life in multiple sclerosis. Acta Neurol Scand 2001; 104: 257–261.
3. Wang JL, Reimer MA, Metz LM, et al. Major depression and quality of life in individuals with multiple sclerosis. Int J Psychiatry Med 2000; 30: 309–317.
4. Lobentanz IS, Asenbaum S, Vass K, et al. Factors influencing quality of life in multiple sclerosis patients: disability, depressive mood, fatigue and sleep quality. Acta Neurol Scand 2004; 110: 6.
5. Mohr DC, Goodkin DE, Likosky W, et al. Treatment of depression improves adherence to interferon beta-1b therapy for multiple sclerosis. Arch Neurol 1997; 54: 531–533.
6. Feinstein A. An examination of suicidal intent in patients with multiple sclerosis. Neurology 2002; 59: 674–678.
7. Sadovnick AD, Remick RA, Allen J, et al. Depression and multiple sclerosis. Neurology 1996; 46: 628–632.
8. Schiffer RB, Caine ED, Bamford KA, et al. Depressive episodes in patients with multiple sclerosis. Am J Psychiatry 1983; 140: 1498–1500.
9. Minden SL, Orav J, Reich P. Depression in multiple sclerosis. Gen Hosp Psychiatry 1987; 9: 426–434.
10. Joffe RT, Lippert GP, Gray TA, et al. Mood disorder and multiple sclerosis. Arch Neurol 1987; 44: 376–378.
11. Beck AT, Steer RA, Brown GK. BDI-II: Beck Depression Inventory Manual, 2nd edn. Boston, MA: Harcourt Brace, 1996.
12. Endicott J, Spitzer RL. A diagnostic interview: the schedule for affective disorders and schizophrenia. Arch Gen Psychiatry 1978; 35: 837–844.
13. DSM-III-R. Diagnostic and Statistical Manual of Mental Disorders, 3rd edn., revised. Washington, DC: American Psychiatric Association, 1987.
14. Kessler RC, Andrews G, Mroczek D, et al. The World Health Organization Composite International Diagnostic Interview Short-Form (CIDI-SF). Int J Methods Psychiatr Res 1998; 7: 171–185.
15. Patten SB, Metz LM, Reimer MA. Biopsychosocial correlates of lifetime major depression in a multiple sclerosis population. Mult Scler 2000; 6: 115–120.
16. Kessler RC, McGonagle KA, Zhao S, et al. Lifetime and 12-month prevalence of DSM-III-R psychiatric disorders in the United States: results from the National Comorbidity Study. Arch Gen Psychiatry 1994; 51: 8–19.

17. Patten SB, Beck CA, Williams JV, et al. Major depression in multiple sclerosis: a population-based perspective. Neurology 2003; 61: 1524–1527.

18. Schiffer RB, Babigian HM. Behavioral disorder in multiple sclerosis, temporal lobe epilepsy, and amyotrophic lateral sclerosis. An epidemiologic study. Arch Neurol 1984; 41: 1067–1069.

19. Stenager EN, Stenager E, Koch-Henriksen N, et al. Suicide and multiple sclerosis: an epidemiological investigation. J Neurol Neurosurg Psychiatry 1992; 55: 542–545.

20. Sadovnick AD, Eisen K, Ebers GC, et al. Cause of death in patients attending multiple sclerosis clinics. Neurology 1991; 41: 1193–1196.

21. Stenager EN, Koch-Henriksen N, Stenager E. Risk factors for suicide in multiple sclerosis. Psychother Psychosom 1996; 65: 86–90.

22. Fredrickson S, Cheng Q, Jiang GX, et al. Elevated suicide risk among patients with multiple sclerosis in Sweden. Neuroepidemiology 2003; 22: 146–152.

23. Myslobodsky M, Lalonde FM, Hicks L. Are patients with Parkinson's disease suicidal? J Geriatr Psychiatry Neurol 2001; 14: 120–124.

24. Stenager EN, Stenager E. Suicide and patients with neurologic diseases. Methodologic problems. Arch Neurol 1992; 49: 1296–1303.

25. Joffe RT, Lippert GP, Gray TA, et al. Personal and family history of affective illness in patients with multiple sclerosis. J Affect Disord 1987; 12: 63–65.

26. Michelson D, Gold PW. Pathophysiologic and somatic investigations of hypothalamic–pituitary–adrenal axis activation in patients with depression. Ann NY Acad Sci 1998; 840: 717–722.

27. Fassbender K, Schmidt R, Mössner R, et al. Mood disorders and dysfunction of the hypothalamic–pituitary–adrenal axis in multiple sclerosis: association with cerebral inflammation. Arch Neurol 1998; 55: 66–72.

28. Then Bergh F, Kumpfel T, Trenkwalder C, et al. Dysregulation of the hypothalamo–pituitary–adrenal axis is related to the clinical course of MS. Neurology 1999; 53: 772–777.

29. Than Bergh F, Kumpfel T, Grasser A, et al. Combined treatment with corticosteroids and moclobemide favors normalization of hypothalamo–pituitary–adrenal axis dysregulation in relapsing–remitting multiple sclerosis: a randomized, double blind trial. J Clin Endocrinol Metab 2001; 86: 1610–1615.

30. Mohr DC, Goodkin DE, Islar J, et al. Treatment of depression is associated with suppression of nonspecific and antigen-specific T(H)1 responses in multiple sclerosis. Arch Neurol 2001; 58: 1081–1086.

31. Kahl KG, Kruse N, Faller H, et al. Expression of tumor necrosis factor-alpha and interferon-gamma mRNA in blood cells correlates with depression scores during an acute attack in patients with multiple sclerosis. Psychoneuroendocrinology 2002; 27: 671–681.

32. Mikova O, Yakimova R, Bosmans E, et al. Increased serum tumor necrosis factor alpha concentrations in major depression and multiple sclerosis. Eur Neuropsychopharmacol 2001; 11: 203–208.

33. Roca CA, Su TP, Elpern S, et al. Cerebrospinal fluid somatostatin, mood, and cognition in multiple sclerosis. Biol Psychiatry 1999; 46: 551–556.

34. Sandyk R, Awerbuch GI. Nocturnal melatonin secreation in multiple sclerosis patients with affective disorders. Int J Neurosci 1993; 68: 227–240.

35. Schott K, Schaefer JE, Richartz E, et al. Autoantibodies to serotonin in serum of patients with psychiatric disorders. Psychiatry Res 2003; 121: 51–57.

36. Benito-León J, Morales JM, Rivera-Navarro J. Health-related quality of life and its relationship to cognitive and emotional functioning in multiple sclerosis patients. Eur J Neurol 2002; 9: 497–502.

37. Janssens AC, van Doorn PA, de Boer JB, et al. Impact of recently diagnosed multiple sclerosis on quality of life, anxiety, depression and distress of patients and partners. Acta Neurol Scand 2003; 108: 389–395.

38. Janssens AC, van Doorn PA, de Boer JB, et al. Anxiety and depression influence the relation between disability status and quality of life in multiple sclerosis. Mult Scler 2003; 9: 397–403.

39. Amato MP, Ponziani G, Rossi F, et al. Quality of life in multiple sclerosis: the impact of depression, fatigue and disability. Mult Scler 2001; 7: 340–344.

40. Chwastiak L, Ehde DM, Gibbons LE, et al. Depressive symptoms and severity of illness in multiple sclerosis: epidemiologic study of a large community sample. Am J Psychiatry 2002; 159: 1862–1868.

41. Smith SJ, Young CA. The role of affect on the perception of disability in multiple sclerosis. Clin Rehabil 2000; 14: 50–54.

42. Bruce JM, Arnett PA. Self-reported everyday memory and depression in patients with multiple sclerosis. J Clin Exp Neuropsychol 2004; 26: 200–214.

43. Freal JE, Kraft GH, Coryell JK. Symptomatic fatigue in multiple sclerosis. Arch Phys Med Rehabil 1984; 65: 135–138.

44. Krupp LB, Alvarez LA, LaRocca NG, et al. Fatigue in multiple sclerosis. Arch Neurol 1988; 45: 435–437.

45. Fisk JD, Pontefract A, Ritvo PG, et al. The impact of fatigue on patients with multiple sclerosis. Can J Neurol Sci 1994; 21: 9–14.

46. Kroencke DC, Lynch SG, Denney DR. Fatigue in multiple sclerosis: relationship to depression, disability, and disease pattern. Mult Scler 2000; 6: 131–136.

47. Giovanni G, Thompson AJ, Miller DH, et al. Fatigue is not associated with raised inflammatory markers in multiple sclerosis. Neurology 2001; 57: 676–681.

48. Ford H, Trigwell P, Johnson M. The nature of fatigue in multiple sclerosis. J Psychosom Res 1998; 45: 33–38.

49. Janardhan V, Bakshi R. Quality of life in patients with multiple sclerosis: the impact of fatigue and depression. J Neurol Sci 2002; 205: 51–58.

50. Mohr DC, Hart SL, Goldberg A. Effects of treatment for depression on fatigue in multiple sclerosis. Psychosom Med 2003; 65: 542–547.

51. Clark CM, Fleming JA, Li D, et al. Sleep disturbance, depression, and lesion site in patients with multiple sclerosis. Arch Neurol 1992; 49: 641–643.

52. Bombardier CH, Blake KD, Ehde DM, et al. Alcohol and drug abuse among persons with multiple sclerosis. Mult Scler 2004; 10: 35–40.

53. Ehde DM, Gibbons LE, Chwastiak L, et al. Chronic pain in a large community sample of persons with multiple sclerosis. Mult Scler 2003; 9: 605–611.

54. Rabins PV, Brooks BR, O'Donnell P, et al. Structural brain correlates of emotional disorder in multiple sclerosis. Brain 1986; 109: 585–597.

55. Pujol J, Bello J, Deus J, et al. Lesions in the left arcuate fasciculus region and depression symptoms in multiple sclerosis. Neurology 1997; 49: 1105–1110.

56. Pujol J, Bello J, Deus J, et al. Beck Depression Inventory factors related to demyelinating lesions of the left arcuate fasciculus region. Psychiatry Res 2000; 99: 151–159.

57. Bakshi R, Czarnecki D, Shaikh ZA, et al. Brain MRI lesions and atrophy are related to depression in multiple sclerosis. Neuroreport 2000; 11: 1153–1158.

58. Benedict RH, Bakshi R, Simon JH, et al. Frontal cortex atrophy predicts cognitive impairment in multiple sclerosis. J Neuropsychiatry Clin Neurosci 2002; 14: 44–51.

59. Di Legge S, Piattella MC, Pozzilli C, et al. Longitudinal evaluation of depression and anxiety in patients with clinically isolated syndrome at high risk of developing early multiple sclerosis. Mult Scler 2003; 9: 302–306.

60. Berg D, Supprian T, Thomae J, et al. Lesion pattern in patients with multiple sclerosis and depression. Mult Scler 2000; 6: 156–162.

61. Benesova Y, Neidermayerova I, Mechl M, et al. The relation between brain MRI lesions and depressive symptoms in multiple sclerosis. Bratisl Lek Listy 2003; 104: 174–176.

62. Zorzon M, de Masi R, Nasuelli D, et al. Depression and anxiety in multiple sclerosis. A clinical and MRI study in 95 subjects. J Neurol 2001; 248: 416–421.

63. Zorzon M, Zivadinov R, Nasuelli D, et al. Depressive symptoms and MRI changes in multiple sclerosis. Eur J Neurol 2002; 9: 491–496.

64. Benedict RH, Weinstock-Guttman B, Fishman I, et al. Prediction of neuropsychological impairment in multiple sclerosis: comparison of conventional magnetic resonance imaging measures of atrophy and lesion burden. Arch Neurol 2004; 61: 226–230.

65. Feinstein A, Roy P, Lobaugh N, et al. Structural brain abnormalities in multiple sclerosis patients with major depression. Neurology 2004; 62: 586–590.

66. Mohr DC, Epstein L, Luks TL, et al. Brain lesion volume and neuropsychological function predict efficacy of treatment for depression in multiple sclerosis. J Consult Clin Psychol 2003; 71: 1017–1024.

67. Walther EU, Hohlfeld R. Multiple sclerosis: side effects of interferon beta therapy and their management. Neurology 1999; 53: 1622–1627.

68. Jacobs LD, Cookfair DL, Rudick RA, et al. Intramuscular interferon beta-1a for disease progression in relapsing multiple sclerosis. The Multiple Sclerosis Collaborative Research Group (MSCRG). Ann Neurol 1996; 39: 285–294.

69. PRISMS (Prevention of Relapses, Disability by Interferon beta-1a Subcutaneously in Multiple Sclerosis) Study Group. Randomized double-blind placebo-controlled study of interferon beta-1a in relapsing/remitting multiple sclerosis. Lancet 1998; 352: 1498–1504.

70. Patten SB, Metz LM. Interferon beta-1a and depression in relapsing–remitting multiple sclerosis: an analysis of depression data from

the PRISMS clinical trial. Mult Scler 2001; 7: 243–248.

71. Paty DW on behalf of the SPIMS (Secondary Progressive Interferon beta-1a MS) Study Group. Results of a 3 year, double-blind, placebo-controlled study of interferon beta-1a (Rebif) in secondary-progressive MS. J Neurol 1999; 246 (Suppl 1): I/15.

72. Vial T, Descotes J. Clinical toxicity of the interferons. Drug Saf 1994; 10: 115–150.

73. The INFbeta- Multiple Sclerosis Study Group. Interferon beta-1b is effective in relapsing–remitting multiple sclerosis. I. Clinical results of a multicenter, randomized, double-blind, placebo-controlled trial. Neurology 1993; 43: 655–661.

74. European Study Group on Interferon beta-1b in Secondary Progressive MS. Placebo-controlled multicenter randomized trial of interferon beta-1b in treatment of secondary progressive multiple sclerosis. Lancet 1998; 352: 1491–1497.

75. Neilley LK, Goodin DS, Goodkin DE, et al. Side effect profile of interferon beta-1b in MS: results of an open label trial. Neurology 1996; 46: 552–554.

76. Feinstein A, O'Connor P, Feinstein K. Multiple sclerosis, interferon beta-1b and depression: a prospective investigation. J Neurol 2002; 249: 815–820.

77. Mohr DC, Likosky W, Dwyer P, et al. Course of depression during the initiation of interferon beta-1a treatment for multiple sclerosis. Arch Neurol 1999; 56: 1263–1265.

78. Borras C, Rio J, Porcel J, et al. Emotional state of patients with relapsing–remitting MS treated with interferon beta-1b. Neurology 1999; 52: 1636–1639.

79. Feinstein A. Multiple sclerosis, disease modifying treatments and depression: a critical methodological review. Mult Scler 2000; 6: 343–348.

80. Patten SB, Fridhandler S, Beck CA, et al. Depressive symptoms in a treated multiple sclerosis cohort. Mult Scler 2003; 9: 616–620.

81. Zephir H, De Seze J, Stojkovic T, et al. Multiple sclerosis and depression: influence of interferon beta therapy. Mult Scler 2003; 9: 284–288.

82. Krupp LB, Rizvi SA. Symptomatic therapy for underrecognized manifestations of multiple sclerosis. Neurology 2002; 58 (8 Suppl 4): S32–S39.

83. Scott TF, Allen D, Price TR, et al. Characterization of major depression symptoms in multiple sclerosis patients. J Neuropsychiatry Clin Neurosci 1996; 8: 318–323.

84. Mohr DC, Likosky W, Bertagnolli A, et al. Telephone-administered cognitive–behavioral therapy for the treatment of depressive symptoms in multiple sclerosis. J Consult Clin Psychol 2000; 68: 356–361.

85. Mohr DC, Boudewyn AC, Goodkin DE, et al. Comparative outcomes for individual cognitive–behavior therapy, supportive–expressive group psychotherapy, and sertraline for the treatment of depression in multiple sclerosis. J Consult Clin Psychol 2001; 69: 942–949.

86. Tesar N, Baumhackl U, Kopp M, et al. Effects of psychological group therapy in patients with multiple sclerosis. Acta Neurol Scand 2003; 107: 394–399.

87. Schiffer RB, Wineman NM. Antidepressant pharmacotherapy of depression associated with multiple sclerosis. Am J Psychiatry 1990; 147: 1493–1497.

88. Scott TF, Nussbaum P, McConnell H, et al. Measurement of treatment response to sertraline in depressed multiple sclerosis patients using the Carroll scale. Neurol Res 1995; 17: 421–422.

89. Barak Y, Ur E, Achiron A. Moclobemide treatment in multiple sclerosis patients with comorbid depression: an open-label safety trial. J Neuropsychiatry Clin Neurosci 1999; 11: 271–273.

90. Ben-Zacharia AB, Lublin FD. Palliative care in patients with multiple sclerosis. Neurol Clin 2001; 19: 801–827.

91. Mattingly G, Baker K, Zorumski CF, et al. Multiple sclerosis and ECT: possible value of gadolinium-enhanced magnetic resonance scans for identifying high-risk patients. J Neuropsychiatry Clin Neurosci 1992; 4: 145–151.

92. Messmer UM, Mancuso ML, Battaglia MA, et al. Peer support groups in multiple sclerosis: current effectiveness and future directions. Mult Scler 2004; 10: 80–84.

93. Sandyk R. Suicidal behavior is attenuated in patients with multiple sclerosis with treatment of electromagnetic fields. Int J Neurosci 1996; 87: 5–15.

94. Patti F, Ciancio MR, Reggio E, et al. The impact of outpatient rehabilitation on quality of life in multiple sclerosis. J Neurol 2002; 249: 1027–1033.

95. Grotenhermen F. Cannabinoids do not reduce objective measurements in muscle spasticity, but people with multiple sclerosis perceive some benefit. Evidence-based Healthcare 2004; 8: 159.

96. Oken BS, Kishiyama, MA, Zajdel D, et al. Randomized controlled trial of yoga and exercise in multiple sclerosis. Neurology 2004; 62: 2058–2084.

12

Traumatic brain injury and depression

Moises Gaviria and Jorge A Franco

INTRODUCTION

Traumatic brain injury (TBI), broadly defined as brain injury from externally inflicted trauma, may result in significant impairment of an individual's physical, cognitive and psychosocial functioning. In the USA, an estimated 1.5–2 million people incur TBI each year, principally as a result of vehicular incidents, falls, acts of violence and sports accidents. The number of people surviving TBI with impairment has increased significantly in recent years, which is attributed to faster and more effective emergency care, quicker and safer transportation to specialized treatment facilities and advances in acute medical management. While the risk of having a TBI is substantial among all age groups, this risk is highest among adolescents, young adults and persons older than 75 years.

Each year, approximately 70 000–90 000 individuals suffer a TBI resulting in a long-term, substantial loss of functioning. The consequences of TBI include a dramatic change in the individual's life course, profound disruption of the family, enormous loss of income or earning potential and large expenses over a lifetime. There are approximately 300 000 hospital admissions annually for persons with mild or moderate TBI, and an additional unknown number of traumatic brain injuries that are not diagnosed but may result in long-term disability. As the cumulative result of post-traumatic brain injuries, an estimated 5.3 million men, women and children are living with a permanent TBI-related disability in the USA today[1,2].

Although TBI may result in physical impairment, the more problematic consequences involve the individual's cognition, emotional functioning and behavior. TBI can cause a heterogeneous group of neuropsychiatric symptoms, because the injuries tend to be diffuse, and rarely limited to one discrete brain region or functional system. Prevalence rates for psychiatric disorders after TBI are generally high. Major depression is the disorder cited most frequently, with rates spanning from 14 to 77% after TBI[3–5]. Other disorders reported with high frequencies are substance abuse, ranging from 5 to 28%[3–6], post-traumatic stress disorder (PTSD) ranging from 3 to 27%[3,7], and other anxiety disorders, ranging from 3 to 28%[3]. These impact interpersonal relationships, school and work. Cognitive–behavioral remediation, pharmacological management, assistive technology, environmental manipulation, education and counseling are among currently used treatments of these sequelae.

Depression in general is an illness that affects almost every aspect of life and interferes with the individual's activities at home as well as at work. Depression has both physical and psychosocial effects, influencing the cognitive, emotional and behavioral domains. Therefore, those who present with depression after a TBI are reported to be less able to follow their medical regimens, experience higher levels of distress and experience more decreased functioning than patients who are not depressed[8,9].

Why is depression so prevalent after TBI? Is depression a direct consequence of the trauma, or is it a co-morbid illness? These are questions that are yet to be answered, but there are some theories, which include pre-existing depression, pre-injury personality disorder, site and extent of central nervous system (CNS) damage, difficulty with social integration after injury and poor family support[8,10,11]. Regardless of its etiology and predisposing factors, depression complicates the recovery process of survivors of TBI. For instance, apathy and cognitive disturbances associated with post-TBI depression may interfere with the rehabilitation process[10], increase the distress of patients and family members, and have a negative impact on the quality of life among survivors of TBI[12].

EPIDEMIOLOGY

Psychiatric disorders are a major cause of disability after TBI. Major depression has been the most studied psychiatric disorder after a TBI. The rates have varied from 14 to 77% for major depression, 2 to 14% for dysthymia and 2 to 17% for bipolar illness. The disparity in report of depressive disorders is the result of multiple factors:

- Many studies have been limited by small unrepresentative samples, while data from other studies may have been derived from community-based clinics or rehabilitation centers.
- The lack of uniformity in the diagnostic criteria of depression and other depression-like disorders associated with TBI contribute to diagnostic errors and yield false-positive and -negative data. Furthermore, the diagnostic criteria listed in the DSM III-R, DSM-IV and DSM-IV-R have yet to be validated in patients with TBI[13,14].
- There is inadequate selection of diagnostic instruments of depression. Many depression inventories contain items that overlap with TBI sequelae, increasing the instrument's sensitivity at the expense of specificity. Furthermore, most of these tests and inventories were created for the non-medical population and have not been validated in post-TBI patients. Most of the studies have relied on rating scales or a relative's report rather than on structured interviews and established diagnostic criteria.
- The time of assessment after injury affects response patterns on tests of depression.

Before the introduction of DMS-III in 1980, the most extensive study on psychiatric disorders after TBI was reported in 1969 by Achte *et al.*[15], who examined 3552 veterans for psychosis with a follow-up of 22–26 years. Since the introduction of DSM-III, Hibbard *et al.* conducted a follow-up study for 8 years using structured psychiatric interviews[16,17].

In 2002, Koponen *et al.* published a 30-year follow-up study of axis I and II psychiatric disorders after TBI[4]. In their study 60 patients were assessed on average, 30 years after TBI. DSM-IV axis I disorders were diagnosed on a clinical basis with the aid of the Schedules for Clinical Assessment in Neuropsychiatry (version 2.1), and axis II disorders were diagnosed with the Structured Clinical Interview for DSM-III-R Personality Disorders. Cognitive impairment was measured with a neuropsychological test battery and the Mini-Mental State Examination.

Of the 60 patients, 29 (48.3%) had had an axis I disorder that began after TBI, and 37 (61.7%) had had an axis I disorder during their lifetimes. The most common novel disorders after traumatic brain injury were major depression (26.7%), alcohol abuse or dependence (11.7%), panic disorder (8.3%) and schizoid personality disorder (6.7%). Nine patients (15.0%) had DSM-III-R organic personality syndrome.

The findings of this study suggest that TBI not only temporarily disturbs brain function but may cause decades-long or even permanent vulnerability to psychiatric disorders in some individuals. TBI seems to make patients particularly susceptible to depressive episodes, delusional disorder and personality disturbances. Major depression was observed at the early stage after TBI, but also after 30-year follow-up. Its occurrence was not significantly related to the severity of the brain injury. The authors concluded that psychiatric evaluation and follow-up is important and should be included in the routine treatment of TBI due to the high rate of psychiatric disorders found in this study.

ETIOLOGY

There are numerous theoretical difficulties in identifying the causal relation between TBI and the increased prevalence of depression. First, neural damage may lead directly to depression; this may occur when the damage affects non-depressive functions such as initiation and conation, as in the case of a patient with frontal lobe damage who appears apathetic. Prigatano and Summers[18] refer to the neuropsychologically mediated behavior disturbances as those that are 'a direct effect of brain damage'. Second, the patient may develop a reactive depression to the experience of the damaging event (e.g. car accident) or the traumatic sequelae (e.g. inability to return to work). Reactive depression may often develop in individuals as their insight into their physical and neuropsychological deficit grows[19]. This post-TBI psychopathology, which is supported by one longitudinal study of post-TBI depression[20–22] may represent a heterogeneous group with superficially similar representations with varying etiologies. The conclusion of this study stated that cases of post-TBI depression with acute onset may be viewed as a direct result of tissue damage, whereas the association between TBI with delayed-onset depression may be more likely to be mediated by psychological and psychosocial variables.

In regards to an understanding of post-TBI depression, the hypothesis formulated by Mayberg[23] and based on functional imaging studies of depression, noted that there is a neural circuit that encompasses the paralimbic frontal and temporal cortex, basal ganglia and amygdala, and controls mood. Furthermore, any disruption of this circuit may cause depression independent of the etiology. An alternative hypothesis formulated by Federoff *et al.*[24] noted that there may be an interruption of biogenic amines containing neurons as they pass through the basal ganglia or frontal subcortical white matter. However, the hypothesis that this left dorsolateral lesion is unduly disruptive of this neurotransmitter pathway fails to explain the association between right hemispheric lesions and depression.

Additional factors such as time since injury, psychosocial stressors pertaining to injury limitations and development of insight are important variables in understanding post-TBI depression. Moldover *et al.*[25] concluded that post-TBI depressive patients represent a heterogeneous group with varying etiological pathways leading to the same diagnosis; one group is perhaps best conceptualized as suffering from secondary or 'neuropsychologically mediated' depression and is most likely to display depressive symptoms in the months immediately following brain injury. A second group of depressed patients is most likely to develop a form of depression peaking further after injury. This primary depression is likely to be associated with events occurring after the injury, such as failure to reintegrate into social roles. These events when dealt with via a depressogenic cognitive framework lend themselves to self-perpetuating cycles of depression and withdrawal. The authors concluded that the conceptualization of depression post-TBI as a clinical heterogeneous condition will help to establish which therapeutic approaches should be applied for whom and when[25].

SUICIDE AFTER TRAUMATIC BRAIN INJURY

A person with a TBI can experience devastating and enduring changes in all aspects of life, including employment, and family and social relationships. Profound emotional responses of anxiety and, in particular, major depression are frequent, and suicidal ideation and suicide attempts after a TBI have been reported[26].

Owing to its comparative rarity, however, there have been only a few studies that have examined the occurrence of completed suicide after TBI. In a review, Harris and Barraclough[27] listed six studies, all performed before 1980, which together suggested increased rates of suicide among populations with brain injury. In by far the largest of these studies, Achté *et al.*[28] found 85 suicide cases among 6498 war veterans with brain injury. In another earlier longitudinal study with 25-year follow-up, Roberts[29] found three cases of suicide among 479 patients with TBI. In more recent long-term follow-up studies, Wilson[30] reported three suicides among 55 patients with TBI, Klonoff and Lage[31] found two suicides among a group of 111 patients from a rehabilitation program, and Tate *et al.*[32] reported eight suicides from among 896 patients also with TBI.

Teasdale and Engberg[33] found increased rates of suicide among patients with relatively severe TBI. The magnitude of the rates found in this study, however, was not of alarming proportions, only being about 1% over a 15-year period. Nonetheless, as Tate *et al.*[32] remarked, 'the small magnitude in no way corresponds to the seriousness of the problem' and awareness of a suicide risk should be present in the assessment of any TBI.

Alarming parallels may be noted between suicidality and brain injury. Several of the key risk factors for TBI are similar to those for suicide; these include: age (usually mid-adolescence to mid-twenties), gender (predominantly males), lower socioeconomic status, presence of drug and alcohol use, and psychological disturbance. Other parallels exist in the reported psychological states of people with suicidal ideation and those with brain injuries. Both populations may experience profound feelings of hopelessness, despair, worthlessness, loss of sense of integrity, and problems with impulsiveness, rigidity and isolation.

Factors that may adversely influence a person with brain injury toward suicide may be seen at a number of levels[34]:

(1) Some direct effects of brain injuries (e.g. cognitive and motor disturbances,

emotional disorders, increased impulsivity, depression, rigidity, hyperactivity) may precipitate suicidal behavior.

(2) Longer-term implications of these effects of brain injuries may result in profound personality changes, which might adversely influence suicidality.

(3) Changes in capabilities and competencies post-injury may increase the likelihood of depression and the risk of suicide for people with TBI.

(4) At a contextual level, brain injury often results in a catastrophic, life-changing event for individuals and their families. Many TBI survivors experience dramatic and permanent changes in work status, role, income, family life, support network and quality of life. This may predispose them to significant depressive reactions and feelings of social isolation, helplessness and hopelessness.

(5) Premorbid social functioning, alcohol use, previous psychiatric problems and family history all influence suicidality.

Given the multiplicity of suicide risk factors for people with brain injury, the need for appropriately targeted and tailored suicide prevention strategies would appear evident. However, Kuipers and Lancaster, in a research project aimed at developing suicide prevention strategies based on the perspective of people with brain injury[35], stressed the importance of recognizing differences between the perspectives of people with TBI compared with their family members, the importance of informal relationships in suicide prevention and that TBI-specific rehabilitation services play an important role in suicide prevention.

ASSESSMENT OF DEPRESSION AFTER TRAUMATIC BRAIN INJURY

Diagnosis of depression in patients who have suffered a TBI offers a challenge to clinicians. It is not an easy task to differentiate between a true mood disorder and other symptoms that are present after damage of the frontal lobes and other structures that regulate mood and emotions and which may result in multiple other disturbances. These disturbances might include diminished concentration and memory, difficulty with problem solving in complex situations, apathy, lack of awareness of cognitive deficits, irritability, lability, disinhibition and outbursts of anger. Moreover, appetite and sleep symptoms are frequently observed after TBI[34].

As seen in Table 12.1, most of the symptoms listed overlap to a large degree with symptoms of depression and, therefore, create a diagnostic dilemma. Except for subjective feelings of guilt, feelings of worthlessness and thoughts of death or dying, the symptoms of a major depressive episode are consistent with many aspects of TBI.

During the assessment process the clinician has a number of DSM-IV diagnostic categories from which to choose in determining an appropriate diagnosis (Table 12.2). Since major depressive disorder and mood disorder due to a general medical condition are the most frequently used categories, they are explained in detail in Tables 12.3 and 12.4[36].

Another important issue in the assessment process of people with TBI is the type of instrument used to diagnose and study depression. Most studies have measured depression by using standardized assessment instruments which were developed for use with the psychiatric population. Standard instruments such as the Beck Depression Inventory (BDI), or the

Table 12.1 Summary of behavioral abnormalities associated with disorders of the frontal lobes

Medial orbitofrontal cortex	*Lateral orbitofrontal cortex*	*Anterior cingulate prefrontal cortex*
Anergy	Irritability	Akinetic mutism
Anhedonia	Tactlessness	Marked apathy
Hyper- or hypophagia	Fatuous euphoria	Psychic emptiness/reduced creativity
Circadian dysfunction	Impulsivity	Poverty of spontaneous speech
Alimentary changes	Undue familiarity	
Depression	Lability	
Dysphoria	Mania	

Table 12.2 Possible DSM-IV-TR diagnosis when assessing depressive symptoms in patients with traumatic brain injury (TBI)

Major depressive disorder
Depressive disorder not otherwise specified
Mood disorder due to TBI with depression features or with a major depression-like episode
Adjustment disorder with depressed mood
Dementia due to head trauma with or without behavioral disturbance
Substance-induced mood disorders

Hamilton Rating Scale for Depression (HAM-D) were created to diagnose depression in non-medical populations, and many of the symptoms listed, such as neurovegetative or cognitive symptoms, may be confused with deficit seen in TBI patients even without depression. This factor may contribute to overdiagnosing depression in the medical population. On the other hand, patients may deny the presence of a depressed mood as part of a general unawareness of deficit or a denial syndrome. This would result in an underdiagnosis of depression.

The goal in assessment is to find an instrument that is both sensitive and specific in diagnosing depression. Only one instrument to date, the Neurobehavioral Functioning Inventory-R (NFI-R), has a depression scale that deals with the problem of measuring depression in persons with TBI by including mostly subjective feelings associated with depression and avoiding somatic and neurological items that contribute to diagnostic confound. The NFI-R is an 83-item scale, rated on a five-point scale. It contains 13 depression items, all of them subjective feeling states[37]. Several recent studies have used the NFI-R to characterize post-injury depression symptoms. For example, Kreutzer *et al.*[37] used the NFI-R to identify DSM-IV major depressive disorder symptoms. They found that 42% of 722

Table 12.3 Major depressive disorder. This is the most studied of these categories. The specific diagnostic criterion lends itself to be readily measured and, therefore, more easily diagnosed (modified from DSM-IV-TR)

In the same 2 weeks, the patient has had five or more of the following symptoms, which are a definite change from usual functioning. Either depressed mood or decreased interest or pleasure must be one of the five.

- *Mood.* For most of nearly every day, the patient reports depressed mood or appears depressed to others.
- *Interests.* For most of nearly every day, interest or pleasure is markedly decreased in nearly all activities (noted by the patient or by others).
- *Eating and weight.* Although not dieting, there is a marked loss or gain of weight (such as 5% in one month) or appetite is markedly decreased or increased nearly every day.
- *Sleep.* Nearly every day the patient sleeps excessively or not enough.
- *Motor activity.* Nearly every day others can see that the patient's activity is agitated or retarded.
- *Fatigue.* Nearly every day there is fatigue or loss of energy.
- *Self-worth.* Nearly every day the patient feels worthless or inappropriately guilty. These feelings are not just about being sick; they may be delusional.
- *Concentration.* Noted by the patient or by others, nearly every day the patient is indecisive or has trouble thinking or concentrating.
- *Death.* The patient has had repeated thoughts about death (other than the fear of dying), suicide (with or without a plan) or has made a suicide attempt.

These symptoms cause clinically important distress or impair work, social or personal functioning.

This disorder is not directly caused by a general medical condition (traumatic brain injury, for example) or the use of substances, including prescription medications.

Unless the symptoms are severe (defined as severely impaired functioning, severe preoccupation with worthlessness, ideas of suicide, delusions or hallucinations or psychomotor retardation), the episode has not begun within 2 months of the loss of a loved one. This item helps to differentiate from an adjustment disorder

outpatients averaging 35.3 months post-injury met the prerequisite number of criterion 'A' depression symptoms. Fatigue (46%), frustration (41%) and poor concentration (38%) where the most commonly cited manifestations of depression. Feeling hopeless, feeling worthless and having difficulty enjoying activities are the three symptoms that chiefly differentiate depressed patients from non-depressed patients. The depression and TBI item overlap of the more traditional depression measures makes the instruments sensitive, but not specific, in diagnosing depression. In short, they 'find' depression in persons with TBI using traditional depressive diagnostic criteria, but they probably overdiagnose depression in this same population[14].

Table 12.4 Mood disorder due to a general medical condition, namely, mood disorder due to traumatic brain injury (TBI) with depression features or with a major depression-like episode

This category is used when a medical condition – in this case TBI – fully accounts for the depressive symptoms (modified from DSM-IV-TR)

- The patient's clinical presentation is dominated by a mood disorder that persists and is characterized by either or both of:
  – Depressed mood or markedly decreased interest or pleasure in nearly all activities, or
  – Mood that is elevated, expansive or irritable.
- History, physical examination or laboratory findings suggest a general medical condition that seems likely to have directly caused these symptoms.
- No other mental disorder (such as adjustment disorder secondary to having a medical disorder) better accounts for these symptoms.
- The symptoms do not occur solely during a delirium.
- These symptoms cause clinically important distress or impair work, social or personal functioning.

Specify whether:

With manic features, mood is mainly elevated or irritable.

With depressive features, mood is mainly depressed, but criteria for a major depressive episode are not fulfilled.

With major depression-like episode, all criteria (other than the general medical condition exclusion) are fulfilled.

With mixed features, manic and depressive symptoms are present in about equal parts.

Having a prior psychiatric disorder is a potent predictor of depression after TBI; therefore, it is important to ask about history of depression in TBI survivors.

Another critical issue in the evaluation process of TBI patients is the time of the assessment. Is the person very early in his recovery and still severely impaired from a cognitive standpoint? Is the person aware of his surroundings and/or deficits? Are the symptoms more consistent with neurological injury and brain-injury recovery? or Is the person being evaluated later in his recovery and is he experiencing significant psychosocial distress or internal distress?[14] Most studies show that the first year after injury is the time during which TBI patients are most vulnerable to the development of axis I disorders – particularly depression and anxiety disorders – suggesting the need for preventive and early interventions for everyone after TBI[3].

TREATMENT

Treatment of post-TBI patients with depression should be global, must include pharmacological and non-pharmacological interventions, and should include not only the patient, but also their family or caregivers. Few studies have tried to address the presence of depression or anxiety disorders in the caregivers of patients with TBI. Given the burden that it implies for families, we can extrapolate from studies in patients with Alzheimer's disease that show that almost 50% of caregivers present a major depressive disorder, suggesting the need for preventive and early interventions in families and caregivers.

Traditional approaches to help TBI patients improve their psychosocial adjustment often include pharmacological interventions combined with psychotherapy, behavioral modification and various other behavioral interventions and teaching strategies[38].

PHARMACOLOGICAL TREATMENT

There are several general principles that should be followed when using medications in patients with psychiatric syndromes following TBI:

- TBI patients are more sensitive to adverse reactions of medications than are patients with no history of TBI.
- Doses of psychotropic medications must be raised and lowered in small increments over protracted periods of time.
- Medications must be given in a manner to enhance the probability of benefit and to reduce the possibility of adverse reactions.
- TBI patients very often take several medications; therefore, drug–drug interactions should be taken into account.
- Owing to frequent changes in symptomatology and clinical status of TBI patients, continuous reassessment is necessary to determine whether or not the medication is required.
- In the presence of depression after TBI, standard guidelines used in the treatment of major depression should be used.

The effectiveness of antidepressant medications is generally comparable between classes and within classes of medication. Therefore, the initial selection of an antidepressant medication will largely be based on the anticipated side-effects, the safety or tolerability of these side-effects for individual patients, patients' preference, quantity and quality of clinical trial regarding the medication, and its cost[39]. In TBI patients, antidepressants with a minimal lowering of seizure threshold, the fewest interactions, and the fewest sedative, hypotensive and anticholinergic side-effects are preferred. Thus, the selective serotonin reuptake inhibitors (SSRIs) are usually the first-line medications prescribed. These would be initiated at low dosages (e.g. *S*-citalopram starting at 5 mg/day, sertraline starting at 25 mg/day, or fluoxetine starting at 10 mg/day)[2]. Other SSRIs such as paroxetine, citalopram and fluvoxamine, as well as the selective norepinephrine reuptake inhibitor (SNRI) venlafaxine, also appear effective[39]. The most common side-effects of SSRIs include headache, gastrointestinal complaints, insomnia, decreased libido and sexual dysfunction.

The most common and disabling antidepressant side-effects in patients with TBI are the anticholinergic effects, which are manifested in impaired attention, concentration and memory. The tricyclic antidepressants have high affinity for muscarinic-receptors and therefore are highly anticholinergic, and should be avoided or used after careful consideration of alternative medications. *S*-Citalopram, fluoxetine, sertraline, citalopram, mirtazapine, trazodone, venlafaxine, nefazodone and bupropion all have minimal or no anticholinergic effect.

If apathy is the predominant symptom, a medication with dopaminergic effect would be preferable. It can be used as a first-line medication, or in combination with a SSRI. Bupropion, amantadine, or a stimulant medication, are some available options.

Psychostimulants and other dopaminergically active agents (e.g. methylphenidate, dextroamphetamine, pergolide and bromocriptine) may modestly improve arousal and speed of information processing, reduce distractibility and improve some aspects of executive function[40]. This group of drugs may also be beneficial as antidepressants. However, the magnitude and temporal course of their therapeutic effect is still a matter of controversy[41]. Stimulants of choice are dextroamphetamine (8–60 mg/day) and methylphenidate (10–60 mg/day). They are usually given twice a day with the last dose at least 6 hours before sleep, in order to prevent initial insomnia. Treatment begins at lower doses, which are then gradually increased. Sustained-release forms may permit once-daily dose schedules. Patients taking stimulants need close medical observation to prevent abuse or toxic effects. The most common side-effects are anxiety, dysphoria, headaches, irritability, anorexia, insomnia, cardiovascular symptoms, dyskinesias or even psychotic symptoms.

Amantadine, a drug with complex pharmacological effects on dopaminergic, cholinergic and *N*-methyl-D-aspartate (NMDA) receptors, might be of some use for the treatment of motivational deficits[42]. It is usually started at low doses (50 mg bid) and gradually increased to 200 mg bid. There is also some empirical evidence of the beneficial effects of cholinesterase inhibitors such as donepezil on cognitive functioning, motivation and general well-being[43]. The dose range is 5–10 mg/day and the more common side-effects are insomnia, diarrhea and dizziness. These are usually transient and may be minimized by a gradual increase in their dosage[44].

Electroconvulsive therapy (ECT) remains a highly effective and underused option for the treatment of depression overall, and it can be used effectively after acute or severe TBI[45].

NON-PHARMACOLOGICAL TREATMENT

Psychosocial, family and occupational interventions are often required. Cognitive behavioral therapy and interpersonal therapy are the psychotherapeutic approaches that have the best documented efficacy in the literature for the specific treatment of major depressive disorder. Patient preference and the availability of clinicians with appropriate training and expertise in the specific approach are also factors in the choice of a particular form of psychotherapy[25]. Most clinicians would agree that the neurocognitive impairments of a given TBI patient greatly influence the psychological intervention that is attempted. For example, patients with severely limited intellectual capacities and significant impulse control problems tend to be treated 'behaviorally' rather than from a psychodynamic or interpersonal perspective. Likewise, patients who struggle with depression and loss of social status but who have the potential to return to work or a productive lifestyle may benefit from psychotherapeutic interventions aimed at improving self-awareness and re-establishing a sense of personal meaning in the face of their losses[44].

Finally, we should not forget that the patient with brain injury must be highly motivated to participate in the counseling process. Simply realizing that things are not going well is not enough, and the patient must be motivated and willing to work to produce change. This applies not only to interpersonal therapy and to supportive psychotherapy, but also to cognitive therapy. In all psychotherapies, this is indexed by regularly keeping appointments, willingness to discuss difficulties resulting from the head injury and willingness to do homework if necessary to promote change[46].

REFERENCES

1. Silver J, McAllister T, Yudofsky S. Textbook of Traumatic Brain Injury. Washington DC: American Psychiatric Publishing, 2004.
2. Yudofsky S, Hales R. Essentials of Neuropsychiatry and Clinical Neurosciences. Washington, DC: American Psychiatric Publishing, 2003.
3. Ashman TA, Spielman LA, Hibbard MR, et al. Psychiatric challenges in the first 6 years after traumatic brain injury: cross-sequential analyses of axis I disorders. Arch Phys Med Rehabil 2004; 85 (4 Suppl 2): S36–S42.
4. Koponen S, Taiminen T, Portin R, et al. Axis I and II psychiatric disorders after traumatic brain injury: a 30-year follow up study. Am J Psychiatry 2002; 159: 1315–1321.
5. Fann JR, Katon WJ, Uomoto JM, Esselman PC. Psychiatric disorders and functional disability in outpatients with traumatic brain injuries. Am J Psychiatry 1995; 152: 493–499.
6. Max W, Mackenzie E, Rice D. Head injuries: cost and consequences. J Head Trauma Rehabil 1991; 6: 76–91.
7. Bryant RA, Marosszeky JE, Crooks J, Gurka JA. Postraumatic stress disorder after severe traumatic brain injury. Am J Psychiatry 2000; 157: 629–631.
8. Underhill AT, Lobello SG, Stroud TP, et al. Depression and life satisfaction in patients with traumatic brain injury: a longitudinal study. Brain Injury 2003; 17: 973–982.
9. Lecrubier Y. The burden of depression and anxiety in general medicine. J Clin Psychiatry 2001; 62: 4–9.
10. Jean-Bay E. The biobehavioral correlates of post traumatic brain injury depression. J Neurosci Nurs 2000; 32: 169–176.
11. Rosenthal M, Christensen BK, Ross TP. Depression following traumatic brain injury. Arch Phys Med Rehabil 1998; 79: 90–103.
12. Corrigan JD, Bogner JA, Mysiw WJ, et al. Life satisfaction after traumatic brain injury. J Head Trauma Rehabil 2001; 16: 543–555.
13. Jorge RE, Robinson RG. Are depressive symptoms specific for a depressed mood in TBI? J Nerv Ment Disord 1993; 181: 91–99.
14. Rogers P. Diagnosis depression in persons with brain injuries: a look at theories, the DSMIV and depression measures. Brain Injury 2003; 17: 889–900.
15. Achte KA, Hillbom E, Alberg V. Psychosis following brain injuries. Acta Psychiatr Scand 1969; 45: 1–18.
16. Hibbard MR, Uysal S, Kepler K, et al. Axis I psychophatology in individuals with traumatic brain injury. J Head Trauma Rehabil 1998; 13: 24–39.
17. Hibbard MR, Uysal S, Kepler K, et al. Axis II psychophatology in individuals with traumatic brain injury. Brain Injury 2000; 14: 45–61.
18. Prigatano GP, Summers JD. Depression in traumatic brain injury patients. In: Robertson MM, Katona CLE, eds.

Depression and Physical Illness. New York: Wiley, 1997: 341–358.

19. Lezak M, O'Brien KP. Longitudinal study of emotional, social and physical changes after traumatic brain injury. J Learn Disabil 1988; 21: 456–463.

20. Jorge RE, Robinson RG, Arndt SU, et al. Comparison between acute- and delayed-onset depression following traumatic brain injury. J Neuropsychiatry Clin Neurosci 1993; 5: 43–49.

21. Jorge RE, Robinson RG, Arndt SU, et al. Depression after mild traumatic brain injury: a 1 year longitudinal study. J Affect Disord 1993; 27: 233–243.

22. Jorge RE, Robinson RG, Starkstein SE, Arndt SU. Depression and anxiety following traumatic brain injury. J Neuropsychiatry Clin Neurosci 1993; 5: 369–374.

23. Mayberg HS. Functional imaging studies in secondary depression. Psychiatric Ann 1994; 24: 643–647.

24. Federoff JP, Starkstein SE, Forrester AW, et al. Depression in patients with acute traumatic brain injury. Am J Psychiatry 1992; 149: 918–923.

25. Moldover J, Goldberg K, Prout M. Depression after traumatic brain injury: a review of evidence for clinical heterogeneity. Neuropsychol Rev 2004; 14: 143–154.

26. Persinger MA. Sense of a presence and suicidal ideation following traumatic brain injury: indications of right-hemispheric intrusions from neuropsychological profiles. Psychol Rep 1994; 75: 1059–1070.

27. Harris EC, Barraclough B. Suicide as an outcome for mental disorders: a meta analysis. Br J Psychiatry 1997; 170: 205–228.

28. Achte KA, Lonnqvist J, Hillbom E. Suicides following war brain injuries. Acta Psychiatr Scand Suppl 1971; 225: 1–94.

29. Roberts AH. Severe Accidental Head Injury: An Assessment of Long-term Prognosis. London: MacMillan, 1979.

30. Wilson BA. Life after brain injury: long term outcome of 101 people seen for rehabilitation 5–12 years earlier. In: Fourez J, Page N, eds. Treatment Issues and Long Term Outcomes. Bowen Hills, Queensland: Australian Academic Press, 1994; 1–6.

31. Klonoff PS, Lage GA. Suicide in patients with traumatic brain injury: risk and prevention. J Head Trauma Rehabil 1995; 10: 16–24.

32. Tate R, Simpson G, Flanagan S, et al. Completed suicide after traumatic brain injury. J Head Trauma Rehabil 2000; 12: 16–28.

33. Teasdale TW, Engberg AW. Suicide after traumatic brain injury: a population study. J Neurol Neurosurg Psychiatry 2001; 71: 436–440.

34. Stein DG. In pursuit of new strategies for understanding recovery from brain damage: problems and perspectives. In: Boll T, Bryant BK, eds. Clinical Neuropsychology and Brain Function: Research, Measurement and Practice. Washington, DC: American Psychological Asssociation, 1994: 9–56.

35. Kuipers P, Lancaster A. Focus on clinical practice, Part II. Developing a suicide prevention strategy on the perspectives of people with brain injury. J Head Trauma Rehabil 2000; 15: 1275–1284.

36. American Psychiatric Association (APA). Diagnostic and Statistical Manual of Mental Disorders, 4th edn., Text revision. Washington, DC: American Psychiatric Association, 2000.

37. Kreutzer JS, Marwitz JH, Seel R, Serio CD. Validation of a neurobehavioral functioning inventory for adults with traumatic brain injury. Arch Phys Med Rehabil 1996; 77: 116–124.

38. American Psychiatric Association. Practice Guidelines for the Treatment of Psychiatric Disorder. Washington, DC: Compendium, 2004.

39. Lee HB, Lyketsos CG, Rao V. Pharmacological management of the psychiatric aspect of TBI. Int Rev Psychiatry 2003; 15: 359–370.

40. Zafonte RD, Lexell J, Cullen N. Possible applications for dopaminergic agents following traumatic brain injury: Part 2. J Head Trauma Rehabil 2001; 16: 112–116.

41. Whyte J, Vacarro M, Grieb-Neff P, Hart T. Psychostimulant use in the rehabilitation of individuals with traumatic brain injury. J Head Trauma Rehabil 2002; 17: 284–299.

42. Meythaler JM, Brunner RC, Johnson A, Novack TA. Amantadine to improve neurorecovery in traumatic brain injury. J Head Trauma Rehabil 2002; 17: 300–313.

43. Cummings JL. Cholinesterase inhibitors: a new class of psychotropic compounds. Am J Psychiatry 2000; 157: 4–15.

44. Freo U, Pizzolato G, Dam M, et al. A short review of cognitive and functional neuroimaging studies of cholinergic drugs: Implications for therapeutic potentials. J Neural Transm 2002; 109: 857–870.

45. Kant R, Coffey CE, Bogyl AM. Safety and efficacy of ECT in patients with head injury: a case series. J Neuropsychiatr Clin Neurosci 1999; 11: 32–37.

46. Bennett T. Individual psychotherapy and minor head injury. Cogn Rehabil 1989; 7: 20–25

Index